2nd Edition
Revised for the 17th Edition IEE Wiring Regulations

Electrical Installations

NVQ and Technical Certificate Book 2

Dave Allan

John Blaus

on behalf of JTL

www.jtltraining.com

www.heinemann.co.uk
✓ Free online support
✓ Useful weblinks
✓ 24 hour online ordering

01865 888118

Heinemann is an imprint of Pearson Education Limited, a company incorporated in England and Wales, having its registered office at Edinburgh Gate, Harlow, Essex, CM20 2JE. Registered company number: 872828

www.heinemann.co.uk

Heinemann is a registered trademark of Pearson Education Limited

First published 2008

12 11 10 09 08
10 9 8 7 6 5 4 3 2 1

British Library Cataloguing in Publication Data
A catalogue record for this book is available from the British Library

ISBN 978 0 435 46705 0

Typeset by HL Studios
Layout by Ken Vail Graphic Design
Original illustrations © Pearson Education, 2008
Cover design by G D Associates
Cover photo/illustration © Pearson Education/Gareth Boden
Printed in the UK by Scotprint

Acknowledgements
Every effort has been made to contact copyright holders of material reproduced in this book. Any omissions will be rectified in subsequent printings if notice is given to the publishers.

Websites
The websites used in this book were correct and up-to-date at the time of publication. It is essential for tutors to preview each website before using it in class so as to ensure that the URL is still accurate, relevant and appropriate. We suggest that tutors bookmark useful websites and consider enabling students to access them through the school/college intranet.

Contents

Acknowledgements

JTL would like to express its appreciation to all those members of staff who contributed to the development of this book, ensuring that the professional standards expected were delivered and generally overseeing the high quality of the final product. Without their commitment and support – as much to each other as to the project, especially during some of the more fraught days – this project would not have been completed successfully. Particular thanks go to Dave Allan, Brian Tucker, Mike Crossley, Nigel Buckhurst, Phil Cunningham and John Langley. Our thanks also to Paul and Rita Hurt for their patience and assistance during the photoshoot and to John Blaus for his help and advice throughout the project.

The author and publishers are very grateful to the following for assistance in the production of this book and permission to reproduce copyright material:

C.K. Tools (Carl Kammerling International Ltd.)
Paul and Rita Hurt
JTL
Gareth Boden
Ginny Stroud-Lewis

Figure 7.39 on page 274 is reprinted from Electrical Installation Work, 5th edition, Brian Scadden, page 147, Copyright (2005), with permission from Elsevier.

Figure 7.41 on page 275 is reprinted from Advanced Electrical Installation Work, 4th edition, Trevor Linsley, page 97, Copyright (2005), with permission from Elsevier.

The authors and publishers would like to thank the following for permission to reproduce photographs:

AAF-Creatas, p119; Alamy Images, p43, p52, p89; Alamy Images/ Daniel Templeton, p1; Alamy Images/Dynamic Graphics /Photis, p279; Art Directors and Trip, p123 (top), p293, p307, p308, p324; Art Directors and Trip/Andrew Lambert, p294; Art Directors and Trip/ H Rogers, p279 (bottom), p310, p314; Corbis/Roger Ressmeyer, p241; Andrew Lambert Photography/Science Photo Library, p251; Ginny Stroud-Lewis, p122, p123 (bottom), p223, p298; Unknown, p113, p159, p167, p168, p205, p245, p252, p254, p260, p262, p274; Photodisc, p90 (top); Shutterstock/Jasena Lukŝa, p90 (bottom); Shutterstock/rorem, p91 (top), Shutterstock/Andrey Borodin, p91 (middle); Shutterstock/stocksnapp, p91 (bottom); Getty Images/Photo library, p95.

All other photos copyright Pearson Education Ltd/Gareth Boden.

IEE 17th Edition Wiring Regulations

Although every effort has been made to ensure the accuracy of information given in terms of compliance with BS 7671: 2008, at the time of printing certain key documents, such as the IEE On-Site Guide and the IEE Guidance Note 3 were unavailable. Therefore please verify information given with these documents as they become available.

Introduction

What is this book?

This book, and its companion volume, Book 1, have been designed with you in mind. They have a dual purpose:

- To lead you through the City & Guilds 2330 Level 2 and Level 3 Certificates in Electrotechnical Technology (Buildings & Structures)
- To provide a future reference book that you will find useful to dip into, long after you have gained your qualifications.

These books have been specially produced to help you achieve the City & Guilds 2330 at Levels 2 and 3 and between them aim to cover the underpinning knowledge requirement of the relevant NVQ qualification. Both these schemes are available throughout England & Wales and are designed to set a quality standard for learning and training in the electrotechnical sector, whether you are a new entrant to the industry or an experienced worker wishing to update your qualifications.

The books' contents underpin the various topics that you will be examined in as part of the Level 2 and 3 components of the City & Guilds 2330 Technical Certificate, with each chapter concluding with job knowledge tests that will allow you to measure your knowledge and understanding of the various topics.

Qualifications

Generally speaking there are two qualifications that should be considered at this point:

1. *National Vocational Qualifications (NVQs)*: The central feature of any NVQ is the National Occupational Standards (NOS) on which they are based. NOS are statements of performance that describe what competent people in a particular occupation are expected to be able to do … think of it as being a bit like a job description. They cover all the main aspects of an occupation, including current best practice, the ability to adapt to future requirements and the knowledge and understanding that underpins competent performance. Therefore, an NVQ is a qualification that is awarded when individuals can demonstrate that they are actually competent to do the job within the workplace. Consequently, the most valid form of 'evidence of ability' is

by someone watching you do the job at your workplace. However, you will still be expected to demonstrate your knowledge and understanding of these tasks and this is normally measured using external written examinations (such as the Technical Certificate) and oral questioning by an assessor. There are actually five levels of qualification within the NVQ system, with level five being the highest. For the electrical contracting industry, it is currently taken that the NVQ Level 3 is the minimum standard that must be attained for the award of electrician status via the Joint Industry Board (JIB).

2. *Technical Certificate*: For this industry, this is the primary means of providing the underpinning job knowledge and understanding that supports the NVQ competencies. Please note that for apprentices, success is required in both the Level 2 and Level 3 component of the City & Guilds 2330.

As a rough guide, within the electrotechnical industry, meeting the requirements of points 1 and 2 above allow an operative to be able to meet the current industry level for recognition with the JIB as a qualified electrician. However, specific requirements for recognition should always be checked with the JIB.

How this book can help you

There are other key features of this book which are designed to help you make progress and reinforce the learning that has taken place, such features are:

- **Photographs**: easy to follow sequences of key operations.
- **Illustrations**: clear drawings, many in colour, showing essential information about complex components and procedures.
- **Margin notes**: short helpful hints to aid you to good practice.
- **Tables, bullet points and flowcharts**: easy to follow features giving information at a glance.
- **On the job scenarios**: typical things that happen on the job: what would you do?
- **Did you know?**: useful information about things you always wondered about.
- **End of section job knowledge checks**: test yourself to see if you have absorbed all the information, are you ready for the real test ?
- **Glossary**: clear definitions and explanations of those strange words and phrases.

Why use this book?

Because it is structured to give you all the basic information required to help you gain the current industry Technical Certificate qualifications and set you on course to an exciting long term career inside a challenging and varied industry … Well done for choosing such a good start!

Chapter 1

Electrical science

OVERVIEW

The work tasks that a competent electrician will undertake are many and varied. However, as well as having practical competence, we also need to know about the operating principles behind the job. The purpose of this chapter is to look at the principles of electrical science and electronics necessary to support electrical and electronic installations. It follows on from the Electrical Science chapter in Book 1 which you may want to refer back to as you work through this chapter. It will cover:

- **Alternating current theory**
- **Transformers**
- **Instruments and measurement**

Alternating current theory

This is one of those subjects that can only be described as theoretical. So you will have to concentrate very carefully and please try to keep awake for this one!

In this section we will be looking at the following areas:

- what is alternating current?
- values of an alternating waveform
- frequency and period
- power factor
- resistance and phasor representation
- inductance
- capacitance
- phasors
- impedance
- resistance and inductance in series (RL)
- resistance and capacitance in series (RC)
- resistance, inductance and capacitance in series (RLC)
- resistance, inductance and capacitance in parallel.
- power in an a.c. circuit
- the power triangle
- three-phase power

What is alternating current?

Alternating current (**a.c.**) is a flow of electrons, which rises to a maximum value in one direction and then falls back to zero before repeating the process in the opposite direction. In other words, the electrons within the conductor do not drift (flow) in one direction, but actually move backwards and forwards.

The journey taken, i.e. starting at zero, flowing in both directions and then returning to zero, is called a cycle. The number of cycles that occur every second is said to be the frequency and this is measured in **hertz** (Hz).

Values of an alternating waveform

If we look at the graph of a sine wave (Figure 1.1), there are several values that can be measured from such an alternating waveform.

Instantaneous value

If we took a reading of induced electromotive force (e.m.f.) from the sine wave at any point in time during its cycle, this would be classed as an instantaneous value.

Average value

Using equally spaced intervals in our cycle (say every 30°) we could take a measurement of current as an instantaneous value. To find the average we would add together all the instantaneous values and then divide by the number of values used. As with the average of anything, the more values used, the greater the accuracy. *For a sine wave only*, we say that the average value is equal to the maximum value multiplied by 0.637. As a formula:

Average current = Maximum (peak) current × 0.637

Peak value

You will remember from Book 1 when the loop in an a.c. generator has rotated for 90° it is cutting the maximum lines of magnetic flux and therefore the greatest value of induced e.m.f. is experienced at this point. This is known as the peak value and both the positive and negative half cycles have a peak value.

Peak to peak value

In Book 1 we said that the maximum value of induced e.m.f., irrespective of direction, is called the maximum or peak value. The voltage measured between the positive and negative peaks is known as the **peak to peak value**. The graph in Figure 1.1 shows this more clearly.

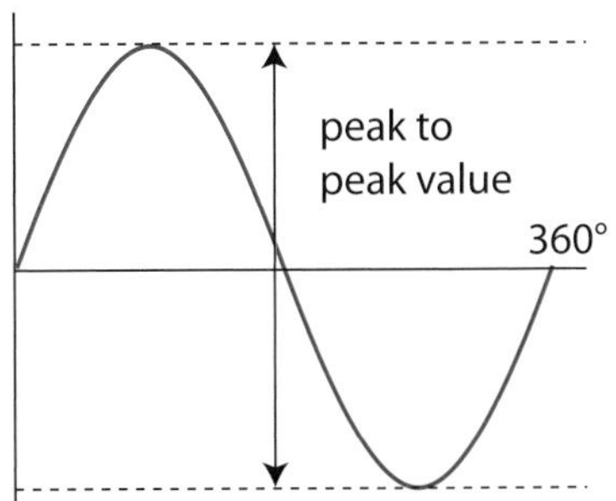

Figure 1.1 Peak to peak value

Root mean square (r.m.s.) or effective value of a waveform (voltage and current)

In Book 1 we showed that in direct current (d.c.) circuits, the power delivered to a resistor is given by the product of the voltage across the element and the current through the element. However, this is only true of the instantaneous power to a resistor in an a.c. circuit.

In most cases the instantaneous power is of little interest, and it is the average power delivered over time that is of most use. In order to have an easy way of measuring power, the r.m.s. method of measuring voltage and current was developed.

The r.m.s. or effective value is defined as being the a.c. value of an equivalent d.c. quantity that would deliver the same average power to the same resistor.

When current flows in a resistor, heat is produced. In Book 1 you saw that, when it is direct current flowing in a resistor, the amount of electrical power converted into heat is expressed by the formulae:

$$P = I^2 \times R \quad \text{or} \quad P = V \times I$$

However, an alternating current having a maximum (peak) value of 1 A does not maintain a constant value (see Figure 1.1). The alternating current will not produce as much heat in the resistance as will a direct current of 1 A. Consider the circuits in Figure 1.2.

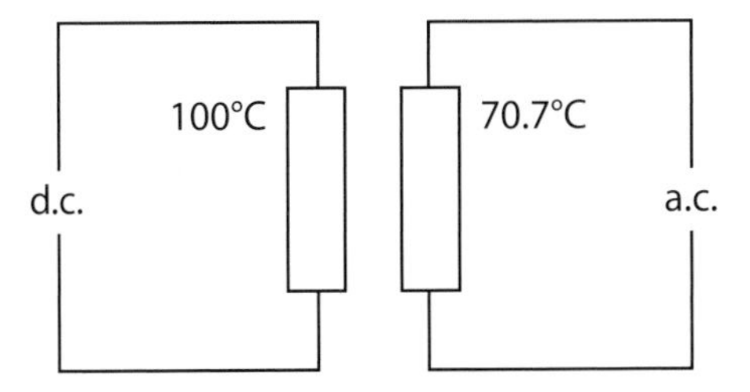

Figure 1.2 d.c. and a.c. circuits

In both the circuits in the figure, the various supplies provide a maximum (peak) value of current of 1 A to a known resistor. However, the heat produced by 1 ampere of alternating current is only 70.7°C compared to the 100°C of heat that is produced by 1 ampere of direct current. We can express this using the following formula:

$$\frac{\text{Heating effect of 1A maximum a.c.}}{\text{Heating effect of 1A maximum d.c.}} = \frac{70.7}{100} = 0.707$$

Therefore, **the effective or r.m.s. value of an a.c. = 0.707 × I_{max}**

where I_{max} = the peak value of the alternating current.

We can also establish the maximum (peak) value from the r.m.s. value with the following formula:

$$I_{max} = I_{r.m.s.} \times 1.414$$

The rate at which heat is produced in a resistor is a convenient way of establishing an effective value of alternating current, and is known as the 'heating effect' method.

An alternating current is said to have an effective value of one ampere when it produces heat in a given resistance at the same rate as one ampere of direct current.

To see where the r.m.s. value sits alongside the peak value, look at the folowing figures.

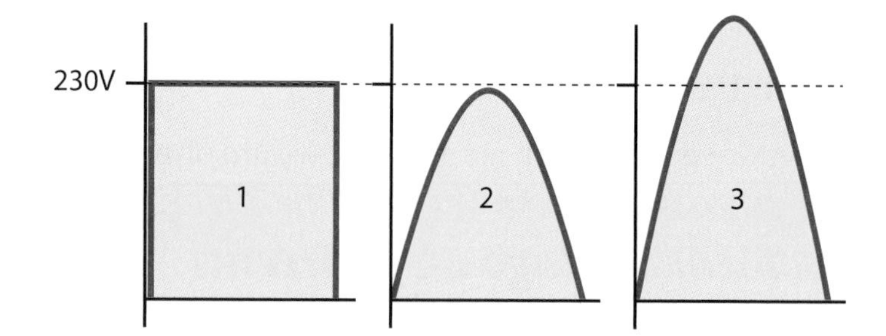

Figure 1.3 Peak value diagrams

The wave in Figure 1.3(1) represents a 230 V d.c. supply, running for a set period of time, with the heating effect produced shown as the shaded area.

The wave in Figure 1.3(2) represents an AC 50 Hz supply in which the voltage peaks at 230V during one half-cycle, with the heating effect produced shown as the shaded area. As the wave only reaches 230V for a small period of time, less heat is produced overall than in the d.c. wave.

The wave in in Figure 1.3(3) represents an increased peak value of voltage to give the same amount of shaded area as in the d.c. example. In this diagram 230 V has become our r.m.s. value.

Important Note: Unless stated otherwise, all values of a.c. voltage and current are given as r.m.s. values.

Frequency and period

Remember from Book 1 that the number of cycles that occur each second is referred to as the frequency of the waveform and this is measured in hertz (Hz). The frequency of the UK supply system is 50 Hz.

In the basic arrangement of the a.c. generator loop, if one cycle of e.m.f. was generated with one complete revolution of the loop over a period of one second, then we would say the frequency was 1 Hz. If we increased the speed of loop rotation so that it was producing five cycles every second, then we would have a frequency of 5 Hz.

Definition

Pole pair – any system consisting of a north and south pole

We can therefore say that the frequency of the waveform is the same as the speed of the loop's rotation, measured in revolutions per second. We can express this using the following equation:

Frequency (f) = Number of revolutions (n) × Number of pole pairs

If we apply this to the simple a.c. generator and rotate the loop at 50 revolutions per second, then:

Frequency = 50 × 1 (there is 1 × pole pair) = 50 Hz

The amount of time taken for the waveform to complete one full cycle is known as the periodic time (T) or period. Therefore, if 50 cycles are produced in one second, one cycle must be produced in a fiftieth of one second. This relationship is expressed using the following equations:

$$\text{Frequency (f)} = \frac{1}{\text{Periodic time}} = \frac{1}{T} \qquad \text{Periodic time (T)} = \frac{1}{\text{Frequency}} = \frac{1}{f}$$

Power factor

When we are dealing with a.c. circuits, we are often looking at the way power is used with particular types of component within the circuit.

Generally speaking, **power factor** is a number less than 1.0, which is used to represent the relationship between the apparent power of a circuit and the true power of that circuit. In other words:

$$\textbf{Power Factor (PF)} = \frac{\textbf{True Power (PT)}}{\textbf{Apparent Power (PA)}} \qquad \text{Or: } \textbf{PF} = \frac{\textbf{PT}}{\textbf{PA}}$$

In terms of units:

$$\textbf{Power Factor (PF)} = \frac{\textbf{Power (W)}}{\textbf{Voltage} \times \textbf{Current (VA)}}$$

Power factor has no units, it is a number. It is also determined by the phase angle, which we will cover shortly.

We will be coming back to this quite frequently over the next few pages, so by the end of the section you will have a good understanding of power factor.

Resistance (R) and phasor representation

Whereas the sine wave is useful, it is also difficult and time consuming to draw. We can therefore also represent a.c. by the use of phasors. A **phasor** is a straight line whose length is a scaled representation of the size of the a.c. quantity and whose direction represents the relationship between the voltage and current, this relationship being known as the phase angle.

To see briefly how we use phasors, let us look at the following diagram, where a tungsten filament lamp has been included as the load.

Circuits like this are said to be **resistive**, and in this type of circuit the values of e.m.f. (voltage) and current actually pass through the same instants in time together. In other words, as voltage reaches its maximum value, so does the current (see Figure 1.4).

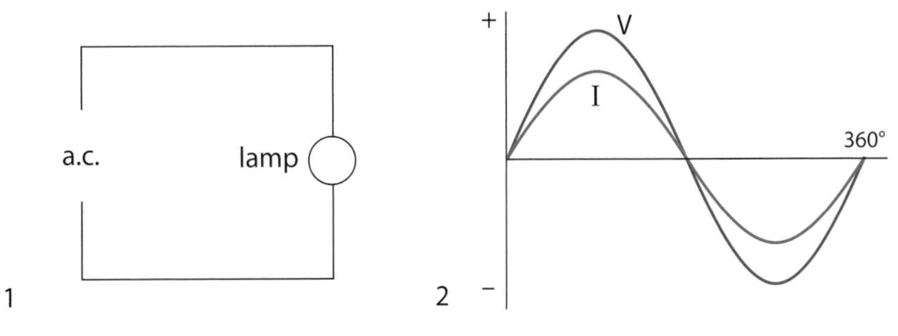

Figure 1.4 Circuit and sine wave diagrams

This happens with all resistive components connected to an a.c. supply and as such the voltage and current are said to be '**in phase**' with each other, or possess a zero phase angle.

The graph in Figure 1.4 shows this when represented by a sine wave. However, we could also show this by using a phasor diagram as shown in Figure 1.5.

Figure 1.5 Phasor diagram for zero phase angle

We can therefore say that a resistive component will consume power and we would carry out calculations as we would for a d.c. circuit (i.e. using $P = V \times I$).

We can also say that resistive equipment (filament lamps, fires, water heaters) use this power to create heat, but such a feature in long cable runs, windings etc. would be seen as unsuitable power loss in the circuit (i.e. using $P = I^2R$).

Inductance (L)

If the load in our circuit were not a filament lamp, but a motor or transformer (something possessing windings), then we say that the load is **inductive**.

The sine wave used to represent this inductive circuit would look like this

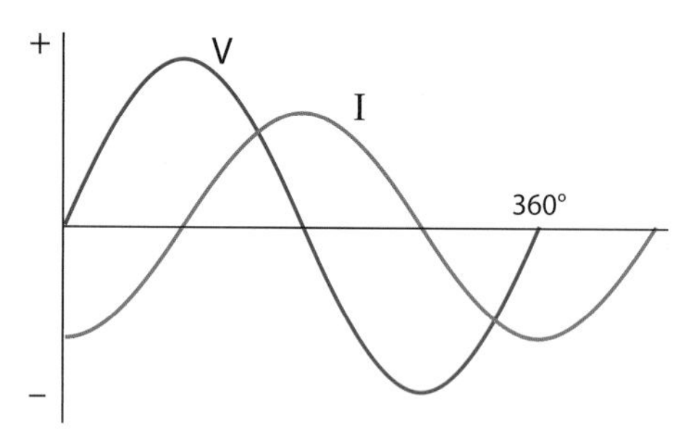

If we represented this as a phasor diagram we end up with

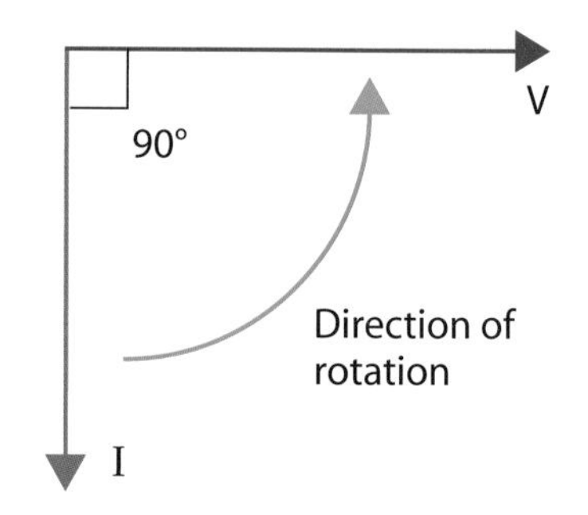

Figure 1.6 Sine wave and phasor diagrams for inductive circuit

With an inductive load the voltage and current become '**out of phase**' with each other. This is because the windings of the equipment set up their own induced e.m.f., which opposes the direction of the applied voltage, and thus forces the flow of electrons (current) to fall behind the force pushing them (voltage). However, over one full cycle, we would see that no power is consumed. When this happens, it is known as possessing a lagging phase angle or power factor.

As voltage and current are no longer perfectly linked, this type of circuit would be given a power factor of less than 1.0 (perfection), for example 0.8.

As we can see from Figure 1.6, the current is lagging the applied voltage by 90°. To make things easier in this exercise, we assumed that the above circuit is purely inductive. However, in reality this is not possible, as every coil is made of wire and that wire will have a resistance. The opposition to current flow in a resistive circuit is resistance.

The limiting effect to the current flow in an inductor is called the **inductive reactance**, which we are able to calculate with the following formula:

$X_L = 2\pi fL$ (Ω)

where:

X_L = inductive reactance (ohms – Ω)

f = supply frequency (hertz – Hz)

L = circuit inductance (henrys – H)

Let us now look at inductive and resistive circuits and see if the current is affected by a lagging phase angle.

To recap, we said that power factor is the relationship between voltage and current and that the ideal situation would seem to be the resistive circuit, where both these quantities are perfectly linked.

In the resistive circuit we know that the power in the circuit could only be the result of the voltage and the current ($P = V \times I$). This is known as the apparent power, and possesses what we call unity power factor, to which we give the value one (1.0).

However, we now know that depending upon the equipment, the true power (actual) in the circuit must take into account the phase angle and will often be less than the apparent power ... but never greater.

True power (in watts) is calculated using the cosine of the phase angle (cos Ø). The formula is:

$P = VI \cos Ø$ (remember we do not have to use the '×' sign)

When there is no phase lag, $Ø = 0$ and $\cos Ø = 1$, a purely resistive circuit. To prove our previous points, let us consider the following.

Example

If we have an inductive load, consuming 3 kW of power from a 230 V supply, with a power factor of 0.7 lagging, then the current (amount of electrons flowing) required to supply the load is:

$$P = V \times I \times \cos Ø$$

Cos Ø = Power Factor. Therefore by transposition:

$$I = \frac{P}{V \times \cos Ø}$$

Or in other words:

$$I = \frac{P}{V \times PF}$$

Therefore:

$$I = \frac{3000}{230 \times 0.7}$$

And so:

$$I = 18.6\,A$$

However, if the same size of load was purely resistive, then $\cos Ø = 0$, thus the power factor would be 1.0, and thus:

$$P = V \times I \times PF$$

Therefore by transposition:

$$I = \frac{P}{V \times PF}$$

Therefore:

$$I = \frac{3000}{230 \times 1}$$

And consequently:

$$I = 13\,A$$

In other words, the lower the power factor of a circuit, then the higher the current will need to be to supply the load's power requirement.

It therefore follows that if power factor is low, then it will be necessary to install larger cables, switchgear etc. to be capable of handling the larger currents. There will also be the possibility of higher voltage drop due to the increased current in the supply cables.

Consequently, local electricity suppliers will often impose a financial fine on premises operating with a low power factor. Fortunately, we have a component that can help. It is called the capacitor.

Capacitance (C)

Simply put, a **capacitor** is a component that stores an electric charge if a potential difference is applied across it.

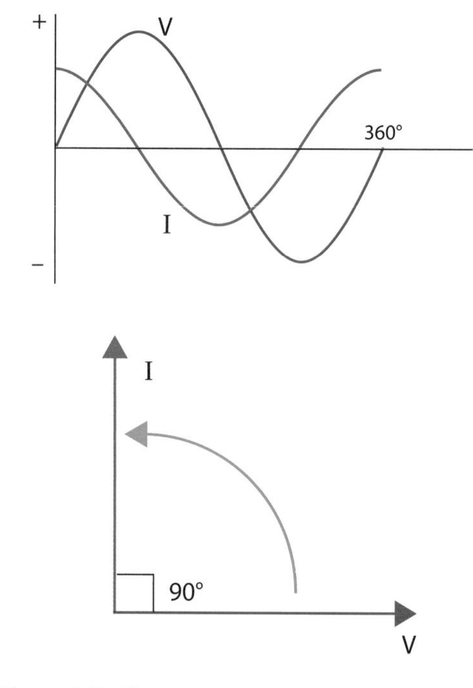

Figure 1.7 Sine wave and phasor diagrams for capacitive circuits

The capacitor's use is then normally based on its ability to return that energy back to the circuit. When a capacitor is connected to an a.c. supply, it is continuously storing the charge and then discharging as the supply moves through its positive and negative cycles. But, as with the inductor, no power is consumed.

This means that in a capacitive circuit, we have a leading phase angle or power factor. The sine wave and phasors used to represent this would look as in Figure 1.7.

As we can see from the previous diagram, the current leads the voltage by 90°. Consequently, the capacitor is able to help because it provides a leading power factor, and therefore if we connect it in parallel across the load, it can help neutralise the effect of a lagging power factor.

The opposition to the flow of a.c. to a capacitor is termed capacitive reactance, which like inductive reactance, is measured in ohms and calculated using the following formula:

$$X_C = \frac{1}{2\pi fC} \ (\Omega)$$

where:

X_C = capacitive reactance (ohms – Ω)

f = supply frequency (hertz – Hz)

C = circuit **capacitance** (farads – F)

Since, in this type of circuit, we have voltage and current but no real power (in watts), the formula of $P = V \times I$ is no longer accurate. Instead, we say that the result of the voltage and current is reactive power, which is measured in reactive volt amperes (VAr).

The current to the capacitor, which does not contain resistance or consume power, is called reactive current.

Well, so far in our attempt to explain power factor, we have looked at a range of different subjects including resistance, inductance, capacitance and also talked about phasor diagrams. As if that wasn't bad enough, some circuits contain combinations of these components.

Sadly, in order to be able to work out these calculations, you will need to use both the trigonometry knowledge you gained in Book 1 and also the following small section regarding the addition of phasors.

Phasors

When sine waves for voltage and current are drawn, the nature of the wave diagram can be based upon any chosen alternating quantity within the circuit. In other words, we can start from zero on the wave diagram with either the voltage or the current.

In electrical science we often need to add together alternating values. If they were 'in phase' with each other, then we would simply add the values together. However, when they are not in phase we cannot do this, hence the need for phasor diagrams.

When we use phasor diagrams the chosen alternating quantity is drawn horizontally and is known as the reference.

When choosing the reference phasor, it makes sense to use a quantity that has the same value at all parts of the circuit. For example in a series circuit the same current flows in each part of the circuit, therefore use current as the reference phasor. In a parallel circuit, the voltage is the same through each branch of the circuit and therefore we use voltage as the reference phasor.

Using the knowledge gained in the previous section, we can now measure all phase angles from this reference phasor.

Our answer, or resultant, is then found by completing a parallelogram. If we use Figure 1.8 as an example, we have been given the values of phasor A and phasor B. Therefore, the result of adding A and B together will be phasor C.

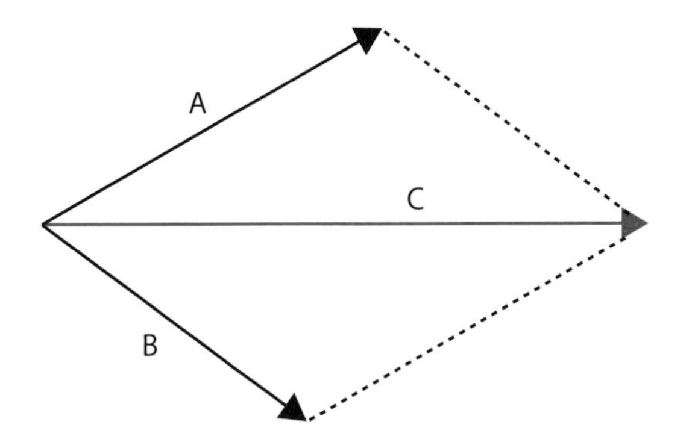

Figure 1.8 Phasor diagram

Impedance

Previously, we have been discussing components with an a.c. circuit. In fact, what those components are actually offering is opposition to the flow of current. By way of a summary, we could say that we now know that:

- The opposition to current in a resistive circuit is called resistance (R) is measured in ohms and the voltage and current are in phase with each other.
- The opposition to current in an inductive circuit is called inductive reactance (X_L), is measured in ohms and the current lags the voltage by 90°.
- The opposition to current in a capacitive circuit is called capacitive reactance (X_C), is measured in ohms and the current leads the voltage by 90°.

However, we also know that circuits will contain a combination of these components. When this happens we say that the total opposition to current is called the **impedance** (Z) of that circuit.

In summary:

- The power consumed by a resistor is dissipated in heat and not returned to the source. This is called the **true power**.
- The energy stored in the magnetic field of an **inductor** or the plates of a capacitor is returned to the source when the current changes direction.
- The power in an a.c. circuit is the sum of true power and reactive power. This is called the **apparent power**.
- **True power is equal to apparent power in a purely resistive circuit** because the voltage and current are in phase. Voltage and current are also in phase in a circuit containing equal values of inductive reactance and capacitive reactance. If the voltage and current are 90 degrees out of phase, as would be the case in a purely capacitive or purely inductive circuit, the average value of true power is equal to zero. There are high positive and negative peak values of power, but when added together the result is zero.
- Apparent power is measured in volt-amps (VA) and has the formula: **P = VI**
- True power is measured in watts and has the formula **P = VI cos Ø**
- **In a purely resistive circuit** where current and voltage are in phase, there is no angle of displacement between current and voltage. The cosine of a zero degree angle is one, and so, the power factor is one. This means that all the energy that is delivered by the source is consumed by the circuit and dissipated in the form of heat.
- **In a purely reactive circuit**, voltage and current are 90 degrees apart. The cosine of a 90 degree angle is zero so the power factor is zero. This means that the circuit returns all the energy it receives from the source, back to the source.
- **In a circuit where reactance and resistance are equal**, voltage and current are displaced by 45 degrees. The cosine of a 45 degree angle is 0.7071, and so, the power factor is 0.7071. This means that such a circuit uses approximately 70 per cent of the energy supplied by the source and returns approximately 30 per cent back to the source.

Definition

Impedance – total opposition to current in a circuit

True or **active power** – the rate at which energy is used

Apparent power – in an a.c. circuit the sum of the true or active power and the reactive power

Resistance and inductance in series (RL)

Consider the following diagram.

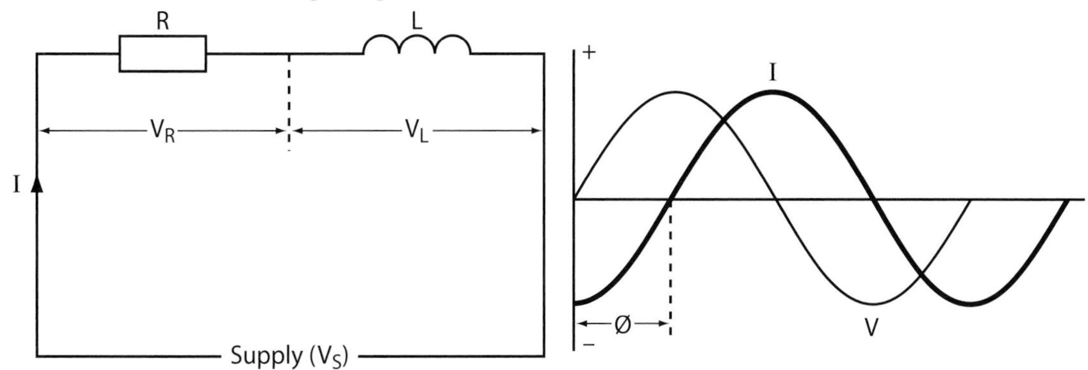

Figure 1.9 Resistor and inductor in series

Here we have a resistor connected in series with an inductor and fed from an a.c. supply. In a series circuit, the current (I) will be common to both the resistor and the inductor, causing **voltage drop** V_R across the resistor and V_L across the inductor.

The sum of these voltages must equal the supply voltage. Here's how to construct a phasor diagram for this circuit.

In a series circuit we know that current will be common to both the resistor and the inductor. It therefore makes sense to use current as our reference phasor. We also know that voltage and current will be in phase for a resistor. Therefore, the volt drop (**p.d.**) V_R across the resistor must be in phase with the current. Also, in an inductive circuit, the current lags the voltage by 90°.

If the current is lagging voltage, then we must be right in saying that voltage is leading the current.

This means in this case that the volt drop across the inductor (V_L) will lead the current by 90°. We can then find the value of the supply voltage (V_S), by completing the parallelogram that was discussed on page nine. When we draw phasors, we always assume that they rotate anti-clockwise and the symbol Ø represents the phase angle.

There are two ways of doing the drawing.

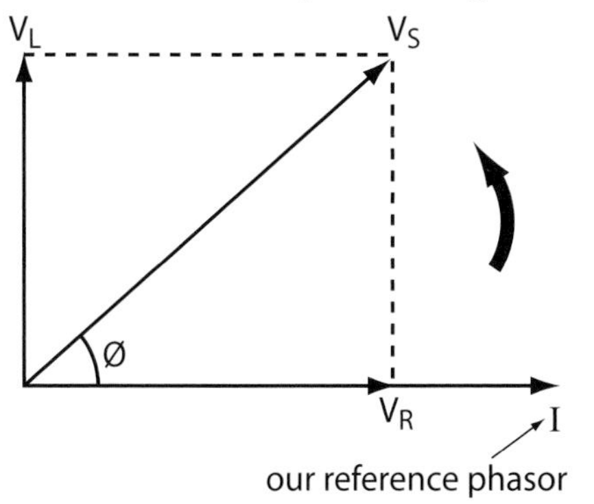

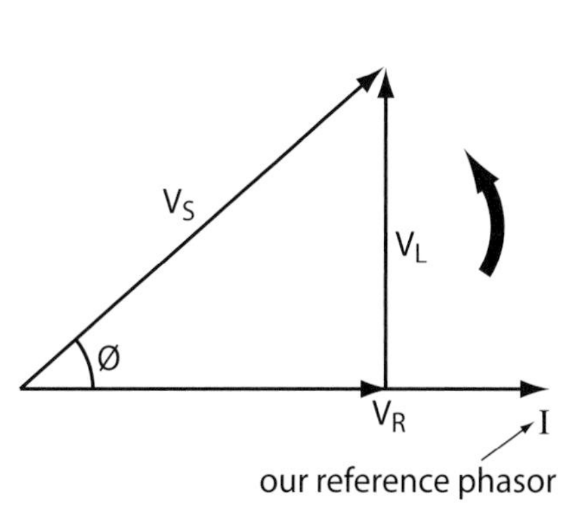

Figure 1.10 Two ways of drawing a phasor diagram

Remember

Because this is an a.c. circuit, you cannot just add the voltages together as you would have done for a d.c. circuit, you need to construct a phasor diagram to work it out

We can see that in the second example, the phasors produce a right-angled triangle. We can therefore use Pythagoras' Theorem to give us the formula:

$$V_S^2 = V_R^2 + V_L^2$$

We can then use trigonometry to give us the different formulae, dependent on the values that we have been given:

$$\cos Ø = \frac{V_R}{V_S} \qquad \sin Ø = \frac{V_L}{V_S} \qquad \tan Ø = \frac{V_L}{V_R}$$

Example 1

A coil of 0.15H is connected in series with a 50 Ω resistor across a 100 V 50 Hz supply. Calculate the following:

a. The inductive reactance of the coil

b. The impedance of the circuit

c. The circuit current

a. Inductive reactance (X_L)

For inductive reactance, we use the formula

$$X_L = 2\pi f L\ (\Omega)$$

Inserting the values, this would give us

$$X_L = 2 \times 3.142 \times 50 \times 0.15 \text{ therefore } X_L = 47.13\ \Omega$$

b. Circuit impedance (Z)

When we have resistance and inductance in series, we calculate the impedance using the following formula:

$$Z^2 = R^2 + X_L^2$$

which becomes

$$Z = \sqrt{R^2 + X_L^2}$$

Figure 1.11 Impedance triangle

In the case of the first formula, isn't this the same as Pythagoras' Theorem for a right-angled triangle ($A^2 = B^2 + C^2$)? We therefore sometimes refer to this as the impedance triangle and it can be drawn for this type of circuit, as shown. Here, the angle (Ø) between sides R and Z is the same as the phase angle between sides R and Z, is the same as the phase angle between current and voltage.

If we therefore apply some trigonometry, the following applies:

$$\cos Ø = \frac{R}{Z} \qquad \sin Ø = \frac{X_L}{Z} \qquad \tan Ø = \frac{X_L}{R}$$

However, using our formula

$$Z = \sqrt{R^2 + X_L^2}$$

then $Z = \sqrt{50^2 + 47.1^2}$ therefore $Z =$ **68.69 Ω**

c. Circuit current (I)

As we are referring to the total opposition to current, we use the formula

$$I = \frac{V}{R} = \frac{100}{68.69} = \mathbf{1.46\ A}$$

Example 2

A coil of 0.159H is connected in series with a 100 Ω resistor across a 230 V 50 Hz supply. Calculate the following:

a. The inductive reactance of the coil

b. The circuit impedance

c. The circuit current

d. The p.d. across each component

e. The circuit phase angle

a. Inductive reactance (X_L)

$X_L = 2\pi f L$ therefore $X_L = 2 \times 3.142 \times 50 \times 0.159 = \mathbf{50\ \Omega}$

b. Circuit impedance (Z)

$Z = \sqrt{R^2 + X_L{}^2}$ therefore $Z = \sqrt{100^2 + 50^2} = \mathbf{111.8\ \Omega}$

c. Circuit current (I)

$I = \frac{V}{Z}$ therefore $I = \frac{230}{111.8} = \mathbf{2.06\ A}$

d. The p.d. across each component (V)

$V_R = I \times R$ therefore $V = 2.06 \times 100 = \mathbf{206\ V}$

$V_L = I \times X_L$ therefore $V = 2.06 \times 50 = \mathbf{103\ V}$

e. Circuit phase angle (Ø)

Using our right-angled triangle:

$\tan Ø = \frac{V_L}{V_R}$ therefore $\tan Ø = \frac{103}{206} = \mathbf{0.5}$

If you then enter 0.5 into your calculator and press the INV key, then press TAN, you should get the number 26.6.

Therefore, the current is lagging voltage by **26.6°**

Resistance and capacitance in series (RC)

Consider the following diagram.

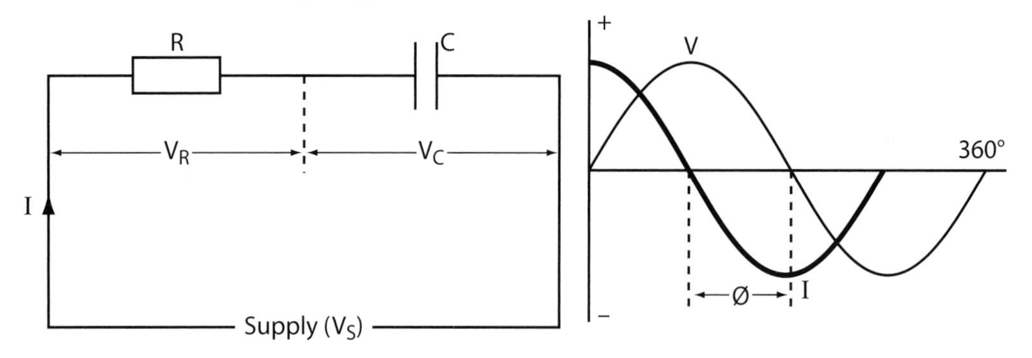

Figure 1.12 A resistor connected in series with a capacitor, fed from an a.c. supply

Here a resistor is connected in series with a capacitor and fed from an a.c. supply. Once again, in a series circuit the current (I) will be common to both the resistor and the capacitor, causing voltage to drop (p.d.) V_R across the resistor and V_C across the capacitor.

As with the Resistance/Inductance (RL) circuit previously, we can take current as the reference phasor. Similarly, the voltage across the resistor will be in phase with that current. Also, if you remember, in a capacitive circuit the current leads the voltage by 90°. Therefore we can say that the voltage across the capacitor will be lagging the current. As before, we can now calculate the supply voltage (V_S) by completion of the parallelograms as follows:

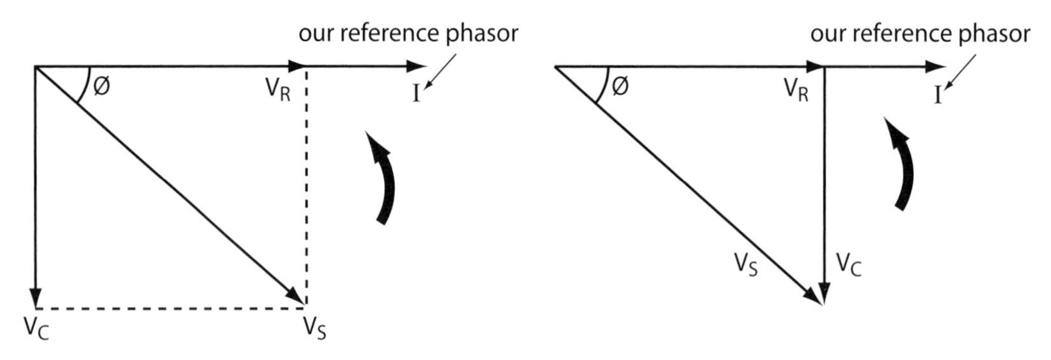

Figure 1.13 Phasor diagrams

As with the inductor, we can apply Pythagoras' Theorem and trigonometry to give us the following formulae:

$$V_S^2 = V_R^2 + V_C^2$$

$$\cos Ø = \frac{V_R}{V_S} \quad \sin Ø = \frac{V_C}{V_S} \quad \tan Ø = \frac{V_C}{V_R}$$

Here's an example:

Example

A capacitor of 15.9μF and a 100 Ω resistor are connected in series across a 230 V 50 Hz supply. Calculate the following:

a. The circuit impedance

b. The circuit current

c. The p.d. across each component

d. The circuit phase angle

a. Circuit impedance (Z)

To be able to find the impedance we must first find the capacitive reactance.

$$X_C = \frac{1}{2\pi fC}$$

however, as the capacitor value is given in μF, we use $X_C = \frac{10^6}{2\pi fC}$

This gives us : $X_C = \frac{10^6}{2 \times 3.142 \times 50 \times 15.9} = \frac{10^6}{4995.78} = \mathbf{200\ \Omega}$

When we have resistance and capacitance in series, we use the following formula:

$Z^2 = R^2 + X_C^2$ which becomes $Z = \sqrt{R^2 + X_C^2}$

therefore $Z = \sqrt{100^2 + 200^2} = \sqrt{50\,000} = \mathbf{224\ \Omega}$

b. Circuit current (I)

$I = \frac{V}{Z}$ therefore $I = \frac{230}{224} = 1.03$ A

c. The p.d. across each component (V)

$V_R = I \times R$ therefore $V_R = 1.03 \times 100 = \mathbf{103\ V}$

$V_C = I \times X_L$ therefore $V_C = 1.03 \times 200 = \mathbf{206\ V}$

d. Circuit phase angle (Ø)

Using our right-angled triangle:

$\tan Ø = \frac{V_C}{V_R}$ therefore $\tan Ø = \frac{206}{103} = \mathbf{2}$

If you then enter 2 into your calculator and press the INV key, then press TAN, you should get the number 63.4.

Therefore, the current is leading voltage by **63.4°**

Resistance, inductance and capacitance in series (RLC)

Consider the following diagram:

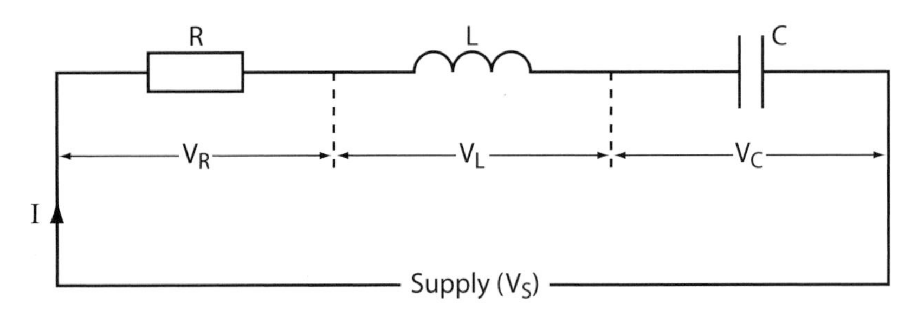

Figure 1.14 Resistor connected in series with an inductor and capacitor, fed from an a.c. supply.

Here we have a resistor connected in series with an inductor and a capacitor then fed from an a.c. supply. This is often referred to as an RLC circuit or a general series circuit. Again, as we have a series circuit, the current (I) will be common to all three components, causing a voltage drop (p.d.) V_R across the resistor, V_L across the inductor and V_C across the capacitor.

Here V_R will be in phase with the current, V_L will lead the current by 90° (because the current lags the voltage) and V_C will lag the current by 90° (because current leads the voltage in a capacitive circuit). Because V_L and V_C are in opposition to each other (one leads and one lags), the actual effect will be the difference between their values, subtracting the smaller from the larger. We can once again calculate V_S by completing a parallelogram as follows.

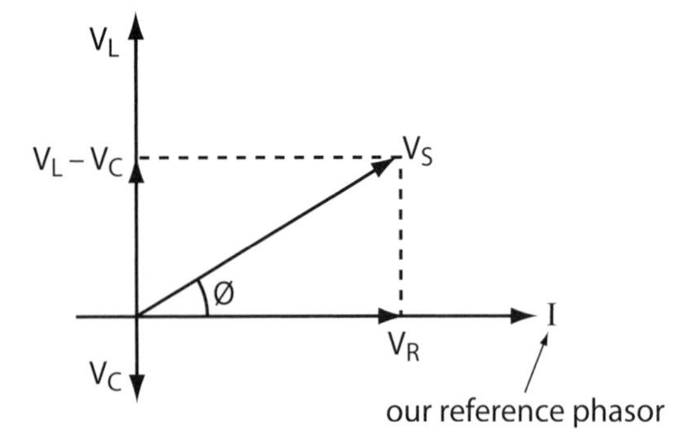

Figure 1.15 Phasor parallelogram

As before, we can now apply Pythagoras' Theorem and trigonometry to give us the following formulae depending on whether V_L or V_C is the larger:

$$V_S^2 = V_R^2 + (V_L - V_C)^2 \quad \text{or} \quad V_S^2 = V_R^2 + (V_C - V_L)^2$$

and because I is the same for each component we get:

$$Z = \sqrt{R^2 + (X_L - X_C)^2} \quad \text{or} \quad Z = \sqrt{R^2 + (X_C - X_L)^2}$$

and finally: $\cos Ø = \frac{V_R}{V_S}$ $\quad \sin Ø = \frac{V_L - V_C}{V_S}$ $\quad \tan Ø = \frac{V_L - V_C}{V_R}$

Example

A resistor of 5 Ω is connected in series with an inductor of 0.02H and a capacitor of 150 μF across a 250 V 50 Hz supply. Calculate the following.

a. The impedance

b. The supply current

c. The power factor

a. Impedance (Z)

In order to find the impedance, we must first find out the relevant values of reactance.

Therefore: $X_L = 2\pi f L = 2 \times 3.142 \times 50 \times 0.02 = \mathbf{6.28\ \Omega}$

$$X_C = \frac{1}{2\pi f C} \quad \text{allowing for microfarads}$$

$$= \frac{10^6}{2 \times 3.142 \times 50 \times 150} = \mathbf{21.2\ \Omega}$$

If you remember, we said that the effect of inductance and capacitance together in series would be the difference between their values. Consequently, this means that the resulting reactance (X) will be found as follows:

$X = X_L - X_C$ or, in this case because X_C is the larger: $X = X_C - X_L$

and therefore: $X = 21.2 - 6.28 = \mathbf{14.92\ \Omega}$

We can now use the impedance formula as follows:

$$Z = \sqrt{R^2 + (X_L - X_C)^2} \quad \text{or} \quad Z = \sqrt{R^2 + (X_C - X_L)^2}$$

which gives us

$$Z = \sqrt{5^2 + 14.92^2} = \mathbf{15.74\ \Omega}$$

Note: X_C is greater than X_L. Therefore we subtract X_L from X_C. Had X_L been the higher, then the reverse would be true. Also, as capacitive reactance is highest, the circuit current will lead the voltage. Had the inductive reactance been the higher, then the current would lag the voltage.

b. Supply current (I)

$$I = \frac{V}{Z} = \frac{230}{15.74} = 14.6\ \text{A}$$

c. Power factor (Ø)

$$\cos Ø = \frac{R}{Z} = \frac{5}{15.74} = 0.32$$

Therefore PF = **0.32 leading**

Resistance, inductance and capacitance in parallel

Consider the following diagram:

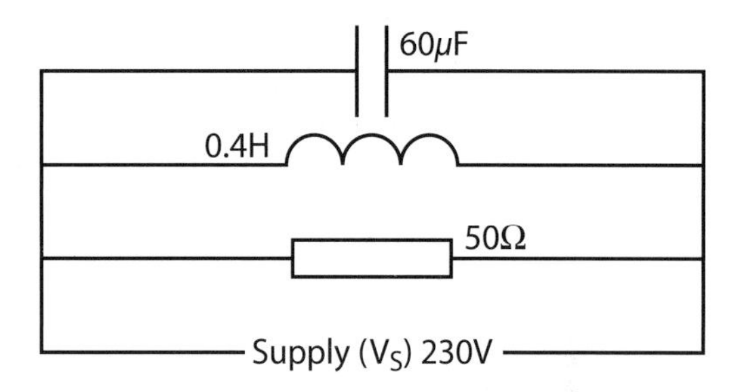

Figure 1.16 Resistor, capacitor and inductor in parallel connected to an a.c. supply

There can obviously be any combination of the above components in parallel. However, to demonstrate the principles involved, we will look at all three connected across and a.c. supply.

As we have a parallel circuit, the voltage (V_S) will be common to all branches of the circuit. Consequently, when we draw our parallelogram we will use voltage as the reference phasor.

In this type of circuit, the current through the resistor will be in phase with the voltage, the current through the inductor will lag the voltage by 90° and the current through the capacitor will lead the voltage by 90°.

Normally, when we carry out calculations for parallel circuits, it is easier to treat each branch as being a separate series circuit. We then draw to scale each of the respective currents and their relationships to our reference phasor, which is voltage.

As with voltage V_L and V_S in the RLC series circuit, the current through the inductor (I_L) and the current through the capacitor (I_C) are in complete opposition to each other. Therefore, the actual effect will be the difference between their two values. We calculate this value by the completion of our parallelogram. But, bear in mind that the bigger value (I_C or I_L) will determine whether the current ends up leading or lagging.

If we use the diagram as the basis for our example, calculate the circuit current and its phase angle relative to the voltage.

To calculate the circuit current and its phase angle relative to the voltage, we must first find the current through each branch:

$$I_R = \frac{V}{R} = \frac{230}{50} = \mathbf{4.6\ A}$$

$$X_L = 2\pi fL = 2 \times 3.142 \times 50 \times 0.4 = \mathbf{126\ \Omega}$$

Therefore $I_L = \frac{V}{X_L} = \frac{230}{126} = \mathbf{1.8\ A}$

$$X_C = \frac{10^6}{2\pi fC} = \frac{10^6}{2 \times 3.142 \times 50 \times 60} = \mathbf{53\ \Omega}$$

Therefore $I_C = \frac{V}{X_C} = \frac{230}{53} = \mathbf{4.3\ A}$

The actual effect will be $I_C - I_L$ which gives 4.3 – 1.8 = **2.5 A**

Now add this to I_R by completing the scale drawing at Figure 1.17. This gives a current of **5.2 A** that is leading voltage by an angle of **28°**.

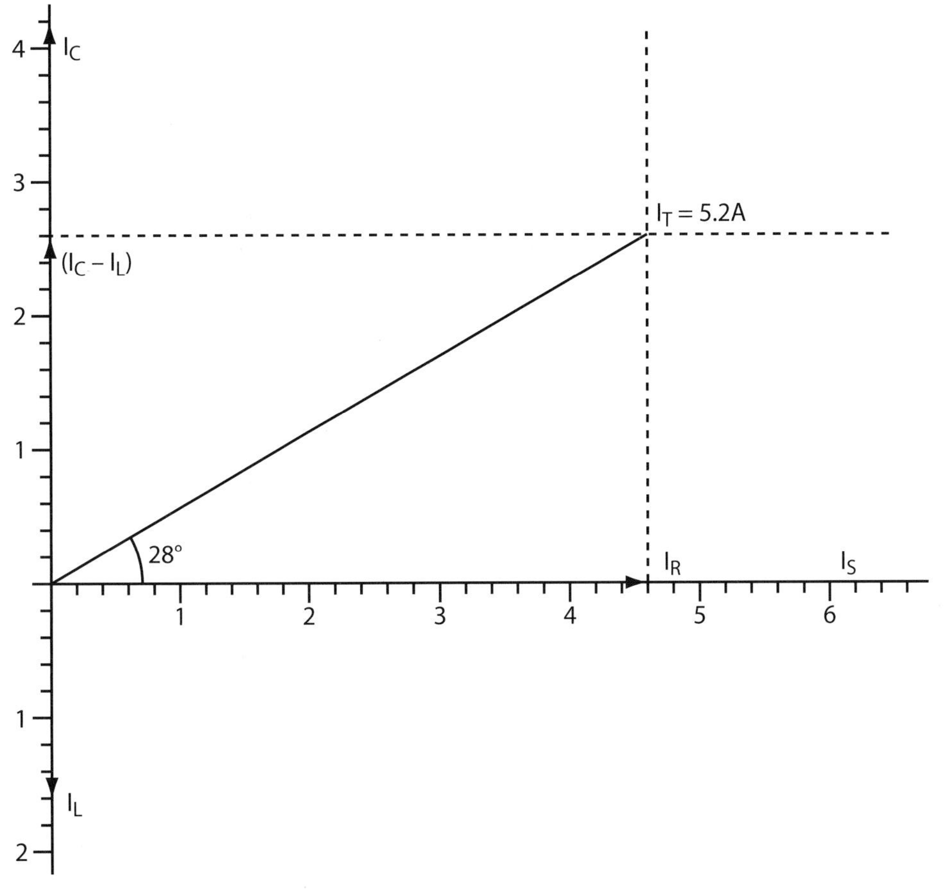

Figure 1.17 Scale drawing

Power in an a.c. circuit

If we were to try to push something against a resistance, we would get hot and bothered as we use up energy in completing the task. When current flows through a resistor, a similar thing happens in that power is the rate of using up energy – in other words, the amount of energy that was used in a certain time.

If a resistor (R) has a current (I) flowing through it for a certain time (t), then the power (energy being used per second) given in watts, can be calculated by the formula:

$P = I^2 \times R$

This power will be dissipated in the resistor as heat and reflects the average power in terms of the r.m.s. values of voltage and current.

What we are effectively saying her is that the average power in a resistive circuit (one which is non-reactive, i.e. doesn't possess inductance or capacitance) can be found by the product of the readings of an ammeter and a voltmeter.

In other words, in the resistive circuit, the power (energy used per second) is associated with that energy being transferred from the medium of electricity into another medium, such as light (filament lamp) or heat (electric fire/kettle). We call this type of power the active power. When we look at the capacitive circuit, we find that current flows to the capacitor, but we have no power.

Look at this wave diagram for voltage (Figure 1.18).

During the first period of the cycle, the voltage is increasing and this provides the energy to charge the capacitor. However, in the second period of the cycle, the voltage is decreasing and therefore the capacitor discharges, returning its energy back to the circuit as it does so. The same is also true of the third and fourth periods.

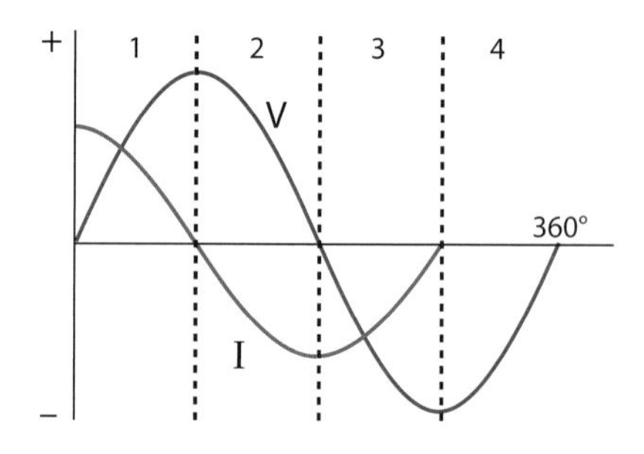

Figure 1.18 Sine wave diagram

This exchange of energy means that we have voltage and current, but no average power and therefore no heating effect. This means that our previous formula ($P = I^2 \times R$) is no longer useful.

We therefore say that the result of voltage and current in this type of circuit is called reactive power and we express this in reactive voltamperes (VAr). Equally, we say that the current in a capacitive circuit, where there is no resistance and no dissipation of energy, is called **reactive current**.

When we come to the inductive circuit we have a similar position. This time, as voltage increases during the first period of the cycle, the energy is stored as a magnetic field in the inductor. This energy will then be fed back into the circuit during the second period of the cycle as the voltage decreases and the magnetic field collapses. In other words, once again the exchange of energy produces no average power (energy used per second).

Circuits are likely to comprise combinations of resistance, inductance and capacitance. In a circuit, which has resistance and reactance, there will be a phase angle between the voltage and current. This relationship has relevance, as power will only be expended in the resistive part of the circuit.

Let's now look at this relationship via a phasor diagram, for a circuit containing resistance and capacitance, remembering that for this type of circuit the current will lead the voltage.

This circuit has two components (resistance and capacitance), so we've drawn two current phasors. In reality, as we already know, these are not currents that will actually flow, but their phasor sum will be the actual current in the circuit I_p(actual).

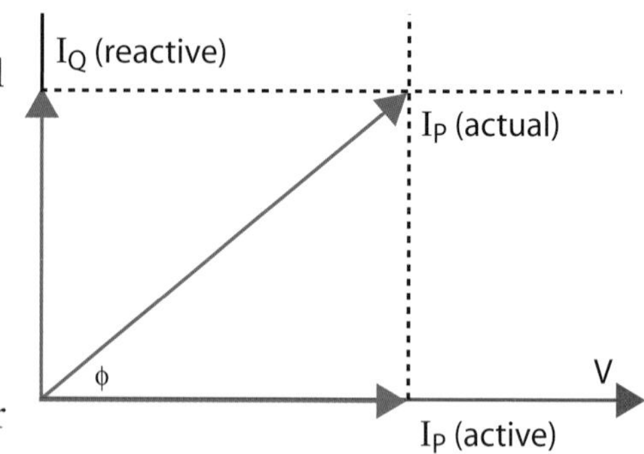

Figure 1.19 Phasor diagram

We also said that in a resistive circuit the voltage and current are in phase, and therefore this section of the current has been represented by the phasor I_p(active). This part of the current is in the active section and we therefore refer to this as the active current.

We then show that part of the current in the reactive section (capacitor) as leading the voltage by 90° and this has been represented by the phasor I_Q(reactive). As stated previously, we refer to this as the reactive current.

We also know that for a resistive circuit, we calculate the power by multiplying together the r.m.s. values of voltage and current ($V \times I$).

Logically, as we know that no power is consumed in the reactive section of the circuit, we can therefore calculate the power in the circuit by multiplying together the r.m.s. value of voltage and the value of current, which is in phase with it (I_p(active)). This would give us the formula:

$$P = V \times I_p(\text{active})$$

But as the actual current will be affected by the reactive current and therefore the phase angle (cos Ø), our formula becomes:

$$P = V \times I_p(\text{active}) \times \cos \varnothing$$

If Ip(active) is zero then Ø = 90° and cos Ø = 0. Therefore P = 0.

We have now established that it is possible in an a.c. circuit for current to flow, but no power to exist.

We also say that the product of voltage and current is power given in watts. However, it would be fair to say that this is not the actual power of the circuit.

The actual (true) power of the circuit has to take on board the effect of the phase angle (cos Ø), the ratio of these two statements being the power factor. In other words:

$$\text{Power Factor } (\cos \varnothing) = \frac{\text{True power (P)}}{\text{Apparent power (S)}}$$

$$= \frac{V \times I_p(\text{actual}) \times \cos \varnothing}{V \times I_p(\text{actual})} = \frac{\text{watts}}{\text{volt-amperes}}$$

To summarise, what we perceive to be the power of a circuit (the apparent power) can also be the true power, as long as we have a unity power factor (1.0). However, as long as we have a phase angle, then we have a difference between apparent power and reality (true power). This difference is the power factor (a value less than unity).

In reality we will use a wattmeter to measure the true power and a voltmeter and ammeter to measure the apparent power.

Remember

Pythagoras' Theorem
The square of the hypotenuse is equal to the sum of the square on the other two sides

The power triangle

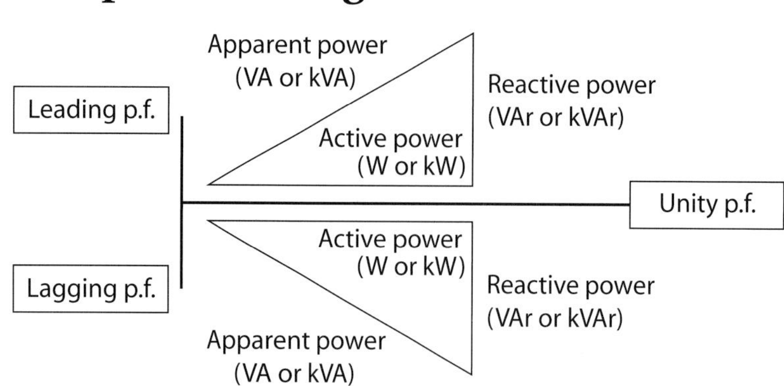

Figure 1.20

We can use Pythagoras' Theorem to help us calculate the different power components within a circuit.

We do so by using Pythagoras' formula as follows:

$$(VA)^2 = (W)^2 + (VAr)^2$$

This is then applied to Figure 1.20, where we have shown both inductive and capacitive conditions.

Example

A resistor of 15 Ω has been connected in series with a capacitor of reactance 30 Ω. If they are connected across a 230 V supply, establish both by calculation and by drawing a scaled power triangle, the following:

(a) the apparent power (b) the true power
(c) the reactive power (d) the power factor.

1.59kVA
1.41kVAr
63.2°
714kW

By calculation

In order to establish the elements of power, we must first find the current. To do this we need to find the impedance of the circuit. Therefore:

$$Z = \sqrt{R^2 + X_C^2} = \sqrt{15^2 + 30^2} = \sqrt{225 + 900} = \sqrt{1125}$$

Therefore $Z = 33.5\,\Omega$

$$I = \frac{V}{Z} = \frac{230}{35.5} = 6.9\text{ A}$$

$$\text{Pf(cos) Ø} = \frac{R}{Z} = \frac{15}{33.5} = 0.45$$

True power $= I \times V \times Pf$
$= 6.9 \times 230 \times 0.45$
$= 714.15\text{ W}$ or 0.714 kW

Apparent power $= I \times V$
$= 6.9 \times 230$
$= 1587\text{ VA}$ or 1.587 kVA

Reactive power $= \sin Ø \times$ apparent power

However, we do not know the value of sin Ø, we must therefore convert cos Ø to an angle and then find the sine of that angle.

cos Ø = 0.45 INV cos of 0.45 = 63.2

sine of 63.2° = 0.89

Therefore Reactive power = sin Ø × apparent power
= 0.89 × 1587
= 1412.43 VAr or 1.412 kVAr

Three-phase power

As we have seen in the previous section, we can find the power in a single-phase a.c. circuit by using the following formula:

$$\text{Power} = V \times I \times \cos Ø$$

Logically, you might assume that for three phase we could multiply this formula by three. Although this is not far from the truth we must remember that this could only apply where we have a balanced three-phase load. We can therefore say, that for any three-phase balanced load, the formula to establish power is:

$$\text{Power} = \sqrt{3} \times (V_L \times I_L \times \cos Ø)$$

However, in the case of an unbalanced load, we need to calculate the power for each separate section and then add them together to get total power.

This was all covered in Electrical Installations Book 1, Chapter 3 and you may like to refresh your memory before working through the three examples that follow.

Example 1

A balanced load of 10 Ω per phase is star connected and supplied with 400 V 50 Hz at unity power factor. Calculate the following:

a. Phase voltage

b. Line current

c. Total power consumed

a. Phase voltage

$$V_L = \sqrt{3} \times V_P$$

Therefore by transposition

$$V_P = \frac{V_L}{\sqrt{3}} = \frac{400}{1.732} = \mathbf{231\ V}$$

b. Line current

$$I_L = I_P$$

and therefore

$$I_P = \frac{V_P}{R_P} = \frac{231}{10} = \mathbf{23.1\ A}$$

c. Total power
In a balanced system

$$\text{Power} = \sqrt{3} \times V_L \times I_L \times \cos Ø$$
$$\text{Power} = 1.732 \times 400 \times 23.1 \times 1 = \mathbf{16\ kW}$$

Example 2

Three coils of resistance 40 Ω and inductive reactance 30 Ω are connected in delta to a 400 V 50 Hz three-phase supply. Calculate the following:

a. The current in each coil

b. Line current

c. Total power

a. Current in each coil
We must first find the impedance of each coil:

$$Z = \sqrt{R^2 + X_L^2} = \sqrt{40^2 + 30^2} = \sqrt{2500} = 50\ \Omega$$

The current in each coil (I_P) can then be found by applying Ohm's Law:

This gives $I_P = \frac{V}{Z} = \frac{400}{50} = \mathbf{8\ A}$

b. Line current
For a delta connected system

$$I_L = \sqrt{3} \times I_P$$

Therefore $I_L = 1.732 \times 8 = \mathbf{13.86\ A}$

c. Total power
We must first find the power factor using the formula:

$$\cos Ø = \frac{R}{Z}$$

This gives us $\cos Ø = \frac{40}{50} = 0.8$
And for a delta system $V_L = V_P$

Therefore we can now use the power formula of :

$$P = \sqrt{3} \times V_L \times I_L \times \cos Ø$$

$$P = 1.732 \times 400 \times 13.86 \times 0.8$$

P = **7682 W** or **7.682 kW**

Example 3

A small industrial estate is fed by a 400 V, three-phase, 4-wire TN-S system. On the estate there are three factories connected to the system as follows:

Factory A taking 50 kW at unity power factor

Factory B taking 80 kVA at 0.6 lagging power factor

Factory C taking 40 kVA at 0.7 leading power factor

Calculate the overall kW, kVA, kVar and power factor for the system.

To clarify, we are trying to find values of P (true power), S (apparent Power) and Q (reactive power). First, we need to work out the situations for each factory.

Factory A
We know that power factor

$$\cos Ø = \frac{\text{True power (P)}}{\text{Apparent power (S)}}$$

We also know that the power factor is 1.0 and that P = 50 kW

Therefore, by transposition:

$$S = \frac{P}{\cos Ø} = \frac{50}{1} = 50 \text{ kVA}$$

And with unity power factor for Factory A, Q = 0

Factory B
Using the same logic, we need to find true power and reactive power.

Therefore $P = \cos Ø \text{ x } S = 0.6 \times 80 \text{ kW} = 48 \text{ kW}$

Reactive component (Q) $= S \times \sin Ø = 80 \times 0.8 = 64$ kVAr

Factory C

$$P = S \times \cos Ø = 40 \text{ kW} \times 0.7 = 28 \text{ kW}$$

$$Q = S \times \sin Ø = 40 \times 0.714 = 28.6 \text{ kVAr}$$

We can now find the **total kW** by addition: 50 + 48 + 28 = **126 kW**

We can find the **total kVAr** as the difference between the reactive power components, the larger one, Factory B, lagging and the smaller one, Factory C, leading:

64 kVAr – 28.6 kVAr = **35.4 kVAr**

We can use Pythagoras' Theorem to find the **total kVA**

$$S = \sqrt{P^2 + Q^2} = \sqrt{126^2 + 35.4^2} = \mathbf{131\ kVA}$$

Consequently, the overall power factor will be:

$$\cos Ø = \frac{P}{S} = \frac{126}{131} = \mathbf{0.96\ lagging}$$

Transformers

The **transformer** is one of the most widely used pieces of electrical equipment and can be found in situations such as electricity distribution, construction work and electronic equipment. Its purpose, as the name implies, is to transform something – the something in this case being the voltage, which can enter the transformer at one level (input) and leave at another (output).

When the output voltage is higher than the input voltage we say that we have a step-up transformer and when the output voltage is lower than the input, we say that we have a step-down transformer.

In this section we will be looking at the following areas:

- mutual inductance
- transformer types
- step-up and step-down transformers.

Mutual inductance

In their operation, transformers make use of an action known as **mutual inductance**. We have looked at inductance in another section. We will revisit the subject here to help you try to understand the concept clearly. Let us look at the following situation. Two coils, primary and secondary, are placed side by side, but not touching each other. The primary coil has been connected to an a.c. supply and the secondary coil is connected to a load, such as a resistor.

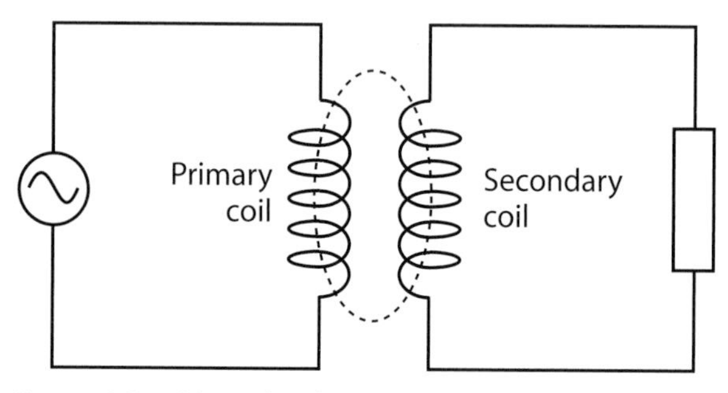

Figure 1.21 Mutual inductance

Did you know?

There is something called self-inductance. This is when a conductor that is formed into a coil, carries an alternating current, the magnetic field created around the coil grows and collapses with the changing current, this will induce into the coils an e.m.f. voltage that is always in opposition to the voltage that creates it

If we now allow current to flow in the primary coil, we know that current flow will create a magnetic field. Therefore, as the current in the primary coil increases up to its maximum value, it creates a changing magnetic flux and, as long as there is a changing magnetic flux, there will be an e.m.f. 'induced' into the secondary coil, which would then start flowing through the load.

This effect, where an alternating e.m.f. in one coil causes an alternating e.m.f. in another coil, is known as mutual inductance.

In transformers we are only really interested in mutual inductance, and it is the rising and falling a.c. current that causes the change of magnetic flux. In other words, **we really need an a.c. supply to allow transformers to operate correctly**.

If the two coils were now wound on an iron core, we would find that the level of magnetic flux is increased and consequently the level of mutual inductance is also increased.

Transformer types

Core-type transformers

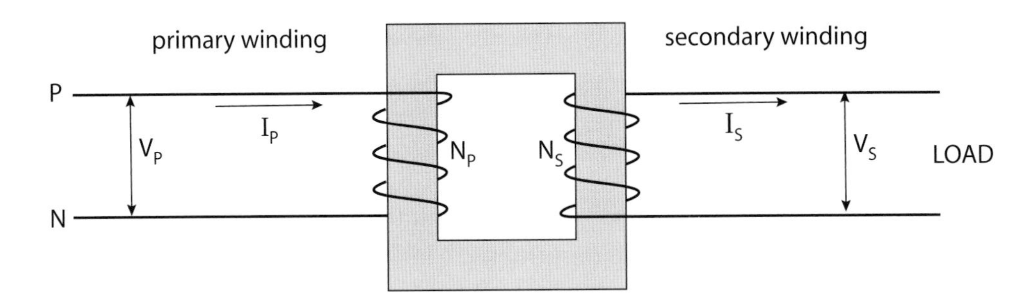

Figure 1.22 Double wound, core type transformer

In Figure 1.22, the supply is wound on one side of the iron core (primary winding) and the output is wound on the other (secondary winding). In other words, double wound means that there is more than one winding.

The number of turns in each winding will affect the induced e.m.f., with the number of turns in the primary being referred to as (Np), and those in the secondary referred to as (Ns). We call this the turns ratio. When voltage (Vp) is applied to the primary winding, it will cause a changing magnetic flux to circulate in the core. This changing flux will cause an e.m.f. (Vs) to be induced in the secondary winding.

Assuming that we have no losses or leakage (i.e. 100 per cent efficient), then power input will equal power output and the ratio between the primary and secondary sides of the transformer can be expressed as follows:

$$\frac{V_P}{V_S} = \frac{N_P}{N_S} = \frac{I_S}{I_P}$$

(where I_P represents the current in the primary winding and I_S the current in the secondary winding).

As we can see from Figure 1.22, a transformer has no moving parts. Consequently, provided that the following general statements apply, it ends up being a very efficient piece of equipment:

- Transformers use laminated (layered) steel cores, not solid metal. In a solid metal core 'eddy currents' are induced which cause heating and power losses.
- Laminated cores, where each lamination is insulated, help to reduce this effect.
- Soft iron with high magnetic properties is used for the core.
- Windings are made from insulated, low resistance conductors. This prevents short circuits occurring either within the windings, or to the core.

The losses that occur in transformers can normally be classed under the following categories.

Remember

No transformer can be 100 per cent efficient. There will always be power losses

Copper losses

> **Definition**
>
> **Hysteresis** – a generic term meaning a lag in the effect of a change of force

Although windings should be made from low resistance conductors, the resistance of the windings will cause the currents passing through them to create a heating effect and subsequent power loss. This power loss can be calculated using the formula:

$$P_C = I^2 \times R \text{ watts}$$

Iron losses

These losses take place in the magnetic core of the transformer. They are normally caused by eddy currents (small currents which circulate inside the laminated core of the transformer) and **hysteresis**. To demonstrate let's say that you push on some material and it bends. When you stop pushing, does it return to its original shape? If it doesn't the material is demonstrating hysteresis. Let's look at this in context.

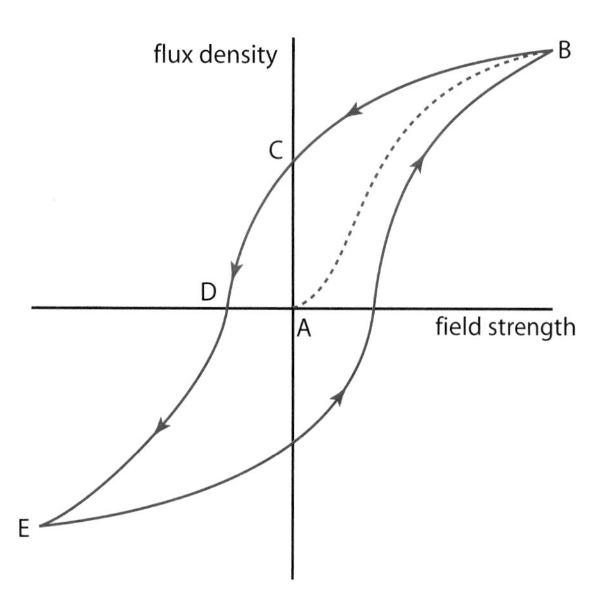

Figure 1.23 Hysteresis loop diagram

Figure 1.23 shows the effect within ferromagnetic materials of hysteresis. Starting with an unmagnetised material at point A; and here both field strength and flux density are zero. The field strength increases in the positive direction and the flux begins to grow along the dotted path until we reach saturation at point B. This is called the initial magnetisation curve.

If the field strength is now relaxed, instead of retracing the initial magnetisation curve, the flux falls more slowly. In fact, even when the applied field has returned to zero, there will still be a degree of flux density (known as the **remanence**) at point C. To force the flux to go back to zero (point D), we have to reverse the applied field. The field strength here that is necessary to drive the field back to zero is known as the **coercivity**. We can then continue reversing the field to get to point E, and so on. This is known as the hysteresis loop.

As we have already said, we can help reduce eddy currents by using a laminated core construction. We can also help to reduce hysteresis by adding silicon to the iron from which the transformer core is made.

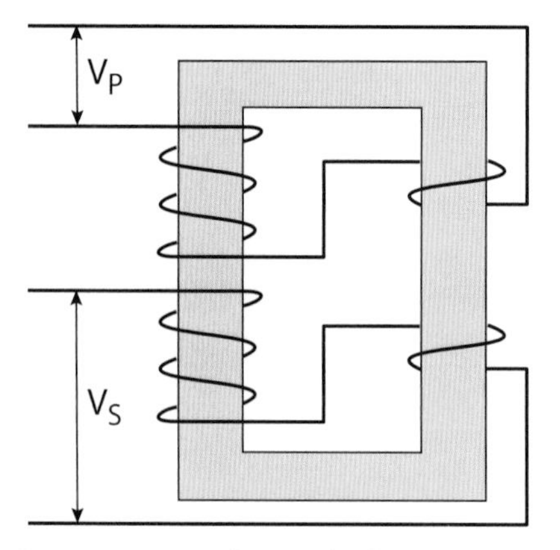

Figure 1.24 Reducing 'leakage'

The version of the double wound transformer that we have looked at so far makes the principle of operation easier to understand.

However, this arrangement is not very efficient, as some of the magnetic flux being produced by the primary winding will not react with the secondary winding and is often referred to as 'leakage'. We can help to reduce this leakage by splitting each winding across the sides of the core (see Figure 1.24).

Shell-type transformers

We can reduce the magnetic flux leakage a bit more, by using a shell-type transformer.

In the shell-type transformer, both windings are wound onto the central leg of the transformer and the two outer legs are then used to provide parallel paths for the magnetic flux.

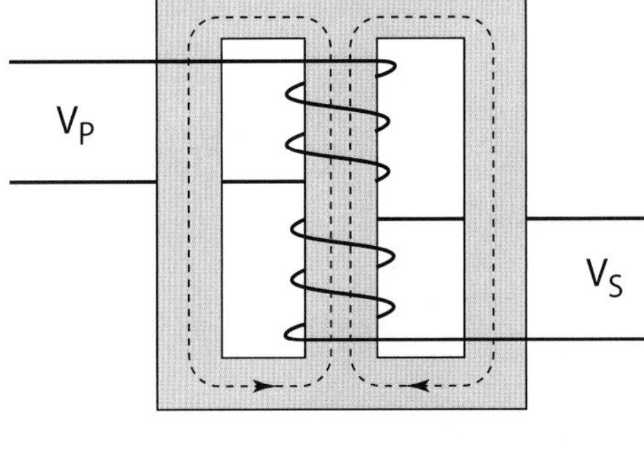

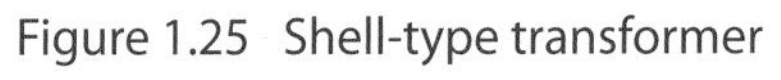

Figure 1.25 Shell-type transformer

The autotransformer

The autotransformer uses the principle of 'tapped' windings in its operation. Remember that the ratio of input voltage against output voltage will depend upon the number of primary winding turns and secondary winding turns (the turns ratio). But what if we want more than one output voltage?

Some devices are supplied with the capability of providing this, such as small transformers for calculators, musical instruments or doorbell systems. Tapped connections are the normal means by which this is achieved. A tapped winding means that we have made a connection to the winding and then brought this connection out to a terminal. Now, by connecting between the different terminals, we can control the number of turns that will appear in that winding and we can therefore provide a range of output voltages.

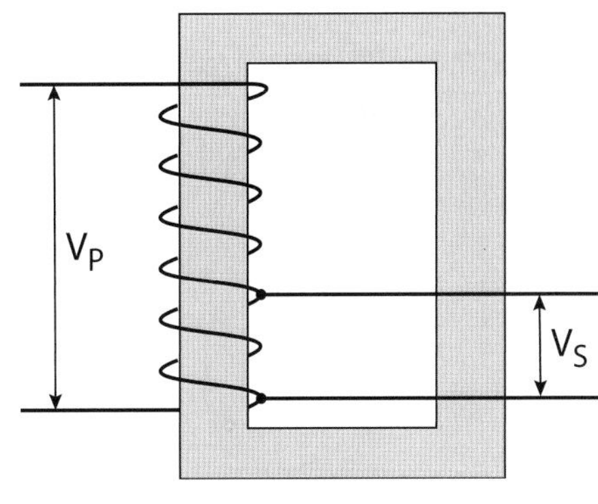

Figure 1.26 Autotransformer

An autotransformer has only one tapped winding and the position of the tapping on that winding will dictate the output voltage.

One of the advantages of the autotransformer is that, because it only has one winding, it is more economical to manufacture. However, on the down side, we have made a physical connection to the winding. Therefore, if the winding ever became broken between the two tapping points, then the transformer would not work and the input voltage would appear on the output terminals. This would then present a real hazard.

Instrument transformers

Instrument transformers are used in conjunction with measuring instruments because it would be very difficult and expensive to design normal instruments to measure the high current and voltage that we find in certain power systems. We therefore have two types of instrument transformer, both being double wound.

The current transformer

Current transformer

The current transformer (c.t.) normally has very few turns on its primary winding so that it does not affect the circuit to be measured, with the actual meter connected across the secondary winding.

Care must be taken when using a c.t. Never open the secondary winding while the primary is 'carrying' the main current. If this happened, a high voltage would be induced into the secondary winding. Apart from the obvious danger, this heat build up could cause the insulation on the c.t. to break down.

The voltage transformer

This is very similar to our standard power transformer, in that it is used to reduce the system voltage. The primary winding is connected across the voltage that we want to measure and the meter is connected across the secondary winding.

Step-up and step-down transformers

Step-up transformers

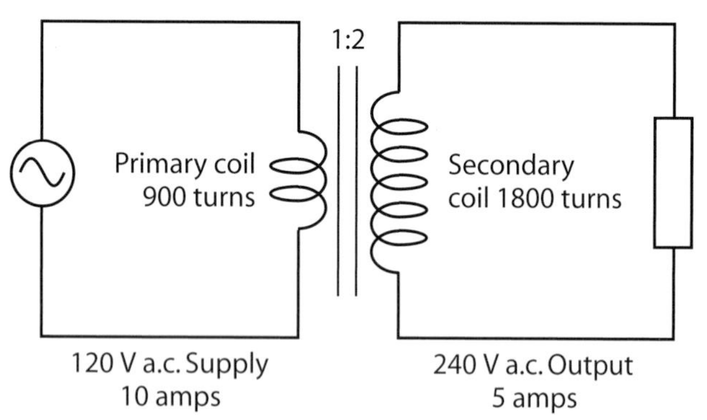

Figure 1.27 Step-up transformer

A step-up transformer is used when it is desirable to step voltage up in value.

The primary coil has fewer turns than the secondary coil. We already know that the number of turns in a transformer is given as a ratio. When the primary has fewer turns than the secondary, voltage and impedance are stepped up. In the circuit shown, voltage is stepped up from 120 V a.c. to 240 V a.c. Since impedance is also stepped up, current is stepped down from 10 amps to 5 amps.

Step-down transformers

2:1

Primary coil
1800 turns

Secondary
coil 900 turns

240 V a.c. Supply
5 amps

120 V a.c. Output
10 amps

Figure 1.28 Step-down transformer

A step-down transformer is used when it is desirable to step voltage down in value.

The primary coil has more turns than the secondary coil. The step-down ratio is 2:1. Since the voltage and impedance are stepped down, the current is stepped up in this case to 10 A.

Safety isolating transformer

Another use of a transformer is to isolate the secondary output from the supply. In a bathroom, the shower socket should be supplied from a 1:1 safety isolating transformer. The output of the transformer has no connection to earth thereby ensuring that output from the transformer is totally isolated from the supply.

In areas of installations where there is an increased risk of electric shock, the voltage is reduced to less than 50 V and is supplied from a safety isolating transformer; we know this as **Separate Extra Low Voltage** (**SELV**) supply.

Transformer ratings

Transformers are rated in kVA (kilovolt-amps). This rating is used rather than watts because loads are not purely resistive. Only resistive loads are measured in watts.

The kVA rating determines the current that a transformer can deliver to its load without overheating. Given volts and amps, kVA can be calculated. Given kVA and volts, amps can be calculated.

The kVA rating of a transformer is the same for both the primary and the secondary. At this point let us try some examples.

Example 1

A transformer having a turns ratio of 2:7 is connected to a 230 V supply. Calculate the output voltage.

When we give transformer ratios, we give them in the order primary then secondary. Therefore in this example we are saying that for every two windings on the primary winding, there are seven on the secondary. If we therefore use our formula:

$$\frac{V_P}{V_S} = \frac{N_P}{N_S}$$

We should now transpose this to get:

$$V_S = \frac{V_P \times N_S}{N_P}$$

However, we do not know the exact number of turns involved. But do we need to, if we know the ratio? Let us find out.

The ratio is 2:7, meaning for every 2 turns on the primary, there will be 7 turns on the secondary. Therefore, if we had 6 turns on the primary, this would give us 21 turns on the secondary, but the ratio of the two has not changed. It is still for every 2 on the primary we are getting 7 on the secondary.

This means we can just insert the ratio rather than the individual number of turns into our formula:

$$V_S = \frac{V_P \times N_S}{N_P} = \frac{230 \times 7}{2} = \mathbf{805\,V}$$

Now, to prove our point about ratios, let us say that we know the number of turns in the windings to be 6 in the primary and 21 in the secondary (which is still giving us a 2:7 ratio). If we now apply this to our formula we get:

$$V_S = \frac{V_P \times N_S}{N_P} = \frac{230 \times 21}{6} = \mathbf{805\,V}$$

The same answer.

Example 2

A single phase transformer, with 2000 primary turns and 500 secondary turns, is fed from a 230V a.c. supply. Find:

(a) the secondary voltage

(b) the volts per turn.

Secondary voltage

$$\frac{V_P}{V_S} = \frac{N_P}{N_S}$$

Using transposition, re-arrange the formula to give:

$$V_S = \frac{V_P \times N_S}{N_P}$$

$$Vs = \frac{230 \times 500}{2000} = \frac{115{,}000}{2000} = \mathbf{57.5\,V}$$

Volts per turn

This is the relationship between the volts in a winding and the number of turns in that winding. To find volts per turn, we simply divide the voltage by the number of turns.

Therefore, in the primary:

$$\frac{V_P}{N_P} = \frac{230}{2000} = \mathbf{0.115\ volts\ per\ turn}$$

In the secondary

$$\frac{V_S}{N_S} = \frac{57.5}{500} = \mathbf{0.115\ volts\ per\ turn}$$

Example 3

A single phase transformer is being used to supply a trace heating system. The transformer is fed from a 230 V 50 Hz a.c. supply and needs to provide an output voltage of 25 V. If the secondary current is 150 A and the secondary winding has 50 turns, find:

(a) the output kVA of the transformer
(b) the number of primary turns
(c) the primary current
(d) the volts per turn.

The output kVA

$$\text{kVA} = \frac{\text{volts} \times \text{amperes}}{1000} = \frac{V_S \times I_S}{1000} = \frac{25 \times 150}{1000} = \mathbf{3.75\ kVA}$$

The number of primary turns

If: $\frac{V_P}{V_S} = \frac{N_P}{N_S}$ then by transposition

$$N_P = \frac{V_P \times N_S}{V_S} = \frac{250 \times 50}{25} = \mathbf{460\ turns}$$

The primary current

If: $\frac{V_P}{V_S} = \frac{I_S}{I_P}$ then by transposition

$$I_P = \frac{V_S \times I_S}{V_P} = \frac{25 \times 150}{230} = \mathbf{16\ A}$$

The volts per turn

In the primary: $\frac{V_P}{N_P} = \frac{230}{460} = \mathbf{0.5\ volts\ per\ turn}$

In the secondary: $\frac{V_S}{N_S} = \frac{25}{50} = \mathbf{0.5\ volts\ per\ turn}$

Example 4

A step-down transformer, having a ratio of 2:1, has an 800 turn primary winding and is fed from a 400 V a.c. supply. The output from the secondary is 200 V and this feeds a load of 20 Ω resistance. Calculate:

(a) the power in the primary winding

(b) the power in the secondary winding.

Well, we know that the formula for power is: $P = V \times I$

And we also know that we can use Ohm's Law to find current: $I = \frac{V}{R}$

Therefore, if we insert the values that we have, we can establish the current in the secondary winding:

$$I_S = \frac{V_S}{R_S} = \frac{200}{20} = \mathbf{10\,A}$$

Now that we know the current in the secondary winding, we can use the power formula to find the power generated in the secondary winding:

$$P = V \times I = 200 \times 10 = 2000 \text{ watts} = \mathbf{2\,kW}$$

We now need to find the current in the primary winding. To do this we can use the formula:

$$\frac{V_P}{V_S} = \frac{I_S}{I_P}$$

However, we need to transpose the formula to find I_P. This would give us:

$$I_P = \frac{I_S \times V_S}{V_P}$$

Which, if we now insert the known values to, gives us:

$$I_P = \frac{I_S \times V_S}{V_P} = \frac{10 \times 200}{400} = \frac{2000}{400} = \mathbf{5\,A}$$

Now that we know the current in the primary winding, we can again use the power formula to find the power generated in the secondary winding:

$$P = V \times I = 400 \times 5 = 2000 \text{ watts} = \mathbf{2\,kW}$$

Instruments and measurement

As electricians we are often responsible for the measurement of different electrical quantities. Some must be measured as part of the inspection and testing of an installation (e.g. Insulation Resistance). However, of the remainder, the most common quantities that we are likely to come across are shown in Table 1.1.

Property	Instrument
Current	Ammeter
Voltage	Voltmeter
Resistance	Ohmmeter
Power	Wattmeter

Table 1.1 Electrical quantities

In this section we will be looking at the following areas:

- measuring current (ammeter)
- measuring voltage (voltmeter)
- measuring resistance (ohmmeter)
- using an ammeter and voltmeter
- loading errors
- meter displays (digital and analogue).

We will look at the different instruments that we use to measure these quantities in the following section.

But before we ever use a meter, we must always ask ourselves:

- Have we chosen the correct instrument?
- Is it working correctly?
- Has it been set to the correct scale?
- How should it be connected?

Measuring current (ammeter)

An ammeter measures current by series connections. Although we know that we can use a multi-meter on site, the actual name for an instrument that measures current is an ammeter. Ammeters are connected in series so that the current to be measured passes through them. The circuit diagram in Figure 1.29 illustrates this. Consequently, they need to have a very low resistance or they would give a false reading.

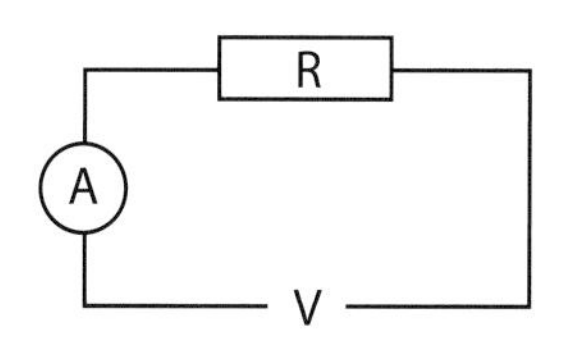

Figure 1.29 Ammeter in circuit

If we now look at the same circuit, but use a correctly set multi-meter, the connections would look as in Figure 1.30.

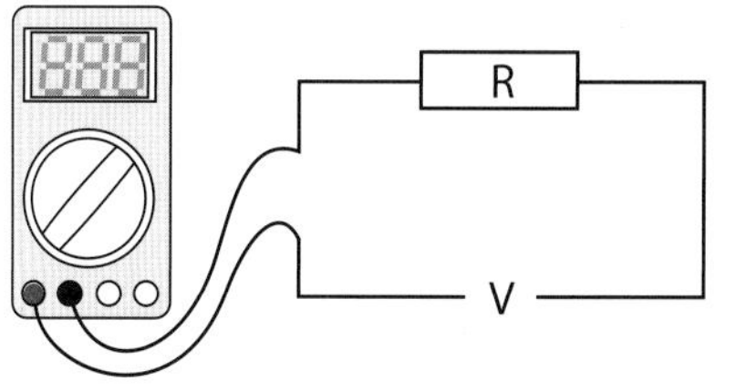

Figure 1.30 Multi-meter in circuit

Measuring voltage (voltmeter)

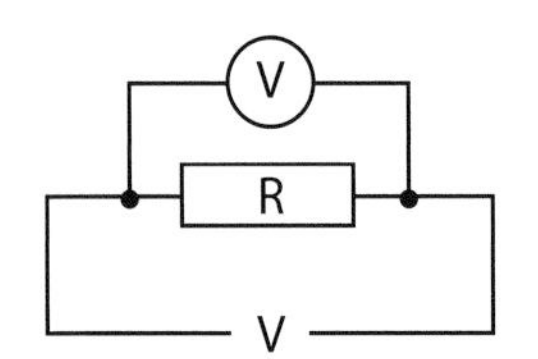

Figure 1.31 Voltmeter in circuit

On site or for general purposes, we can use a multi-meter to measure voltage, but the actual device used for measuring voltage is called a voltmeter. It measures the potential difference between two points (for instance, across the two connections of a resistor). The voltmeter must be connected in parallel across the load or circuit to be measured as shown in Figure 1.31.

If we now look at the same circuit, but use our correctly set multi-meter, the connections would look as in Figure 1.32.

The internal resistance of a voltmeter must be very high if we wish to get accurate readings.

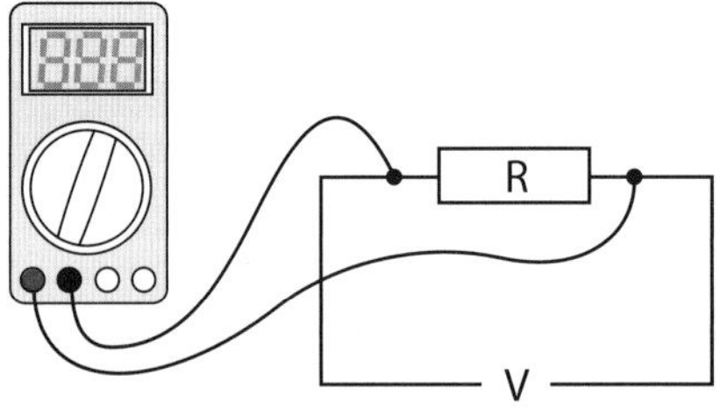

Figure 1.32 Multi-meter measuring voltage

Measuring resistance (ohmmeter)

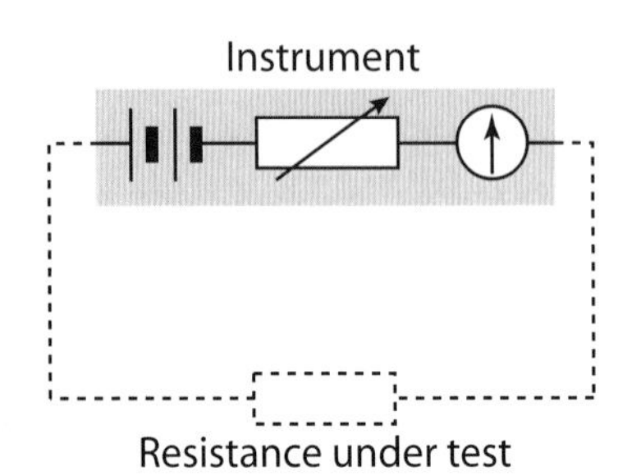

Figure 1.33 Ohmmeter

There are many ways in which we can measure resistance. However, in the majority of cases we do so by executing a calculation based upon instrument readings (ammeter/voltmeter) or from other known resistance values. Once again we can use our multi-meter to perform the task, but we refer to the actual meter as being an ohmmeter.

The principle of operation involves the meter having its own internal supply (battery). The current, which then flows through the meter, must be dependent upon the value of the resistance under scrutiny.

However, before we start our measurement, we must ensure that the supply is off and then connect both leads of the meter together and adjust the meter's variable resistor until full-scale deflection (zero) is reached.

The dotted line in Figure 1.33 indicates the circuit under test. All other components are internal to the meter.

Using an ammeter and voltmeter

If we consider the application of Ohm's Law, we know that we can establish resistance by the following formula:

$$R = \frac{V \text{ (voltmeter)}}{I \text{ (ammeter)}}$$

If we therefore connect an ammeter and voltmeter in the circuit, then by using a known supply (battery) we can apply the above formula to the readings that we get.

In practice, we invariably find that we need a long 'wandering' test lead to perform such a test. Where this is the case, we must remember to deduct the value of the test lead from our circuit resistance.

Loading errors

Connecting a measuring instrument to a circuit will invariably have an effect on that circuit. In the case of the average electrical installation, this effect can largely be ignored. But if we were to measure electronic circuits, the effect may be drastic enough to actually destroy the circuit.

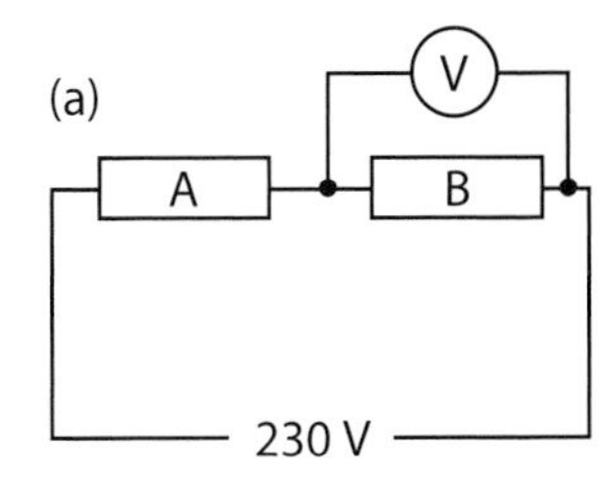

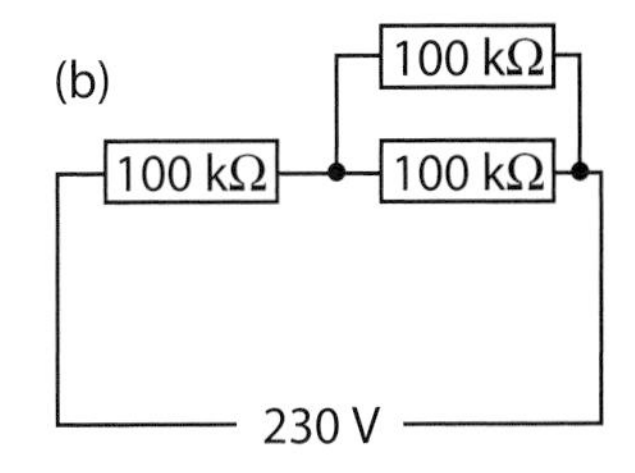

Figure 1.34 Loading errors

We normally refer to these errors introduced by the measuring instrument as loading errors, and below we will look at an example of this.

Let's assume in (a) that we have been asked to measure the voltage across resistor B in the circuit, where both items have a resistance of 100 kΩ. Note that in this arrangement, the potential difference (p.d.) across each resistor will be 115V.

Our instrument, a voltmeter, has an internal resistance of 100 kΩ and would therefore be connected as in the diagram.

In effect, by connecting our meter in this way, we are introducing an extra resistance in parallel into the circuit. We could, therefore, now draw our circuit as in Figure 1.34(b).

Now, if we find the equivalent resistance of the parallel branches:

$$\frac{1}{R_t} = \frac{1}{R_1} + \frac{1}{R_2} = \frac{1+1}{100} = \frac{2}{100}\text{ k}\Omega$$

Therefore:

$$R_t = \frac{100}{2} = 50\text{ k}\Omega$$

And then if we treat the parallel combination as being in series with 100 kΩ resistor A then total resistance:

$$R = R_A + R_t = 100\text{ k}\Omega + 50\text{ k}\Omega = 150\text{ k}\Omega$$

If you applied Ohm's Law again, we would find we now have a current of 0.00153 A (1.53 mA) flowing in the circuit. This in turn would give a p.d. across each resistor of 76 V and not the 115 V we would expect to see.

The problem is caused by the amount of current that is flowing through the meter, and this is normally solved by using meters with a very high resistance (as resistance restricts current). So you now see why we said earlier that voltmeters must have a high internal resistance in order to be accurate.

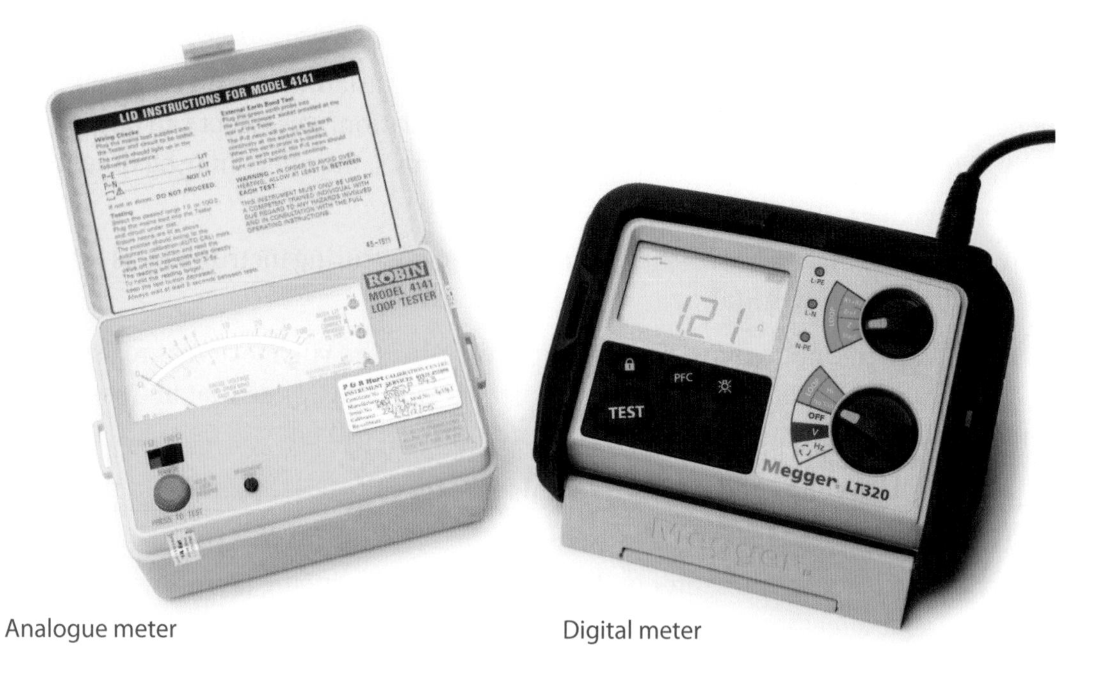

Analogue meter Digital meter

Meter displays

There are two types of display, **analogue** and **digital**. Analogue meters have a needle moving around a calibrated scale, whereas digital tend to show results, as numeric values, via a liquid crystal or LED display.

For many years electricians used the analogue meter and indeed it is still common for continuity/insulation resistance testers to be of the analogue style. However, their use is slowly being replaced by the digital meter. Most modern digital meters are based on semiconductors and consequently have very high impedance, making them ideal for accurate readings and good for use with electronic circuits.

Generally speaking, as they have no moving parts they are also more suited to rugged site conditions than the analogue style.

What sort of meter will I need?

Although instruments exist to measure individual electrical quantities, most electricians will find that a cost effective solution is to use a digital multi-meter similar to the one pictured. Meters like this are generally capable of measuring current, resistance and voltage, both in a.c. and d.c. circuits and across a wide range of values.

Is the meter working correctly?

It is important to check that your meter is functioning correctly and physically fit for the purpose. There are commercially available proving units that will help. However, on site it is more normal to use a known supply as the means of proving voltage measurement and the shorting/separating of leads to prove operation.

Measuring power

In a d.c. circuit it is possible to measure the power supplied by using a voltmeter and an ammeter. Then by using the formula **P = V × I** we can arrive at the power in Watts.

However, in an a.c. circuit, this method only produces the apparent power, a figure that will not be accurate unless we have unity power factor. This is because components such as capacitors and inductors will cause the current to lead or lag the voltage.

For a single-phase a.c. circuit we therefore use the formula:

P = V × I × cos Ø

This indicates how we would connect measuring instruments to establish power. For a single-phase resistive load, we can use the ammeter and voltmeter, but for a circuit containing capacitance or inductance we must use a wattmeter.

Figure 1.35 indicates how a wattmeter is connected into a single-phase circuit. Just like Ohm's Law, where to calculate power we need to know the values of voltage and current, a wattmeter also requires values of voltage and current.

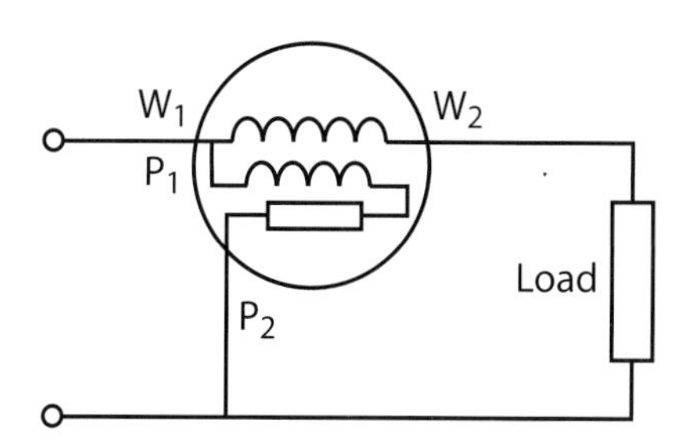

Figure 1.35 Wattmeter connected to load

The instrument shown in Figure 1.35 has a current coil, indicated between W_1 and W_2, that is wired in series with the load, and a voltage coil, shown between P_1 and P_2, wired in parallel across the supply.

This can be used to measure the power in a single-phase circuit or in a balanced three-phase load, where the total power will be equal to three times the value measured on the meter.

Measuring power in a three-phase, four-wire balanced load – one-meter method

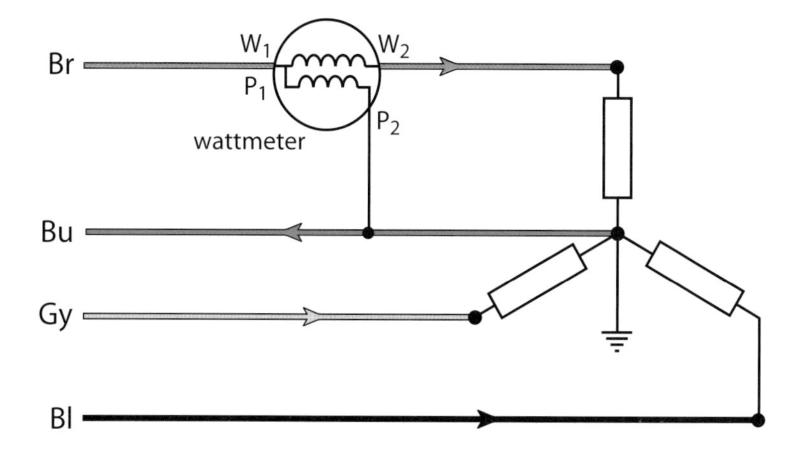

Figure 1.36 One-meter method

Measuring power in a three-phase balanced load – two-wattmeter method

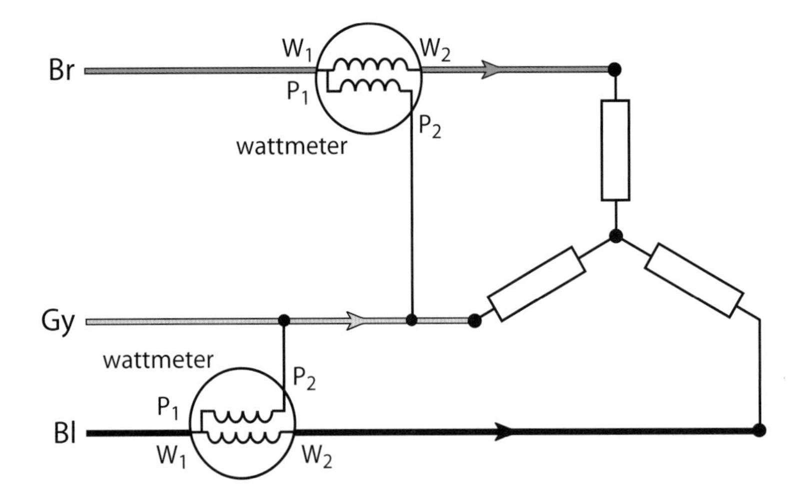

Figure 1.37 Two-meter method

In the two-wattmeter method, the total power is found by adding the two values together. At unity power factor the instruments will read the same and be half of the total load. For other power factors the instrument readings will be different, the difference in the reading could then be used to calculate the power factor.

Measuring power in an unbalanced three-phase circuit – three-wattmeter method

Here the total power will be the sum total of the readings on the three-wattmeters.

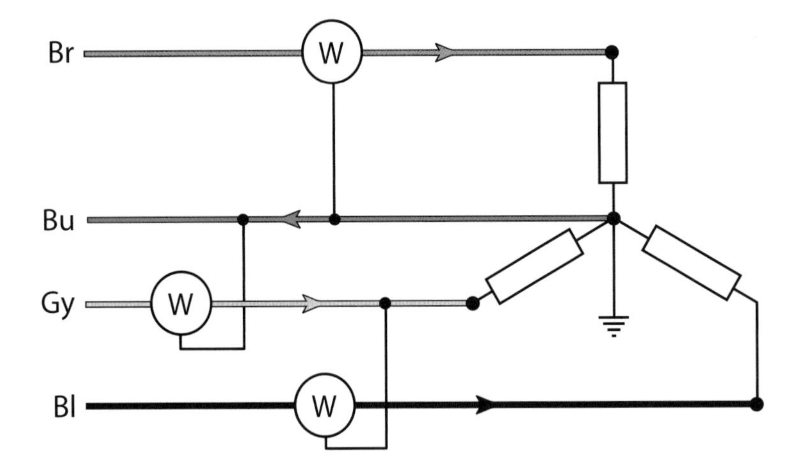

Figure 1.38 Three-phase circuit

This allows the power to be measured in a situation where the load is unbalanced, such as the three-phase supply to a large building where it is impossible to balance the load completely.

Measuring power factor

To measure power factor, there are a number of purpose made instruments available. All of these meters include both voltage and current measurement in circuits. Most designs are based on the clamp meter idea where the meter clamps around the current-carrying conductor. However, additional leads are then required to connect the meter across the supply. Many meters are also combined with the ability to measure all aspects of power, i.e. kW, kVA and kVAr.

The circuit diagram illustrates the connections of a power factor measuring instrument.

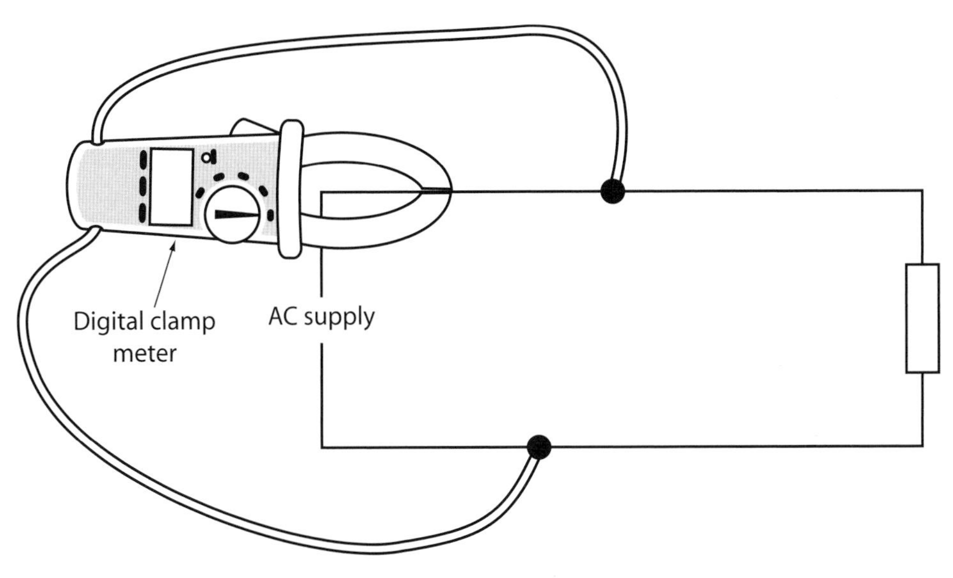

Figure 1.39 Circuit diagram with digital clamp meter

There is also an alternative method. The calculation to establish power factor is to divide true power by the apparent power (VA). As we know that a wattmeter will give us the true power, by adding an ammeter and a voltmeter to our circuit we can establish the apparent power (VA) and therefore establish the power factor. Figure 1.40 shows this arrangement.

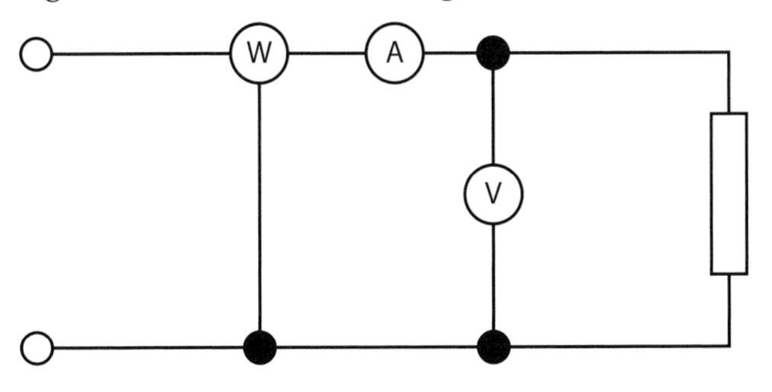

Figure 1.40 Circuit with wattmeter, ammeter and voltmeter connected

Knowledge check

1. A single-phase transformer with 2000 primary turns and 500 secondary turns is fed from a 230 volt a.c. supply. What will the secondary voltage and volts per turn in the secondary be?
2. What is the highest value in a sinusoidal wave form?
3. To calculate the average value of a sinusoidal voltage or current we must multiply the maximum value by what?
4. What is the correct formula for inductive reactance?
5. What is the correct formula for calculating impedance in a circuit containing resistance and inductance in series?
6. Power factor can be expressed as what?
7. Define reactive current.
8. Draw and label the power triangle.
9. On what principle to transformers work?
10. Power factor is the ratio of true power against what?
11. What is a c.t. used to measure?
12. From what is SELV derived?

Chapter 2

Earthing and protection

OVERVIEW

The purpose of earthing is to connect together all metalwork (other than that which is intended to carry current) to earth, so that dangerous potential differences cannot exist either between different metal parts, or between metal parts and earth. The definition given in BS 7671 for earthing is 'the act of connecting the exposed conductive parts of an installation to the main earthing terminal of that installation'.

By using the correct earthing procedures (and correct lightning protection measures), danger to life and the risk of fire to property can be greatly reduced.

This chapter will look at the following areas:

- **Purpose of earthing**
- **Earthing systems**
- **Lightning protection**
- **Electric shock**
- **Automatic disconnection of supply**
- **Residual current devices**
- **The earth-fault loop path**
- **Overcurrent protection**
- **Cable selection**

Purpose of earthing

By connecting to earth all metalwork not intended to carry current, a path is provided for leakage current which can be detected and interrupted by fuses, circuit breakers and residual current devices. Figure 2.1 illustrates the earth return path from the consumer's earth to the supply earth.

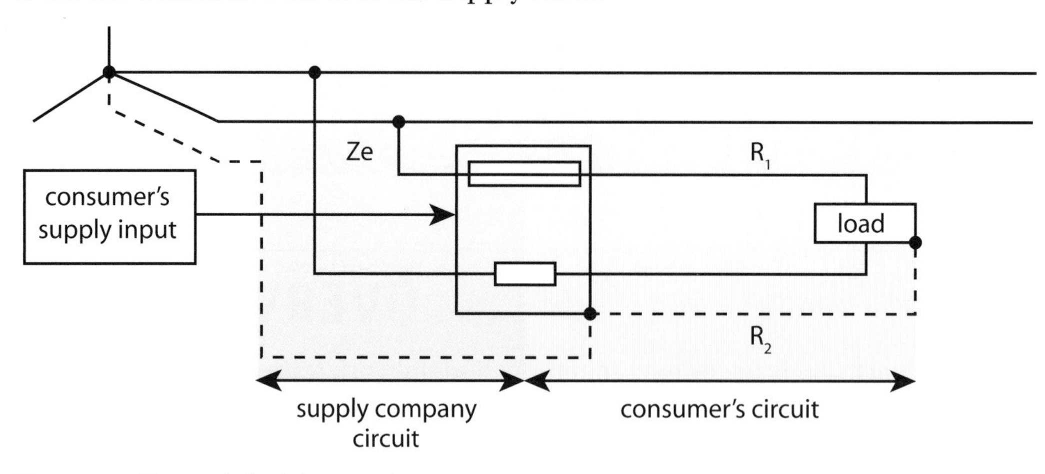

Figure 2.1 The earth-fault loop path

The connection at the consumer's earth can be by means of either an earth electrode at the building where the earth is required or may be in the form of a cable which runs back to the generator or transformer and is then connected to an earth point.

Because the transformer or generator at the point of supply always has an earth point, a circuit is formed when earth-fault currents are flowing. If these fault currents are large enough they will operate the protective device, thereby isolating the circuit. The star point of the secondary winding in a three-phase four-wire distribution transformer is connected to the earth to maintain the neutral at earth potential.

Results of an unearthed appliance

A person touching the appliance shown, which is live due to a fault, completes the earth circuit and receives an electric shock.

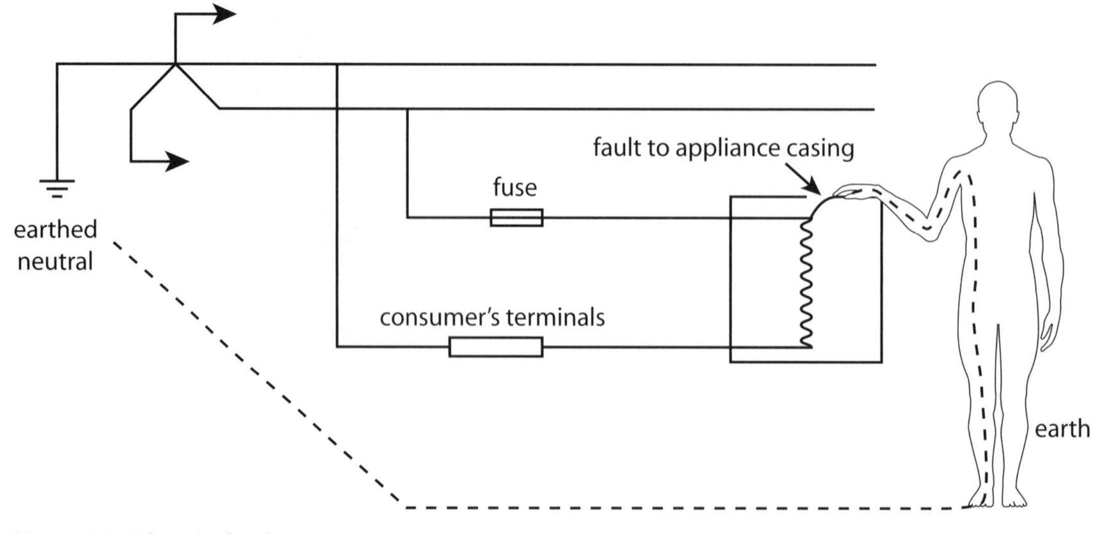

Figure 2.2 Electric shock

Results of a bad earth

A bad earth circuit, that is one with too large a resistance, can sometimes have more disastrous effects than having no earth at all. This is shown in the illustration, where the earth-fault circuit has a high resistance mainly due to a bad contact at point A.

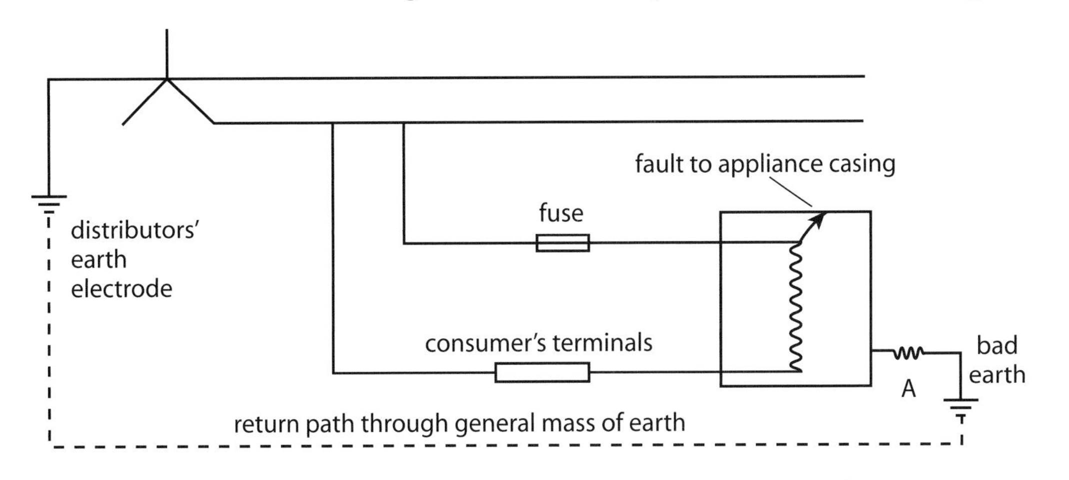

Figure 2.3 Bad earth path

The severity of shock will depend mainly upon the surroundings, the condition of the person receiving the shock and the type of supply. When the current starts to flow, the high resistance connection will heat up and this could be a fire hazard. Also, because the current flowing may not be high enough to blow the fuse or trip the circuit breaker, the appliance casing remains live.

Results of a good earth path

A good earth path, that is a low resistance one, will allow a high current to flow. This will cause the protective device to operate quickly, thereby isolating the circuit and giving protection against electric shock.

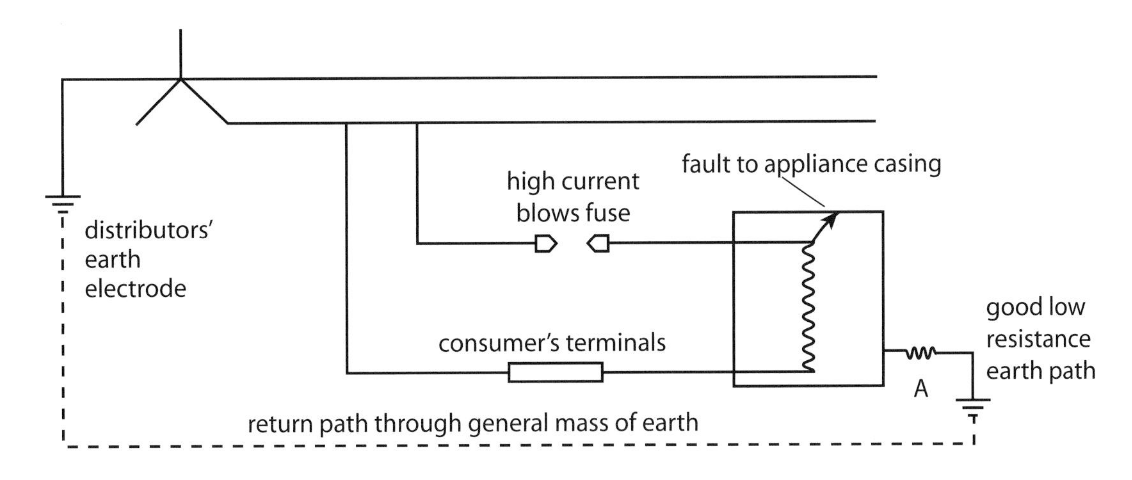

Figure 2.4 Good earth path

Earth-fault loop impedance

The path made or followed by the earth fault current is called the earth-fault loop or phase-earth loop. It is termed **impedance** because part of the circuit is the transformer or generator winding, which is inductive. This inductance, along with the resistance of the cables to and from the fault, makes up the impedance.

1. The circuit protective conductor (cpc) within the installation
2. The consumer's earthing terminal and earthing conductor
3. The earth return path, which can be either by means of an electrode or via the cable armouring
4. The path through the earthed neutral point of the transformer and the transformer winding (or generator winding)
5. The line conductor.

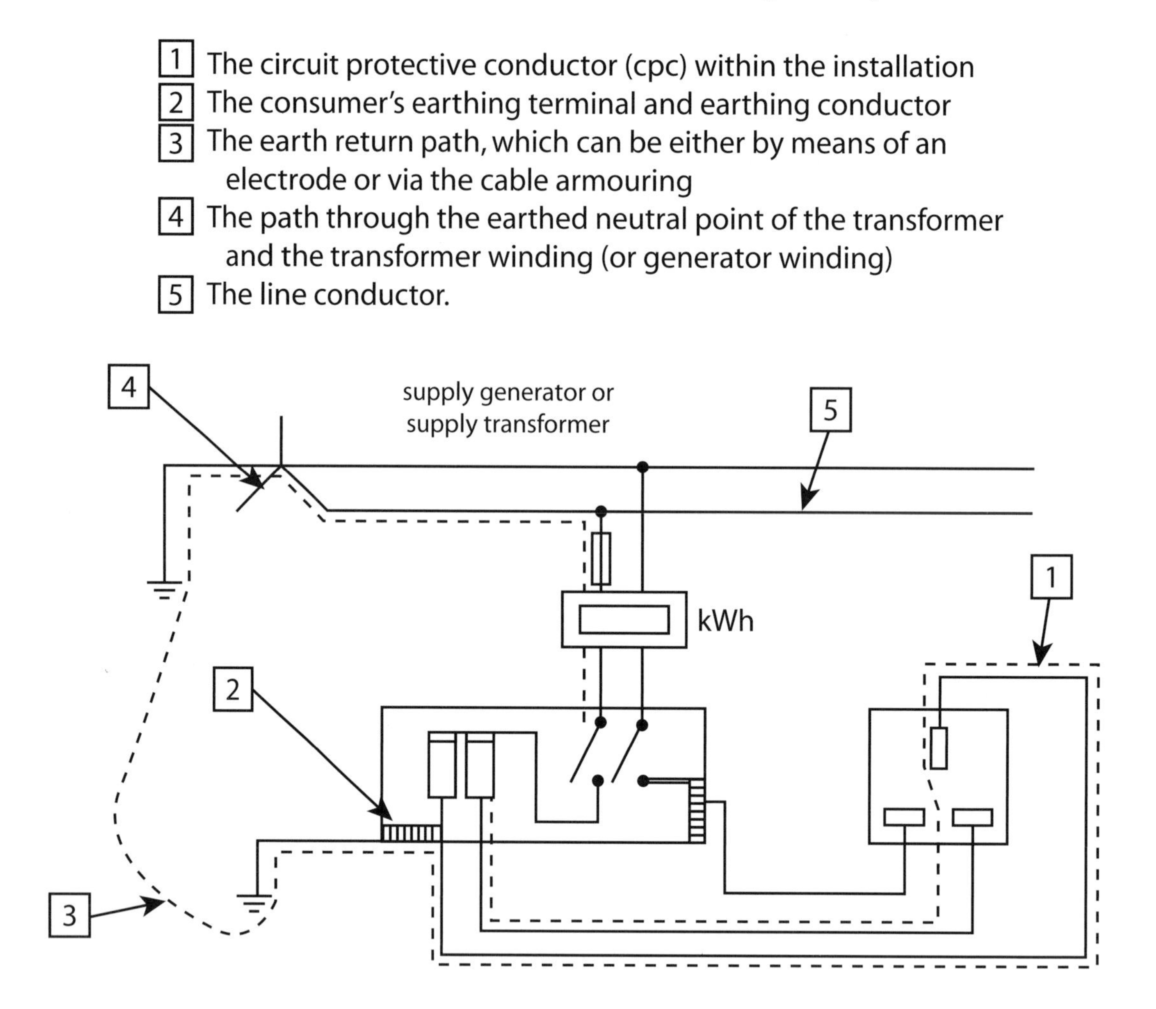

Figure 2.5 Earth-fault loop impedance

Earthing systems

TT system

The letter T when used in this way of describing wiring systems comes from the Latin word 'terra', which means earth. So the first letter T in this instance means that the supply is connected directly to earth at the source. This could be the generator, or transformer at one or more points.

The second letter T means that the exposed metalwork of the installation is connected to the earth by a separate earth electrode. The only connection between these two points is the general mass of earth (soil etc.) as shown.

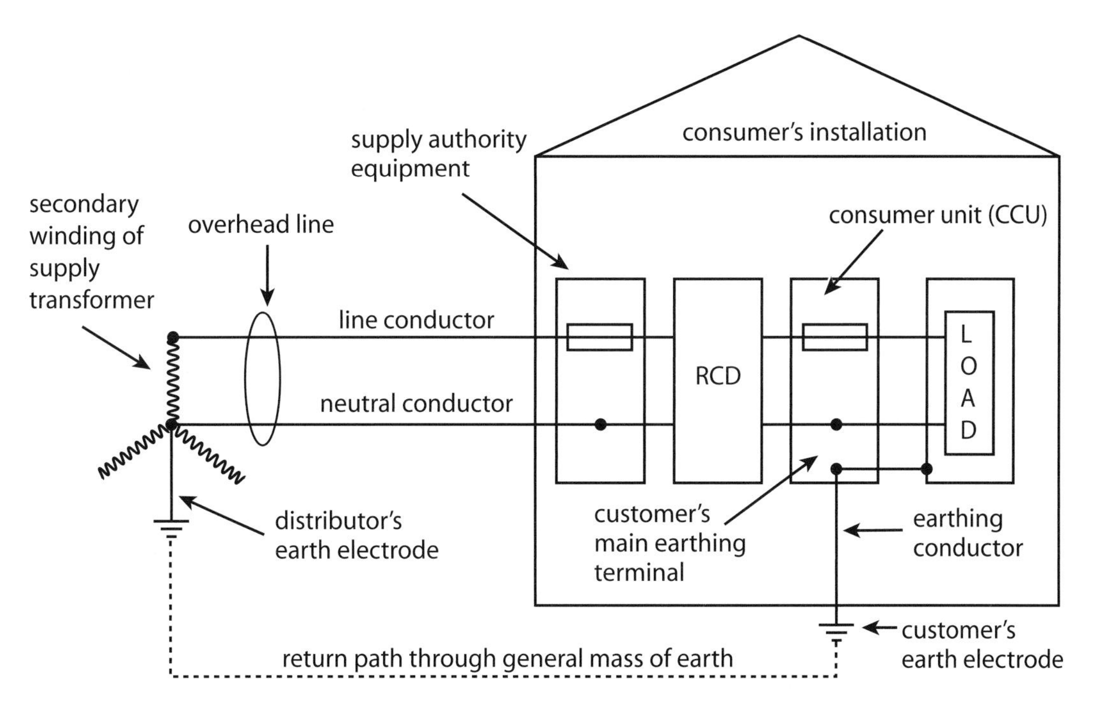

Figure 2.6 Earthing with customer's earth electrode and ground earth return path

When a fault to earth occurs on this system, the earth-fault current will flow around this circuit from the fault, through the earth and transformer windings, along the line conductor and back to the fault position.

In this system the earth may have a very high value of impedance, for example 2300 ohms. From Ohm's Law we can see that the current flow would be:

$$I = \frac{V}{R} = \frac{230}{2300} = 0.1\ \text{A}$$

This current flow of 0.1 A, that is 100 mA, is sufficient to be fatal to people, so protection against even very small currents is vital to prevent danger from electric shock. Regulation 411.5.2 states that a residual current device (RCD) or an overcurrent protective device or both should protect a TT system, with the RCD being preferred. Figure 2.7 illustrates the earth connection.

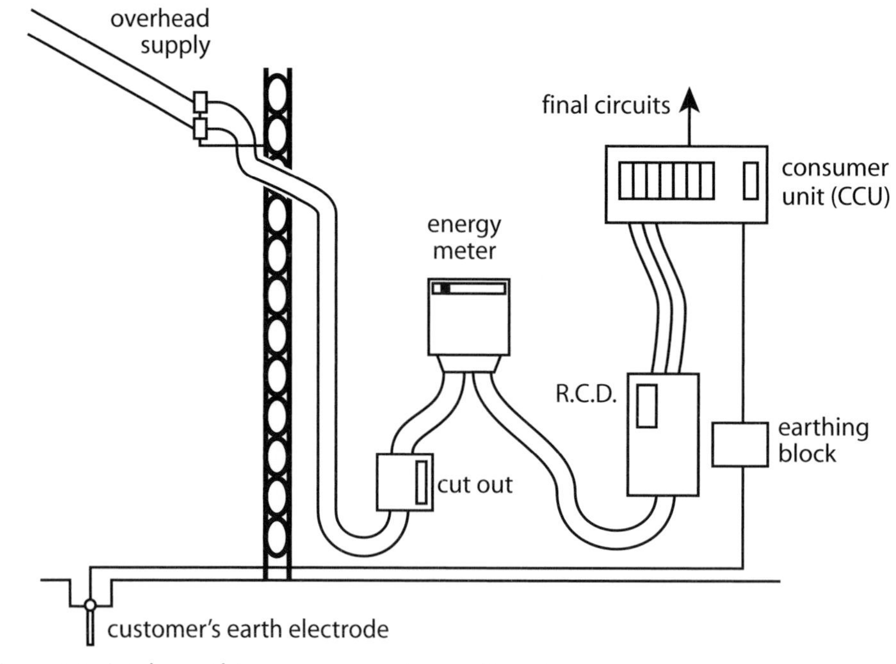

Figure 2.7 Intake earthing arrangements on a TT system

TN-S system

- T means that the supply is connected directly to the earth at one or more points.
- N means the exposed metalwork of the installation is connected directly to the earthing point of the supply.
- S means a separate conductor is used throughout the system from the supply transformer all the way to the final circuit to provide the earth connection.

Figure 2.8 illustrates this system.

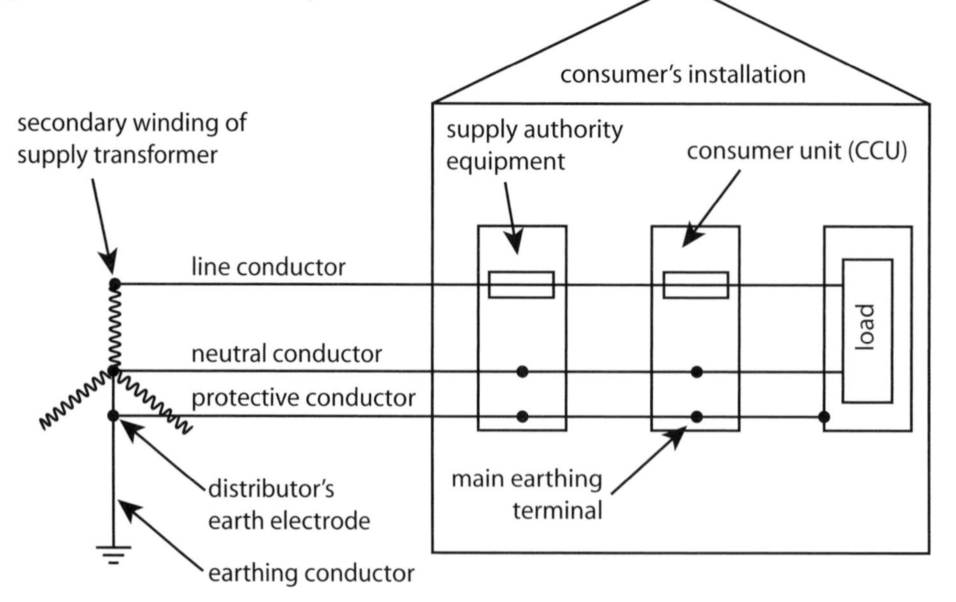

Figure 2.8 TN-S earthing system with metallic earth return path

This earth connection is usually through the sheath or armouring of the supply cable and then by a separate conductor within the installation. As a conductor is used throughout the whole system to provide a return path for the earth-fault current, the return path should have a low value of impedance. Figure 2.9 illustrates the intake earthing arrangements for a TN-S system.

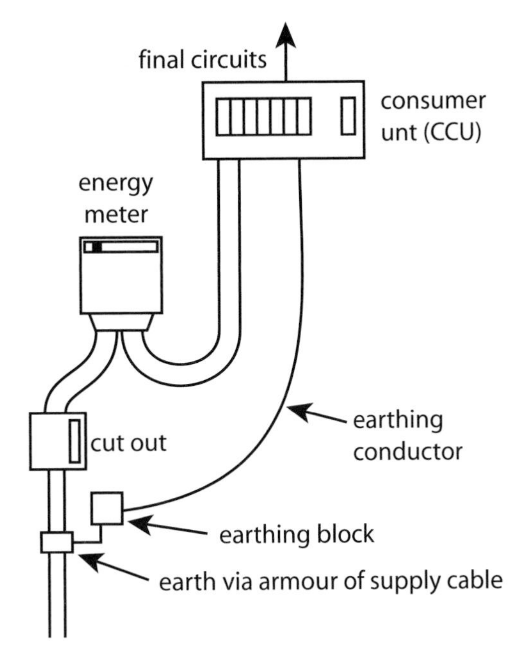

Figure 2.9 TN-S earth connection

TN-C-S system

- T means the supply is connected directly to earth at one or more points.
- N means the exposed metalwork of the installation is connected directly to the earthing point of the supply.
- C means that for some part of the system (generally in the supply section) the functions of neutral conductor and earth conductor are combined in a single common conductor.
- S means that for some part of the system generally in the installation, the functions of neutral and earth are performed by separate conductors.

Figures 2.10 and 2.11 illustrate this system.

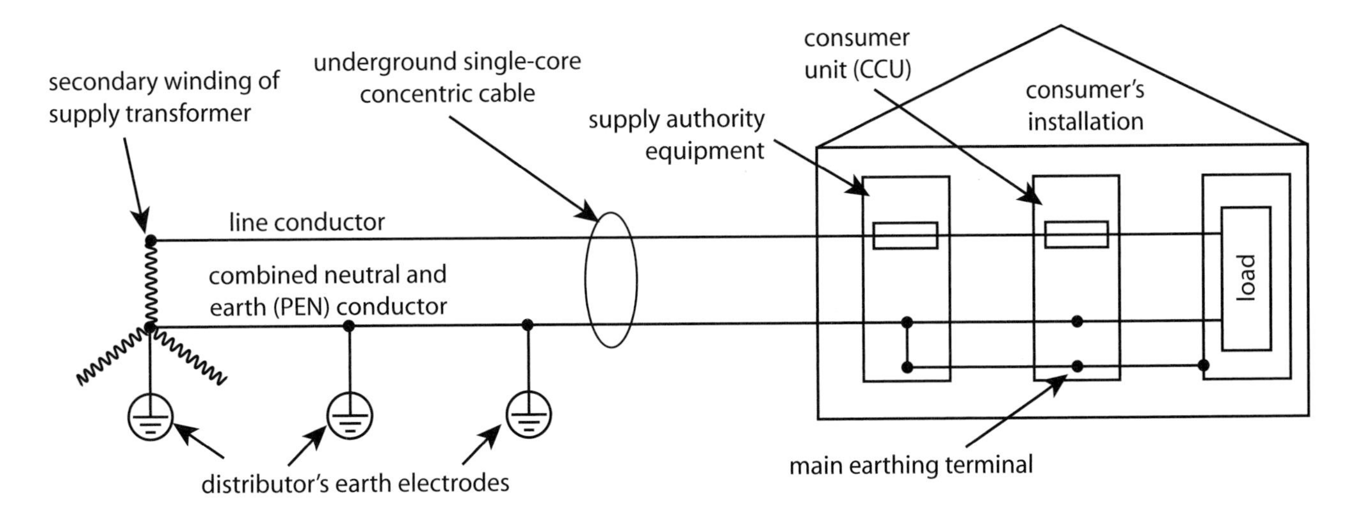

Figure 2.10 TN-C-S, protective multiple earthing (PME) system

In a TN-C-S system, the supply uses a common conductor for both the neutral and the earth. This combined conductor is commonly known as the Protective Earthed Neutral (PEN) or also sometimes as the Combined Neutral & Earth (CNE) conductor.

In such a system the supply PEN is required to be earthed at several points. This type of system is also known as Protective Multiple Earthing (PME) and this will be discussed more fully in the next section.

However, commonly, as shown on page 49, this effectively means that the distribution system is TN-C and the consumer's installation is TN-S; this combination therefore giving us a TN-C-S system.

The intake earthing arrangements for a typical TN-C-S system are shown in Figure 2.11.

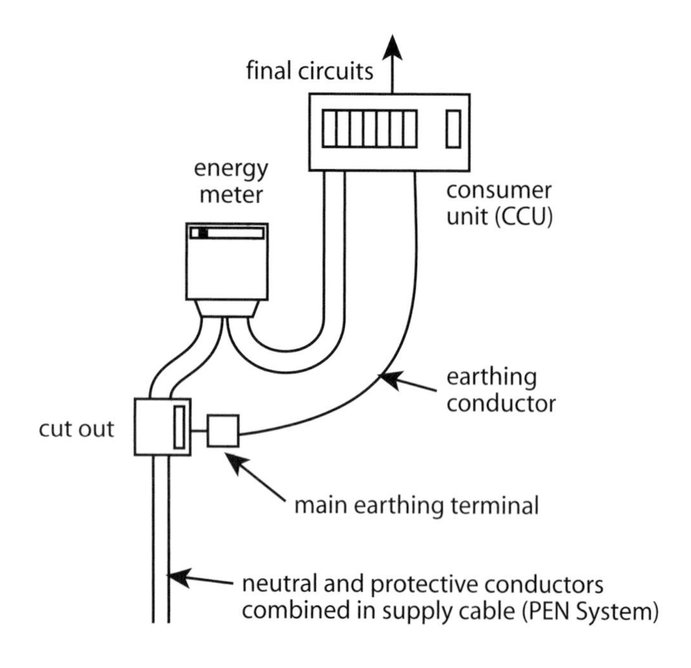

Figure 2.11 TN-C-S intake earthing arrangements

Protective Multiple Earthing (PME)

As shown in the previous section, this system of earthing is used on TN-C-S systems. PME is an extremely reliable system of earthing and is becoming the most commonly used distribution system in the UK today.

With the PME system, the neutral of the incoming supply is used as the earth point and all cpc's connect all metal work, in the installation to be protected, to the consumer's earth terminal.

Consequently, all line to earth faults are converted into line to neutral faults, which ensures that under fault conditions a heavier current will flow, thus operating protective devices rapidly.

However this increase in fault current may produce two hazards:

1. The increased fault current results in an enhanced fire risk during the time the protective device takes to operate.
2. If the neutral conductor ever rose to a dangerous potential relative to earth, then the resultant shock risk would extend to all the protected metalwork on every installation that is connected to this particular supply distribution network.

Because of these possible hazards, certain conditions are laid down before a PME system is used. These include the following:

- PME can only be installed by the supply company if the supply system and the installations it will feed meet certain requirements.
- The neutral conductor must be earthed at a number of points along its length. It is this action that gives rise to the name 'multiple' earthing.
- The neutral conductor must have no fuse or link etc. that can break the neutral path.
- Where PME conditions apply, the main equipotential bonding conductor shall be selected in accordance with the neutral conductor of the supply and Table 54.8 of BS 7671.

Other systems

TN-C system

In this system the neutral and protective functions are combined in a single conductor throughout the system. This system is relatively uncommon and its use is restricted for specific situations. There must be no metallic connection between this system and supply company equipment.

IT system

This system **must not** be connected to the supply company's system. It is a special system used in quarries, telephone exchanges and some industrial processes etc. The consumers must provide their own connection to earth.

FAQ

Q **Why are there different supply systems?**

A This is just recognition that in the past, before nationalisation of power supplies, there were independent suppliers of electricity, each with differing earthing arrangements to meet the requirements of their particular networks. TN-S was the most common but there were TN-C and TN-C-S systems as well. In the main all new-build projects are TN-C-S and as area supply systems are updated they get converted to TN-C-S. Remote and rural areas are supplied as TT systems as on overhead lines but more and more of these are being converted to TN-C-S. Whether there will ever be a time when all the supply systems are the same is uncertain.

Lightning protection

Lightning conductor connection

The protection of buildings and structures against lightning involves connecting various conductive parts to earth, hence its inclusion in this chapter. British Standard Code of Practice 326 (1965) and BS 6651 (1990) are the documents that cover the requirements for lightning protection and this type of installation is very specialised and is not something you would normally be involved with; however, an appreciation of the main requirements is useful.

The discharge current created when lightning strikes can be as high as 200,000 amps, which produces electrical, mechanical and thermal effects.

- The electrical effects happen when the lightning conductor's potential with respect to earth can become very high and 'flash-over' to other metalwork in the structure itself.
- The mechanical effects are caused by air pressure waves when the lightning first strikes or by the high discharge currents flowing to earth, which can cause large mechanical forces to be exerted on to the conductor, its fixings and supports.
- The thermal effects are caused by the rise in temperature of the lightning conductor. When the lightning strike happens a high value of current flows but only for a short time and the effect on the system is normally negligible.

The lightning protective system consists of a network of conductors (copper or aluminium) positioned along the roof and/or walls of a structure at specified distances, bonded to other parts of the structure such as radio and TV masts, and ultimately connected to a common point discharging to earth. The maximum resistance of this network of conductors should not exceed 10 ohms.

Electric shock

BS EN 61140 is a basic safety standard that applies to the protection of persons and livestock. It states that the fundamental rule of protection against electric shock is that hazardous-live-parts shall not be accessible, and that accessible conductive parts shall not be hazardous-live, either in use without a fault, or in single fault conditions.

Therefore BS 7671 states that we must provide the following two levels of protection:

- **Basic protection** – protection in use without a fault.
- **Fault protection** – protection under fault conditions.

In affording these levels of protection, BS 7671 Regulation 410.3.2 requires us to apply a protective measure. That protective measure shall consist of:

- an appropriate combination of provision for basic protection and an independent provision for fault protection or;

- an enhanced protective provision (e.g. reinforced insulation) that provides both basic and fault protection.

However, before we look at these measures we should become familiar with the following definitions:

- **Exposed-conductive-part** – these are those conductive parts of an installation that can be touched and which are not normally live, but could become live under fault conditions. Examples of these are metal casings of appliances such as kettles and ovens or wiring enclosures such as steel conduit, trunking or tray plate.
- **Extraneous-conductive-part** – these are those conductive parts within a building that do not form part of the electrical installation, but could become live under fault conditions. Examples of these are copper water and gas pipes and metal air conditioning system ductwork.

In each part of an installation, one or more protective measures shall be applied, bearing in mind the external influences. Generally, the following measures are permitted:

- Automatic Disconnection of Supply (ADOS)
- Double or reinforced insulation
- Electrical separation from one item of current using equipment
- Extra Low Voltage (SELV and PELV).

In electrical installations Automatic Disconnection of Supply will probably be the most commonly used protective measure as it provides both basic and fault protection.

The requirements for basic and fault protection are given in the respective regulations for each of the protective measures. However, the following information is given as a guide. The particular section or regulation within BS 7671 is shown in brackets.

Automatic Disconnection of Supply (411)

Automatic Disconnection of Supply (ADOS) is a protective measure where:

- Basic protection is given by basic insulation of live parts (416.1), barriers and enclosures (416.2) or, where appropriate, by obstacles (417.2), placing out of reach (417.3).
- Fault protection is given by protective earthing and protective equipotential bonding (411.3.1) and ADOS in case of a fault (411.3.2).

Be aware that protection by obstacles and placing out of reach is basic protection only. They are also only to be used in installations that are under the control or supervision of a skilled person (417.1 and 410.3.5).

However, we must provide additional protection (411.3.3) on a.c. systems for socket outlets with a rated current not exceeding 20 A for use by ordinary persons and for general use. This additional protection is afforded by the use of a 30 mA RCD.

There is an exception to this requirement. This is where the socket outlets are to be used under the supervision of a skilled or instructed person or where the socket is labelled or identified for use with a particular piece of equipment only.

Double (supplementary) or reinforced insulation (412)

These are protective measures where:

- basic protection is given by basic insulation and fault protection by supplementary insulation; or
- both basic and fault protection are given by reinforced insulation between live parts and accessible parts.

Where this protective measure is to be the only protective measure (i.e. the whole circuit consists entirely of equipment with double or reinforced insulation) then the circuit must be under effective supervision in normal use to make sure nothing can affect the protection.

This means that it cannot be used in the context of a circuit where the user could change a piece of equipment without authorisation and therefore cannot be used where socket outlets, light support couplers or cable couplers would be used (412.1.3).

Electrical separation (413)

This is a protective measure where:

- Basic protection is given by basic insulation of live parts or by barriers and enclosures (416); and
- Fault protection is given by simple separation of the separated circuit from other circuits and earth.

To comply with fault protection requirements, the voltage of the separated circuit shall not exceed 500 V (413.3.2) and it is recommended that such circuits are in separate wiring systems (413.3.5) and no exposed conductive part shall be connected to either the protective conductor or exposed conductive parts of other circuits, or to Earth (413.3.6).

Extra Low Voltage provided by SELV or PELV (414)

This is a protective measure consisting of either SELV or PELV, where both basic and fault protection are given when:

- the nominal voltage cannot exceed the upper limit of Band 1 voltage (414.1.1), i.e. 50 V a.c. or 120 V d.c.
- the supply is taken from one of the sources listed in Regulation 414.3 and the conditions of Regulation 414.4 are fulfilled.

Regulation 414.3 states the following sources of supply may be used:

- A safety isolating transformer.

- A source of current providing a degree of safety equivalent to that of the safety isolating transformer (e.g. generator with windings providing equivalent isolation).
- An electro-chemical source (e.g. a battery).
- A source independent of a higher voltage circuit (e.g. a diesel driven generator).
- Certain electronic devices where it has been guaranteed that even in the case of a fault, the voltage on the outgoing terminals cannot exceed the values, or is immediately reduced to values, specified in Regulation 414.1.1.

Additional Protection (415)

In addition to the four protective measures covered, additional protection may be specified. This is particularly so under certain external influences and in some special locations. Two methods are given:

- Residual Current Devices (RCD) operating at 30 mA within 40 ms (415.1.1).
- Supplementary equipotential bonding (415.2).

Please note that an RCD is not recognised as the sole means of protection and therefore doesn't remove the need to apply one of the four protective measures (411 to 414). However, supplementary equipotential bonding is considered as an addition to fault protection.

The protective measures and their structure are summarised in Table 2.1.

<table>
<tr><th colspan="8">Protective measure</th></tr>
<tr><td colspan="2">Automatic disconnection of supply (411)</td><td colspan="2">Double or reinforced insulation (412)</td><td colspan="2">Electrical separation (413)</td><td colspan="2">Extra Low Voltage (414)</td></tr>
<tr><td>Basic</td><td>Fault</td><td>Basic</td><td>Fault</td><td>Basic</td><td>Fault</td><td>Basic</td><td>Fault</td></tr>
<tr><td>Gives basic protection by basic insulation of live parts or by barriers/ enclosures</td><td>Gives fault protection by protective earthing, protective equipotential bonding and ADS in case of fault</td><td>Gives basic protection by basic insulation</td><td>Gives fault protection by supplementary insulation</td><td rowspan="2">Provides basic protection by one of two methods in 416, i.e. basic insulation of live parts of barriers/ enclosures</td><td rowspan="2">Provides fault protection by simple separation of the separated circuit from other circuits and Earth</td><td></td><td></td></tr>
<tr><td colspan="2">There is an additional protection requirement by use of an RCD with socket outlets not exceeding 20 A</td><td colspan="2">Gives basic and fault protection by reinforced insulation between live and accessible parts</td><td colspan="2">Gives basic and fault protection by limiting voltage to the upper limit of Band 1 using a supply such as an isolating transformer with circuits in accordance with Regulation 411</td></tr>
</table>

Table 2.1 Summary of protective measures

Residual current devices

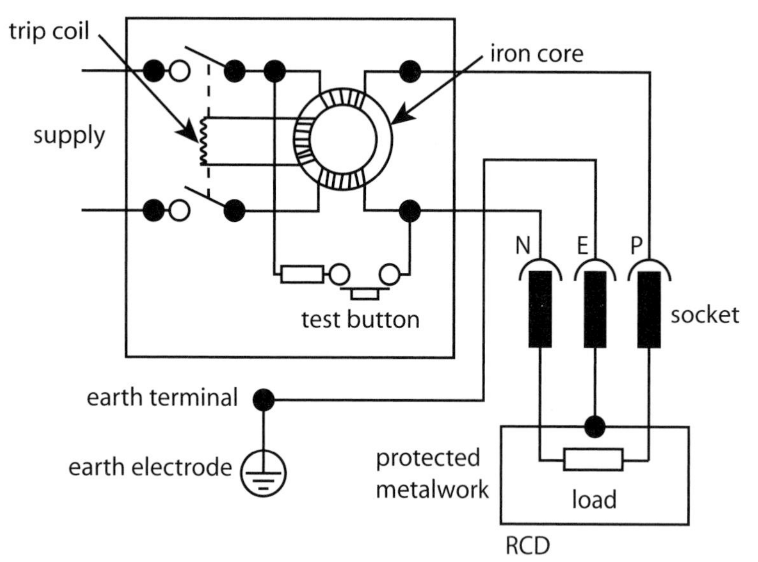

Figure 2.12 Fault detector coil

Residual current devices (RCDs) are a group of devices providing a modern approach to the enhancement of safety in electrical systems. They provide extra protection to people and livestock by reducing the risk of electric shock. Although RCDs operate on small currents, there are circumstances where the combination of operating current and high earth-fault loop impedance could result in the earthed metalwork rising to a dangerously high potential.

The regulations draw attention to the fact that if the product of operating current (A) and earth-fault loop impedance (Ω) exceeds 50V, the potential of the earthed metalwork will be more than 50V above earth potential and hence dangerous. This situation must not be allowed to occur.

The current taken by the load is fed through two equal and opposing coils wound onto a common iron core. Under normal safe working conditions, the current flowing in the live conductor into the load will be the same as that returning via the neutral conductor from the same load. When this happens, the phase and neutral conductors produce equal and opposing magnetic fluxes in the iron core contained within the RCD, resulting in no voltage being induced into the trip coil.

If more current flows in the phase side than the neutral side (or less returns via the neutral because the current has 'leaked' to earth within the installation) an out of balance flux will be produced which will induce a voltage into the fault detector coil. The fault detector coil then energises the trip coil and opens the double pole (DP) switch, due to the residual current produced by the induced voltage in the trip coil.

Functional testing of an RCD's operation should be carried out by operating the test button at regular intervals. Initial and periodic inspection and testing procedures additionally require the RCD to be tested using the appropriate instrument. RCDs are available for single and three phase applications.

Practical application of RCDs (Regulation 531.2):

- An RCD shall disconnect all line conductors of the circuit at substantially the same time.
- The residual operating current of the RCD shall comply with Regulation 411 relative to the type of earthing system.
- The RCD shall be selected and circuits sub-divided so that any normally occurring earth leakage (such as with some computers) doesn't cause nuisance tripping of the RCD.

- The use of a 30 mA RCD with a circuit having a protective conductor is not deemed sufficient to afford fault protection.
- An RCD must be positioned so that its operation isn't affected by the magnetic fields of other equipment.
- Where an RCD is used for fault protection with, but separately from, an overcurrent device, then it must be capable of withstanding (without damage) any stresses likely to be encountered in the event of a fault on the load side of the point at which it is installed.
- RCDs shall be selected and installed so that where required, if two or more RCDs are connected in series and discrimination is required to avoid danger, that discrimination is achieved.
- Where an RCD can be operated by someone other than a skilled or instructed person, it shall be selected and installed so that its settings cannot be altered without using a tool and the result of any alteration being clearly visible.

Sensitivity rated residual operating current	Level of protection	Applications
10 mA	Personnel	High risk areas: schools, colleges, workshops, laboratories. Areas where liquid spillage may occur
30 mA	Personnel	Domestic, commercial and industrial
100 mA	Personnel fire	Only limited personnel protection Excellent fire protection
300 mA	Fire	Commercial and industrial

Table 2.2 Overcurrent protection sensitivity

Residual Current Circuit Breaker with Overload Protection (RCBO)

An RCBO is essentially a marriage between a Miniature Circuit Breaker (MCB) and an RCD, as it comprises both of these components. Consequently, such a device offers protection against the effects of earth leakage, overload and short-circuit currents, while reducing the number of outgoing ways required within a distribution board.

Normally groups of circuits are protected by an RCD, and therefore all circuits would be interrupted under fault conditions, which can cause inconvenience. This device has the advantage of allowing earth-fault protection to be restricted to a single circuit, thus ensuring that only the circuit with the fault is interrupted.

A standard range of 10 A/30 mA, 16 A/30 mA, 20 A/30 mA, 32 A/30 mA and 40 A/30 mA are available from most manufacturers.

The earth-fault loop path

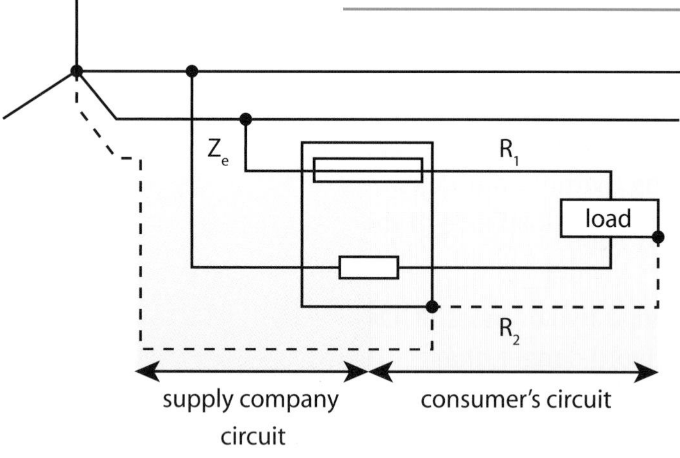

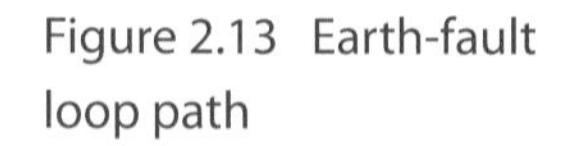
Figure 2.13 Earth-fault loop path

On a full earth-fault, the overcurrent protective device will operate, causing the automatic disconnection of the supply. On a low-level earth leakage fault, the overcurrent protective device will not detect the fault. With a faulty or disconnected earth-fault loop path, metalwork may become live under fault conditions.

An RCD will detect any imbalance between line and neutral conductors, as any difference is assumed to be flowing to earth. This earth 'leakage' will cause the RCD to operate at its rated residual operating current.

Z_s is the sum of Z_e (the supply earth loop impedance) plus R_1 (the resistance of the circuit's phase conductor) and R_2 (the resistance of the circuit protective conductor), i.e.:

$$Z_s = Z_e + (R_1 + R_2)$$

$$\text{Fault current } I_f = \frac{U_o}{Z_s}$$

where U_o is the nominal supply voltage.

Note: This is also known as the Prospective Short Circuit Current (PSSC) I_{SC}. In situations where the value of Z_s is too high to achieve rapid disconnection on a fault then:

(i) Cable conductor size may be increased to reduce the Z_s value.

(ii) RCDs may be installed in combination with overcurrent protection devices and will provide a safe solution.

Earth-fault leakage currents will then be detected and the supply automatically disconnected.

The value of Z_s governs the flow of Earth-Fault Loop Current, the operation of earth-fault protective devices and their circuit disconnection times.

From Tables 41.2 and 41.3 and Appendix 3 Figures 3.3A, 3.5 and 3.6 of BS 7671 it can be seen that:

- A BS 88 32A installed in a circuit of $Z_S = 1.04$ ohms will operate (at a fault current of 220 A) in 0.4 seconds on a 230 V system.
- A BS EN 60898 32A Type C MCB installed in a circuit of $Z_S = 0.72$ ohms will operate (at fault current of 320 A) in 0.1–5.0 seconds on a 230 V system.
- A BS EN 60898 32A Type D MCB installed in a circuit of $Z_S = 0.36$ ohms will operate (at fault current of 640 A) in 0.1–5.0 seconds on a 230 V system.

Regulations state that the protective device protecting a circuit must operate within a time of:

- 0.2 seconds for a final circuit not exceeding 32 A in a TT system
- 0.4 seconds for a final circuit not exceeding 32 A in a TN system
- 1 second for final circuits exceeding 32 A in a TT system

- 5 seconds for final circuits exceeding 32 A in a TN-S system
- 1 second for distribution circuits in a TT system
- 5 seconds for distribution circuits in a TN system.

Overcurrent protection

An overcurrent is simply a current that is in excess of the rated value or the current carrying capacity of a conductor and is the result of either an overload or a fault current.

An **overload** is defined as an overcurrent occurring within a circuit that is electrically sound. In other words a circuit that is carrying more current than it was designed to.

This may be as the result of faulty equipment such as a stalled motor or might occur when many pieces of equipment are added to the circuit, such as when several 13 A plugs are connected to an adaptor and then plugged into a 13 A socket outlet.

Fault current is the other category and this can arise from two circumstances, a short circuit or an earth fault. These may be as the result of a connection error or perhaps a nail has been driven into an energised cable.

A short circuit is defined as being when there is a fault of negligible impedance between live conductors. An earth fault current is defined as being a fault of negligible impedance between a line conductor and an exposed conductive part or protective conductor.

It should be noted the terms short circuit and earth fault have been used only to illustrate these concepts. BS 7671 no longer recognises both variations, referring to them instead under the overall title of fault currents.

Protection against overcurrent is dealt with in Chapter 43 of BS 7671 and is afforded by what we refer to as protective devices. These are devices designed to disconnect the supply automatically, such as fuses or MCBs, and therefore in this section we will look at the following:

- BS 3036 rewireable fuses
- BS 1361/1362 cartridge fuses
- BS 88 HBC fuses
- Type 'D' and Neozed fuses
- Miniature Circuit Breakers
- Fusing factor
- Fuse and circuit breaker operation
- Prospective Short Circuit Current
- Discrimination

BS 3036 rewireable fuses

Early rewireable fuses had a very low short circuit capacity and were very dangerous when operating under fault conditions because the fuse element melts and splashes the melted copper around and can cause fires. Later rewireable fuses incorporated asbestos to protect the fuse holder during the fusing period, thus reducing the risk of fire from scattering hot metal when rupturing.

BS 3036 rewirable fuse

The rewireable fuse consists of a fuse, holder, a fuse element and a fuse carrier. The holder and carrier are made of porcelain or bakelite. The circuits for which this type of fuse is designed have a colour code, marked on the fuse holder:

- 5 A white
- 15 A blue
- 20 A yellow
- 30 A red
- 45 A green.

This type of fuse was very popular in domestic installations although, with the exception of old installations, it is not normally used because it has some serious disadvantages.

Disadvantages of rewirable fuses	Advantages of rewirable fuses
• Easily abused when the wrong size of fuse wire is fitted **Fusing factor** of around 1.8 – 2.0, which means they aren't guaranteed to operate until up to twice the rated current is flowing. As a result cables protected by them must have a larger current carrying capacity • Precise conditions for operation cannot be easily predicted • Do not cope well with high short circuit currents • Fuse wire can deteriorate over time • Danger from hot scattering metal if the fuse carrier is inserted into the base when the circuit is faulty.	• Low initial cost • Can easily see when the fuse has blown • Low element replacement cost • No mechanical moving parts • Easy storage of spare fuse wire

Table 2.3 Advantages and disadvantages of rewireable fuses

Nominal current of fuse wire (A)	Nominal diameter of wire (mm)
3	0.15
5	0.2
10	0.35
20	0.5
20	0.6
25	0.75
30	0.85
45	1.25
60	1.53
80	1.8
100	2

Table 2.4 Size of tinned copper wire for use in semi-enclosed fuses

BS 1361/1362 cartridge fuses

The cartridge fuse consists of a porcelain tube with metal end caps to which the element is attached. The tube is then filled with granulated silica. The BS 1362 fuse is generally found in domestic plug tops used with 13 A BS 1363 domestic socket outlets. There are two common fuse ratings available, the 3 A, which is for use with appliances up to 720 watts (radios, table lamps, electric blankets) and the 13 A fuse which is used for appliances rated over 720 watts (irons, kettles, fan heaters, electric fires, lawn mowers, toasters, refrigerators, washing machines and vacuum cleaners). There are, however, other sizes available (which are more difficult to come across). These are 1, 5, 7 and 10 amp sizes. The BS 1361 fuse is normally found in distribution boards and at main intake positions.

BS 1361 and 1362 fuses

Disadvantages of cartridge fuses	Advantages of cartridge fuses
• They are more expensive to replace than rewireable fuses • They can be replaced with an incorrect size fuse (plug top type only) • The cartridge can be shorted out with wire or silver foil in extreme cases of bad practice • It is not possible to see if the fuse has blown • They require a stock of spare fuses to be kept	• They have no mechanical moving parts • The declared rating is accurate • The element does not weaken with age • They have a small physical size and no external arcing, which permits their use in plug tops and small fuse carriers • They have a low fusing factor around 1.6–1.8 • They are easy to replace

Table 2.5 Advantages and disadvantages of cartridge fuses

BS 88 high breaking capacity (HBC) fuses

The HBC fuse is a sophisticated variation of the cartridge fuse and is normally found protecting motor circuits and industrial installations. It consists of a porcelain body filled with silica, a silver element and lug-type end caps. Another feature is the indicating bead, which shows when the fuse element has blown. It is a very fast acting fuse and can discriminate between a starting surge and an overload.

These types of fuses would be used when an abnormally high prospective short circuit current exists.

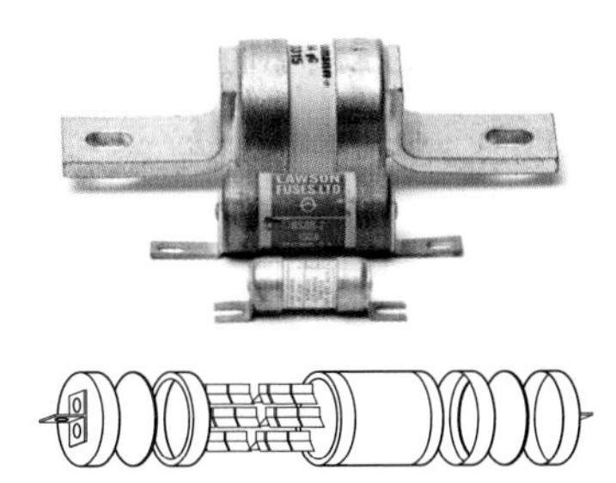
A sectional view of a typical BS 88 HBC fuse

Disadvantages of BS 88 fuses	Advantages of BS 88 fuses
• They are very expensive to replace • Stocks of these spares are costly and take up space • Care must be taken when replacing them, to ensure that the replacement fuse has the same rating and also the same characteristics as the fuse being replaced	• They have no mechanical moving parts • The element does not weaken with age • Operation is very rapid under fault conditions • It is difficult to interchange the cartridge, since different ratings are made to different physical sizes

Table 2.6 Advantages and disadvantages of BS 88 fuses

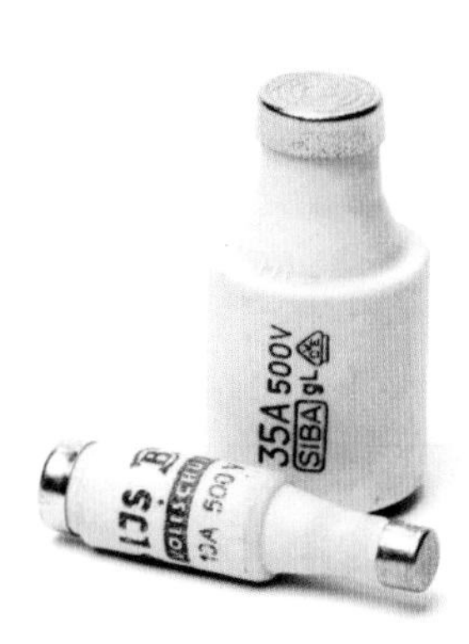

Neozed fuse

Type 'D' and Neozed fuses

Both these fuses are manufactured in Germany and have been developed to European testing regulations where all European testing authorities have approved them. The Neozed is the successor to the 'D' type fuse. You may in the course of your work come into contact with either type of fuse.

Miniature Circuit Breakers (MCBs)

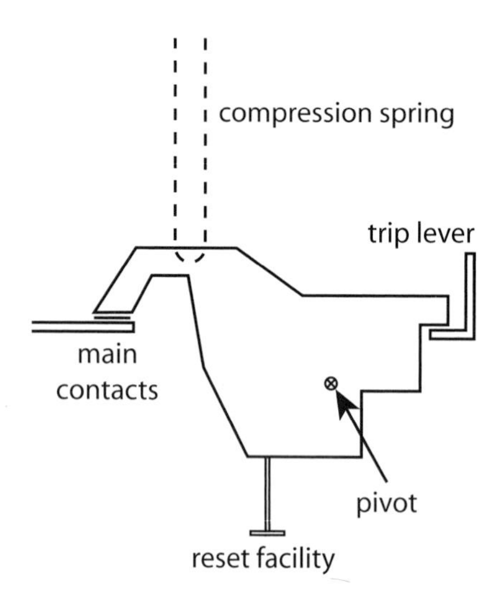

Figure 2.14 Compression spring

Due to improved design and performance, the modern MCB now forms an essential part of the majority of installations at the final distribution level. From about 1970 the benefits of current limiting technology have been incorporated into MCBs, thus providing the designer/user with predictable high performance overcurrent devices.

The circuit breaker is an automatic switch which opens when excess current flows in the circuit. The switch can be closed again when the current returns to normal because the device does not damage itself during normal operation. The contacts of a circuit breaker are closed against spring pressure, and held closed by a latch arrangement. A small movement of the latch will release the contacts, which will open quickly against the spring pressure to break the circuit.

The circuit breaker is so arranged that normal currents will not affect the latch, whereas excessive currents will move it to operate the breaker. There are two basic methods by which overcurrent can operate or 'trip' the latch.

Thermal tripping

The load current is passed through a small heater coil wrapped around a bi-metal strip inside the MCB housing; the heat created depends on the current it carries. This heater is designed to warm the bi-metal strip either directly (the current passes through the bi-metal strip which in effect is part of the electrical circuit) or indirectly (where a coil of current carrying conductor is wound around the bi-metal strip) and the excess current warms the bi-metal strip.

The bi-metal strip is made of two different metals, normally brass and steel (brass expanding more than steel). These two dissimilar metals are securely riveted or

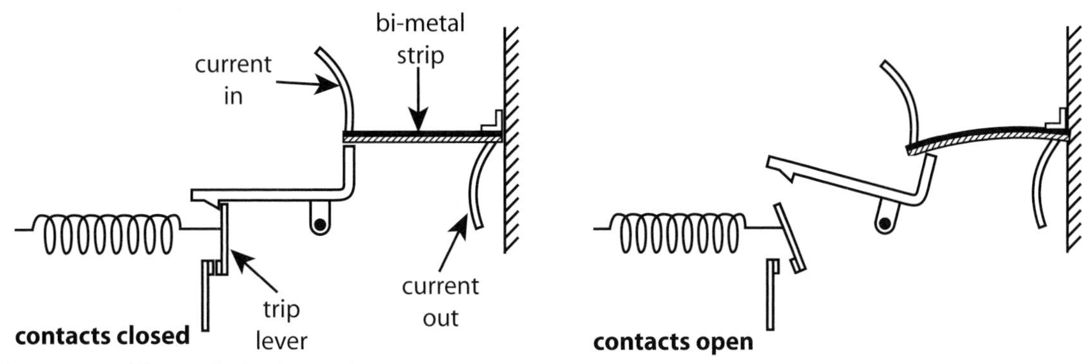

Figure 2.15 Thermal tripping action

welded together along their length. The rate of expansion of the two metals is different so that when the strip is warmed, it will bend and will trip the latch.

The bi-metal strips are arranged so that normal currents will not heat the strip to tripping point.

However, if current increases beyond the rated value, the heater increases in temperature and thus the bi-metal strip is raised in temperature, trips the latch and opens the contacts.

Magnetic tripping

The principle used here is the force of attraction, which can be set up by the magnetic field of a coil carrying the load current. At normal current the magnetic field is not strong enough to attract the latch, but overload currents operate the latch and trip the main contacts.

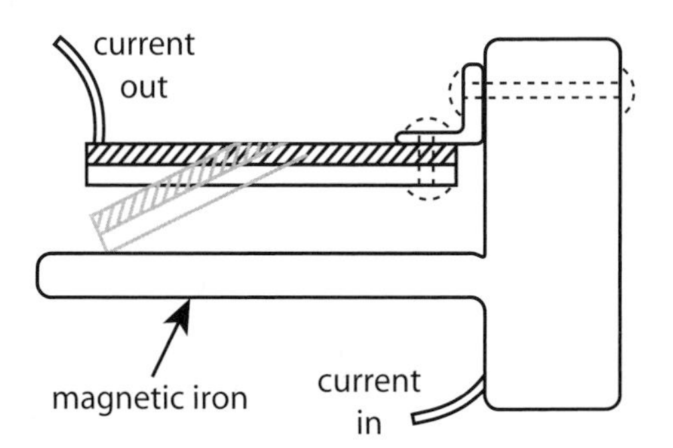

Figure 2.16 A simple attraction type of magnetic trip

The magnetic field is set up by a current in the flexible strip that attracts the strip to the iron. This releases the latch. This is often used in miniature circuit breakers and combined with a thermal trip.

The **oil dashpot** solenoid type is used on larger circuit breakers. The time lag is adjustable by varying the size of the oil-escape hole in the dashpot position. Current rating is adjustable by vertical movement of the plunger.

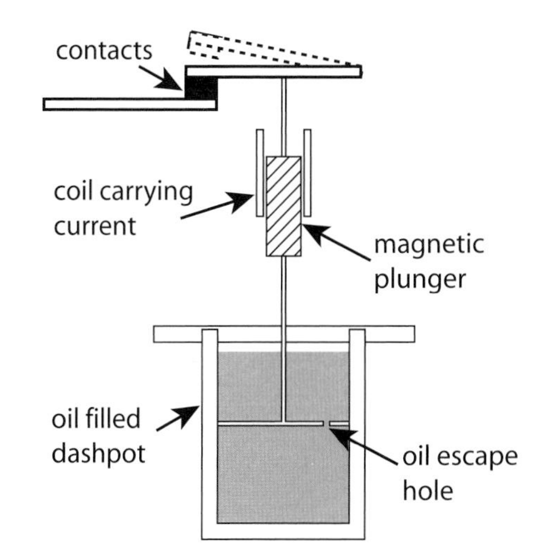

Figure 2.17 Oil filled dashpot

Combined tripping

As we have discussed, circuit breakers use two basic methods in their operation. There is always some time delay in the operation of a thermal trip, as the heat produced by the load current must be transferred to the bi-metal strip. Thermal tripping is therefore best suited to small overloads of comparatively long duration.

Magnetic trips are fast acting for large overloads or short circuits. The two methods are often combined to take advantage of the best characteristics of each.

They are then classified according to their instantaneous tripping current, i.e. the current at which they will operate within 100 ms.

Current circuit breakers (BS EN 60898) are referred to as types B, C and D (there is no type A to avoid confusion with Amperes), however you may come across older ones (BS 3871) that were referred to as types 1, 2, 3 and 4.

If you were to take for example a 20 A MCB of Types 1, 2, 3, and B, C, D and compare their instantaneous tripping currents, you would find that this will vary depending on the type of MCB chosen, and consequently this indicates that a 20 A MCB will trip instantaneously at different fault levels.

Guidance to the circuit breakers of BS 3871 or BS EN 60898 is given in Table 2.7 below, where I_n is the value of current that can be carried indefinitely by the device:

MCB type	Instantaneous trip current	Application
1 B	2.7 to 4 × I_n 3 to 5 × I_n	Domestic and commercial installations having little or no switching surge.
2 C 3	4.0 to 7.0 × I_n 5 to 10 × I_n 7 to 10 × I_n	General use in commercial/industrial installations, where the use of fluorescent lighting, small motors etc. can produce switching surges that would operate a Type 1 or B circuit-breaker. Type C or 3 may be necessary in highly inductive circuits such as banks of fluorescent lighting.
4 D	10 to 50 × I_n 10 to 20 × I_n	Suitable for transformers, X-ray machines, industrial welding equipment etc., where high in-rush currents may occur.

Table 2.7 MCB selection

As an example from this table, a 16 A 'Type C' MCB will trip instantly at 160 A (10 × I_n). However, we can also see that care must be taken when replacing an MCB to ensure that the same type is used if the level of protection is to be maintained.

Disadvantages of MCBs	Advantages of MCBs
• They have mechanical moving parts • They are expensive • They must be regularly tested • Ambient temperature can change performance	• They have factory set operating characteristics which cannot be altered • They will maintain transient overloads and trip on sustained overloads • Easily identified when they have tripped • The supply can be quickly restored

Table 2.8 Advantages and disadvantages of MCBs

Fusing factor

It is evident that each of the protective devices discussed in the previous section provide different levels of protection, e.g. rewireable fuses are slower to operate and less accurate than MCBs. In order to classify these devices it is important to have some means of knowing their circuit breaking and 'fusing' performance. This is achieved for fuses by the use of a fusing factor.

$$\text{Fusing factor} = \frac{\text{Fusing current}}{\text{Current rating}}$$

This is the ratio of the fusing current, which is the minimum current that will cause the fuse to blow and the stated current rating of the fuse or MCB (which is the maximum current that the fuse can sustain without blowing). Fusing currents can be found in Appendix 3 of BS 7671. These tables are logarithmic and the scales increase by factors of ten, not uniformly as may be expected, and therefore the interpretation of these scales will require some practice. The rating of the fuse is the current it will carry continuously without deterioration.

Fusing factors for the above devices can generally be grouped as follows:

- BS 3036 1.8–2.0
- BS 88 1.25–1.7
- BS 1361 1.6–1.9
- MCBs up to 1.5.

The higher the fusing factor, the less accurate – and therefore less reliable – the device selected will be. You may, while looking at fuses, have noticed a number followed by the letters kA stamped onto the end cap of an HBC fuse or printed onto the body of a BS 1361 fuse. This is known as the breaking capacity of fuses and circuit breakers. When a short circuit occurs, the current may, for a fraction of a second, reach hundreds or even thousands of amperes.

The protective device must be able to break or make such a current without damage to its surroundings by arcing, overheating or the scattering of hot particles.

The breaking capacities of MCBs are indicated by an 'M' number e.g. M6. This means that the breaking capacity is 6 kA or 6,000 A. The breaking capacity will be related to the prospective short circuit current.

Fuse and circuit breaker operation

In this part the following symbols will be referred to:

I_z	this is the maximum current carrying capacity of a cable for continuous service, under the particular installation conditions concerned (Appendix 4 of BS 7671)
I_n	this is the nominal current or current setting of the device protecting the circuit against overcurrent (Appendix 3 of BS 7671)
I_2	this is the operating current, i.e. the fusing current or tripping current for the conventional operating time (Appendix 3 of BS 7671)
I_b	this is the design current of the circuit, i.e. the current intended to be carried by the circuit in normal service
I_{ef}	earth fault current
I_{sc}	short circuit current
t	time in seconds
k	material factor found from Tables 43.1 and 54.2 to 54.6 of BS 7671
s	conductor cross-sectional area in mm^2.

Consider a protective device (fuse or MCB) rated at 20 A. This value of current can be carried indefinitely by the device and is known as its nominal current setting, I_n. The value of current that will cause the device to operate, I_2, will be larger than I_n and will be dependent on the device's fusing factor.

This fusing factor figure, when multiplied by the nominal setting I_n, will give the value of operating current I_2.

For fuses to BS 1361, BS 88 and circuit breakers to BS 3871 and BS/EN 60898, the fusing factor has been approximated to 1.45. Therefore our 20 A device would operate when the current reached 29 A (1.45×20).

BS 7671 Regulation 433.1.1 requires co-ordination between conductor and protective device when an overload occurs such that:

(i) The nominal current or setting (I_n) of the protective device is not less than the design current (I_b) of the circuit, therefore $I_b \leq I_n$ (the symbol $\leq$ means less than or equal to). For example, if the circuit design current was 20 A then the protective device would need to have a rating of at least 20 A.

(ii) The nominal current or current setting (I_n) of the protective device does not exceed the lowest value of the current carrying capacities (I_z) of any of the conductors in the circuit. Therefore $I_n \leq I_z$.

The formulae can be combined so that $I_b \leq I_n \leq I_z$.

In simple terms this means that the design current of the circuit must be less than or equal to the protective device rating, which in turn must be less than or equal to the current carrying capacity of the cable.

(iii) The current (I_2) causing effective operation of the protective device does not exceed 1.45 times the lowest of the current carrying capacities (I_z) of any of the conductors of the circuit.

Therefore:

$$I_2 \leq I_z \times 1.45$$

We now have:

$$I_2 = I_n \times 1.45 \text{ and } I_2 \leq I_z \times 1.45$$

We can now say:

$$I_n \times 1.45 \leq I_z \times 1.45$$

We can now remove the 1.45 from the equation, as it is common to both the left hand side (LHS) and the right hand side (RHS) of the equation.

Therefore, this leaves:

$$I_n \leq I_z.$$

This means that the current rating of the protective device must be less than or equal to the current carrying capacity of the cable.

This is only true for situations where BS 1361, BS 88 fuses and circuit breakers complying with BS 3871 and BS/EN 60898 are in use.

Now considering our 20 A device, if the cable is rated at 20 A then condition (ii) is satisfied. The fusing factor is 1.45; therefore condition (iii) is also satisfied, because:

$I_2 = I_n \times 1.45$ which is $20\,A \times 1.45 = 29\,A$

which is the same as $1.45 \times 20\,A$ (I_z) cable rating, because:

$1.45 \times 20\,A = 29\,A$ (rating of the cable).

Problems with BS 3036 rewireable fuses

This type of fuse may have a fusing factor as high as 2. Using the previous example we would see that:

$I_n \times 2 \leq I_z \times 1.45$ would not be true.

In order to comply with condition (iii) I_n should be less than or equal to 0.725 I_z. (This figure is derived from $1.45 \div 2 = 0.725$.)

For example if a cable is rated at 20 A, then I_n for a BS 3036 device should be less than or equal to:

$0.725 \times 20 = 14.5\,A$

As the fusing factor is 2, the operating current I_z would now be:

$2 \times 14.5 = 29\,A$

which conforms to condition (iii) if the fuse size was 14.5 A. However, if the design current required was the same as in the previous example and hence the fuse size (20 A) was the same, then this condition would not apply – the cable size (and hence its rating) would need to be increased. Be wary of using this type of fuse!

If all these requirements are met then this will ensure that the conductor insulation is undamaged when an overload occurs. These factors will be further explored in the cable selection section.

Prospective Fault Current (PFC)

We said in the introduction to this section that there are two types of overcurrent, as the result of overload or fault conditions. BS 7671 Regulation 434 addresses protection against fault current. Additionally, Regulation 435 requires that there is coordination of protection between the overload and fault conditions.

The effects of a 'short circuit' current are:

- **Thermal** – which can cause melting of conductors and insulation, fire or the alteration of the properties of materials
- **Mechanical** – large magnetic fields can build up, resulting in distortion of conductors and breaking of supports and insulators

For fault current protection BS 7671 requires that protective devices are capable of withstanding the short circuit current (the device breaking capacity) at the point at which they are installed (434.5.1) and disconnect quickly enough to prevent damage to the cables (434.5.2). Therefore the fault current shall be determined at every relevant point of the installation by calculation, measurement or enquiry (434.1).

We also said that a fault current is the result of a fault of negligible impedance between live conductors. This means that if two live conductors are touching each

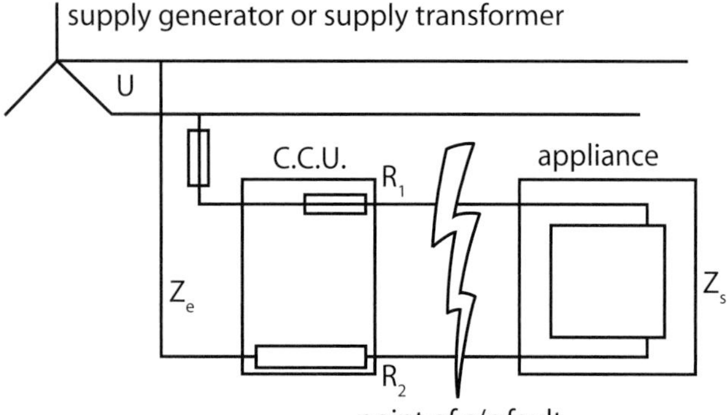

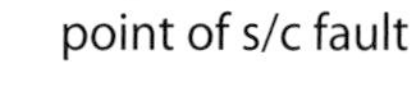

Figure 2.18 Short circuit loop impedance

other, then the resistance/impedance of that joining is so low that it can be ignored. As a result, if there is no resistance then the only opposition to the size of current is the conductors and, as the resistance of conductors is generally low, we can see that the level of current could be very high.

The highest fault currents will arise when the line conductors (this includes the neutral) of a three phase system short together, e.g. brown to grey and the highest value measured will usually be at the origin.

If we look at Figure 2.18, we can see a short circuit between live conductors (line and neutral) and, in this case, the final circuit protective device should operate first.

- U = nominal supply voltage
- Z_e = loop impedance external to the installation
- R_1 = the resistance of the installation line conductor to the fault point
- R_2 = the resistance of the installation return conductor to the fault point
- Z_s = total fault loop impedance,

  $Z_s = Z_e + (R_1 + R_2)$.

Note the return can be via either the neutral or another phase depending on the supply arrangement.

Example

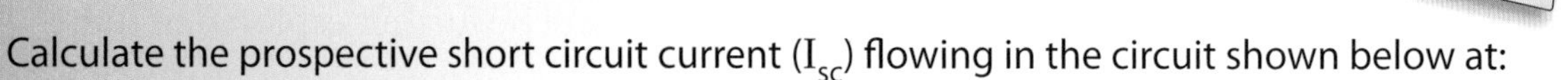

Calculate the prospective short circuit current (I_{sc}) flowing in the circuit shown below at:

(a) the customer consumer unit supply terminals

(b) the terminals of the load given the conditions shown.

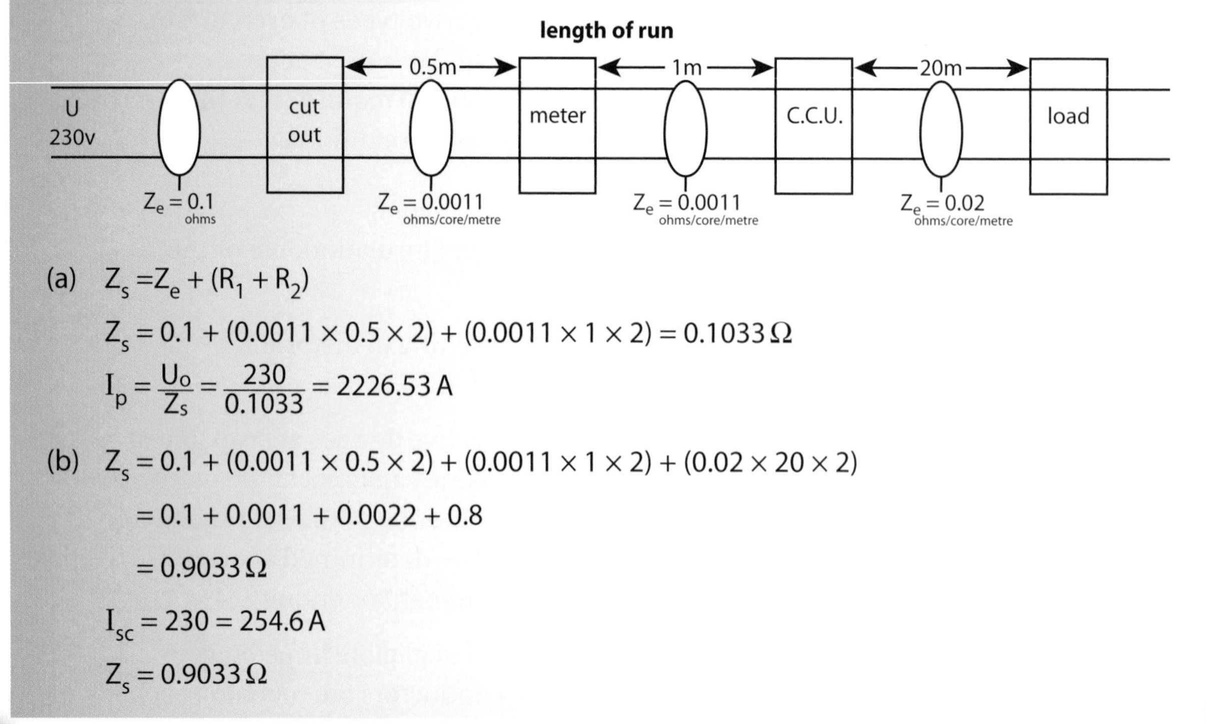

(a) $Z_s = Z_e + (R_1 + R_2)$

$Z_s = 0.1 + (0.0011 \times 0.5 \times 2) + (0.0011 \times 1 \times 2) = 0.1033\,\Omega$

$I_p = \frac{U_o}{Z_s} = \frac{230}{0.1033} = 2226.53\,A$

(b) $Z_s = 0.1 + (0.0011 \times 0.5 \times 2) + (0.0011 \times 1 \times 2) + (0.02 \times 20 \times 2)$

$= 0.1 + 0.0011 + 0.0022 + 0.8$

$= 0.9033\,\Omega$

$I_{sc} = 230 = 254.6\,A$

$Z_s = 0.9033\,\Omega$

Discrimination

In both large and small installations, there is usually a series of fuses and/or circuit breakers between the incoming supply and the electrical outlets. The relative rating of the protective devices used will decrease the nearer they are located to the current using equipment.

Ideally when a fault occurs, only the device nearest the fault operates, thus ensuring minimum disruption to other circuits not associated with the fault. **Discrimination** is said to have taken place when the smaller rated local device operates before the larger device.

Discrimination is generally only a problem when a system uses a mixture of devices. Obviously a particular type of fuse will discriminate against a similar type of fuse if it is of larger rating.

Discrimination is also known as the co-ordination between fuses.

Figure 2.19 relates to a 15 A semi-enclosed rewireable fuse serving a final sub-circuit, backed up by a 30 A MCB. When the fault current approaches 200 amps, the MCB will operate quicker than the fuse. Here discrimination has failed.

Simply because protective devices have different ratings, it cannot be assumed that discrimination will be achieved. This is especially the case where a mixture of different types of device is used. However, as a general rule when fuses are used in series in an installation a 2:1 ratio with the lower-rated devices will be satisfactory.

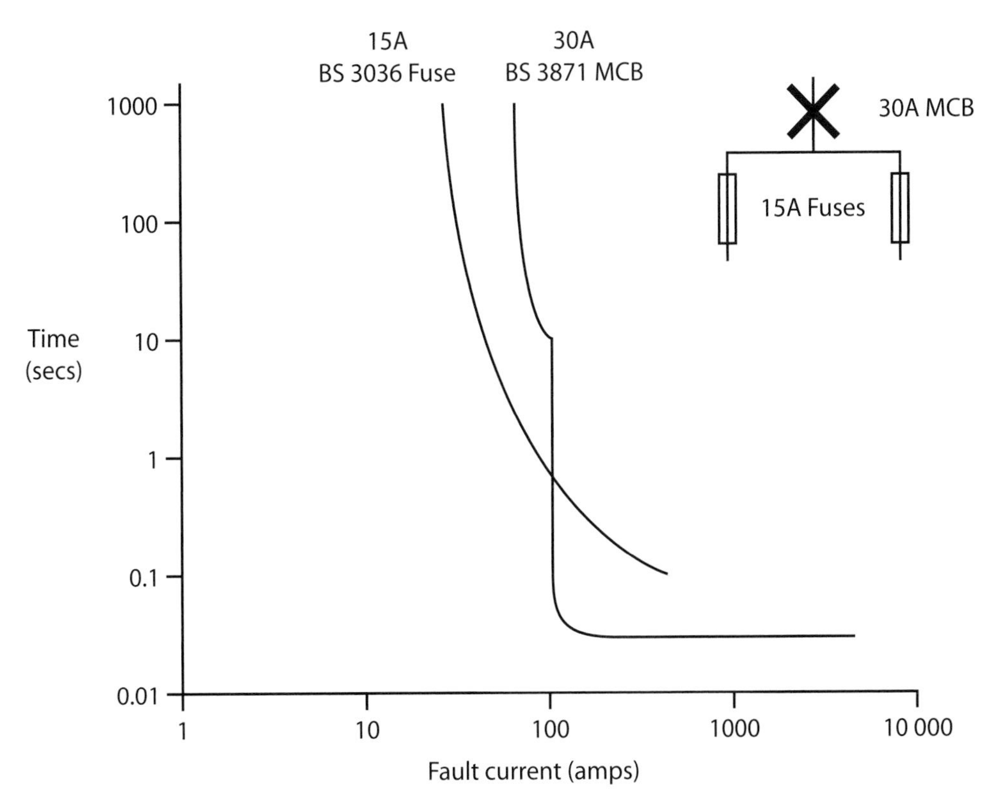

Figure 2.19 Discrimination

Cable selection

Cable selection can be defined as the 'rules' that you must follow when deciding which cable to choose for any installation.

This section will consider the following areas used in cable selection:

- external influences
- design current (I_b)
- rating of the protective device (I_n)
- reference methods
- rating factors
- application of rating factors
- voltage drop
- shock protection
- thermal constraints
- diversity.

External influences

Appendix 5 of BS 7671 lists the 'Classification of External Influences', where each condition of external influence is designated with a code that comprises a group of two capital letters and a number.

The first letter relates to the general category of the external influence (e.g. the Environment). The second letter then relates to the circumstances or nature of that external influence (e.g. the presence of Water). Finally, the number relates to the specific degree or level of that condition within each external influence (e.g. Splashes).

To explain this, let's look at the influences mentioned in the above paragraph and, using BS 7671 Table 5, convert them into a code.

General category of influence	Environment	A
Nature of the influence	Water	D
Level of that influence	Splashes	3

Consequently the designated code for these conditions would be: AD3

Working in the opposite direction and covering each of the three General Categories of external influence, here are three codes and their explanation:

AA4

A	Environment
AA	Environment – Ambient Temperature
AA4	Environment – Ambient Temperature – in the range –5°C to +40°C

BC1

B	Utilisation
BC	Utilisation – Contact with earth
BC1	Utilisation – Contact with earth – None

CA2

C Construction of buildings

CA Construction of buildings – Materials

CA1 Construction of buildings – Materials – Combustible

Once you have decided on the type of cable suitable for the environmental conditions, you must choose the size of conductor to be used. The first factor you need to consider is design current.

Design current (I_b)

You will have to calculate this. It is the normal resistive load current in the circuit. The following formulae apply to single and three phase supplies:

Single phase supplies:

$$U_o = 230\,V$$

$$I_b = \frac{Power}{U_o}$$

Where U_o is the line voltage to earth (supply voltage).

Three phase supplies:

$$U_o = 400\,V$$

$$I_b = \frac{Power}{\sqrt{3} \times U_o}$$

In a.c. circuits, the effects of either highly inductive or highly capacitive loads can produce a poor power factor (PF) (inductive and capacitive loads will be explained later). You will have to allow for this. To find the design current, you may need to use the following equations:

Single phase circuits:

$$I_b = \frac{Power}{U_o \times PF}$$

Three phase circuits:

$$I_b = \frac{Power}{\sqrt{3} \times U_o \times PF}$$

Where PF is the power factor of the circuit concerned.

Rating of the protective device (I_n)

When you have worked out the design current of the circuit (I_b), you must next work out the current rating or setting (I_n) of the protective device.

Regulation 433.1.1 of BS 7671 states that current rating (I_n) must be no less than the design current (I_b) of the circuit.

The reason for this is that the protective device must be able to pass enough current for the circuit to operate at full load, but without the protective device operating and disconnecting the circuit.

Protective devices come in standard values, for example 13 A for plug tops. You can find these values in the *IEE Wiring Regulations*, Tables 41.2, 41.3 and 41.4 or in Appendix 3.

The table you use to select the value of your protective device will depend on the type of equipment or circuit to be supplied and the requirements for disconnection times.

Installation and Reference Methods

This section has vastly increased in size over the previous version of BS 7671. Within Appendix 4 of BS 7671 there are now two tables that should be referred to – 4A1 and 4A2.

As a concept, the method of installation affects the chosen cable's ability to get rid of any heat generated in normal operation. This therefore affects its current carrying capacity.

It is not practical to calculate the current ratings for every single installation method and many of these would have the same rating. Therefore a range of current ratings have been calculated covering all of the installation methods within Table 4A2 and these are known as Reference Methods.

As a process, an electrican should firstly look at 4A1 (see table 2.9) and establish whether the proposed installation method will be allowable for the type of cable that needs to be installed.

Conductors and cables		Installation method							
		Without fixings	Clipped direct	Conduit systems	Cable trunking systems*	Cable ducting systems	Cable ladder, cable tray, cable brackets	On insulators	Support wire
Bare conductors		np	np	np	np	np	np	P	np
Non-sheathed cabled		np	np	P[1]	P[1]	P[1]	np[1]	P	np
Sheathed cables (including armoured and mineral insulated)	Multi-core	P	P[1]	P	P	P	P	n/a	P
	Single core	n/a	P	P	P	P	P	n/a	P

* = including skirting trunking
[1] = non-sheathed

P = permitted
np = not permitted
n/a = not applicable, or not normally used in practice

Table 2.9 Schedule of installation methods in relation to conductors and cables (Table 4A1)

For example, say we want to install a PVC/PVC twin and earth cable, of a yet to be determined size, by clipping it directly to a wall. By looking at the table we can see that this installation method is acceptable (where circled).

Now that we know this, we have to use Table 4A2 to establish the specific installation method and resultant reference method that will govern the current carrying capacity of our cable.

As can be seen from Table 2.10 (where circled) which shows an extract of BS 7671 Table 4A2, our specific installation method is 20 and this is classed as falling within reference method C.

Please note that 4A2 includes specific information relating to flat twin and earth cables in thermal insulation.

Table 4A2 (continued)			
Installation method			**Reference method to be used to determine current-carrying capacity**
Number	**Examples**	**Description**	
13	TV ICT 13	Non-sheathed cables in skirting trunking	B
14	TV ICT 14	Multicore cable in skirting trunking	B
15		Non-sheathed cables in conduit or single-core or mulitcore cable in architrave	A
16		Non-sheathed cables in conduitor or single-core or mulitcore cable in window frames	A
(20)		Single-core or multicore cables: • fixed on (chipped direct), or spaced less than 0.3 × cable diameter from a wooden or masonry wall	(C)
21		Single-core or multicore cables: • fixed directly under a wooden or masonry ceiling	B Higher than standard ambient temperatures may occur with this installation method.
22		Single-core or multicore cables: • spaced from a ceiling	E, F or G Higher than standard ambient temperatures may occur with this installation method.

Table 2.10 BS 7671 Table 4A2

Rating factors (C)

You, as the 'designer' of the installation, need to know the five rating factors, know where they are required and then apply these to the nominal rating of the protection (I_n).

Rating factor	Tables for rating factor values	Symbol
Ambient temperature	Tables 4B1 and 4B2	C_a
Grouping factors	Tables 4C1, 4C2, 4C3 and 4B3	C_g
Thermal insulation	Regulation 523.7	C_i
BS 3036 fuse	0.725 Regulation 433.1.3	C_c
Mineral insulated cable	0.9 Table 4G1A	N/A

Table 2.11 Tables for correction factor values

Ambient temperature factor (C_a)

This is the temperature of the surroundings of the cable, often the temperature of the air in a room or building in which the cable is installed. When a cable carries current, it gives off heat. Therefore the hotter the surroundings of the cable, the more difficult it is for the cable to get rid of this heat. But if the surrounding temperature is low, then the heat given off could be easily let out and the cable could carry more current. Cables must give off this heat safely or they could be damaged and there is a risk of a fire.

You can find the correction factor for **ambient temperature** in Tables 4B1 and 4B2 of BS 7671.

These have been revised since the 16th Edition. Table 4B1 is applicable to cables in free air where 30°C is assumed to be the normal level of ambient temperature. Therefore, 30°C has a value in the table of 1.0 (no effect) and the table then has rating factors for ambient temperatures from 25°C to 95°C.

Table 4B2 is applicable to underground cables where 20°C is assumed to be the normal level of soil temperature. Therefore, 20°C has a value in the table of 1.0 (no effect) and the table then has rating factors for ambient temperatures from 10°C to 80°C.

Note that an ambient temperature below 30°C (e.g. 25°C) will give a correction factor that will allow your choice of cable to be improved. Because a lower ambient temperature will allow the cable to 'give off' heat more easily and safely therefore you could reduce the size of cable. Should a cable run through areas of different ambient temperature, a correction factor should be applied based on the highest temperature only.

Grouping factor (C_g)

As a concept, when cables are grouped together and are touching each other, it is more difficult for them to dissipate any heat they may be producing in operation. Ideally, we should keep them separated, but if this is impossible then you must apply a grouping factor. These are given in Tables 4C1 and 4C2 of BS 7671. A rating factor need not be applied if the horizontal clearance between cables is greater than twice their diameter.

These tables also have revised guidance notes attached to them, which in the case of Table 4C2 (cables laid directly in the ground) can involve rating factors specified within Table 4B3 (soil thermal resistivity other than 2.5 K.m/W).

The rating factors given within Appendix 4 of BS 7671 are based upon groups of 'similar and equally loaded cables'. But what if the group contains cables of differing sizes? For a group of different sized non-sheathed/sheathed cables within conduit, trunking or ducting systems, the following formula can now be applied to calculate the group rating factor:

$$F = \frac{1}{\sqrt{n}}$$

where F is the group rating factor and n is the number of circuits.

Thermal insulation factor (C_i)

Thermal insulation is used in the construction industry within walls or ceilings. However, its effect is the same as wrapping yourself in a thick fur coat on a hot summer day, in that it reduces a cable's ability to dissipate heat.

Therefore Regulation 523.7 of BS 7671 states that:

- for a cable installed in a thermally insulated wall or above a thermally insulated ceiling, where the cable has one side in contact with the thermally conductive surface, current carrying capacities are given in Appendix 4 of BS 7671.
- for a single cable that is totally covered by thermal insulation for **more** than 0.5 m and where no more precise information is available, the current carrying capacity shall be taken as half the value of that same cable when clipped to a surface and open (Reference Method C).
- for a single cable that is totally covered by thermal insulation for **less** than 0.5 m, the current carrying capacity shall be reduced as appropriate depending on the size of the cable, its length in the insulation and the thermal properties of that insulation. Derating factors are therefore given in BS 7671 Table 52.2 for conductor sizes up to 10 mm^2. These are shown in Table 2.12.

Length in insulation (mm)	De-rating factor
50	0.88
100	0.78
200	0.63
400	0.51

Table 2.12 Cable rating (Table 52.2 BS 7671)

Protective device factor (C_c)

Regulation 433.1.2 states that where overload protection is given by either a fuse to BS 88 or 1361, an MCB to BS EN 60898 or an RCBO to BS EN 61009, then the requirements for coordination of conductor and protective device in Regulation 433.1.1 have been complied with. The rating factor for C_c is therefore given as 1.0.

But if the protective device is a rewireable fuse to BS 3036, due to their potentially poor performance we always have a rating factor of C_c equals 0.725.

However, if we have a cable that is installed either in a duct in the ground or is buried directly in the ground, then we apply a rating factor of C_c equals 0.9.

If we therefore had a situation where a cable buried directly in the ground was protected by a BS 3036 fuse, both rating factors would apply and our new rating would be:

$C_c = 0.725 \times 0.9$ and therefore $C_c = 0.653$.

Mineral insulated cable

Table 4G1A (BS 7671) states that, for bare cables (i.e. no PVC outer covering) exposed to touch, the tabulated values should be multiplied by 0.9.

Table 4G2A states that no correction factor for grouping need be applied.

Application of rating factors

You will have seen that nearly all of these factors have a numerical value of less than 1. We also know that the cable must be able to carry current of an equal or greater value than the rating of the fuse that protects the cable.

Therefore, the purpose of applying these factors is to make sure that the cable's conductor is large enough to carry the current without too much heat being generated.

If the cable's ability to give off heat is reduced by external conditions, then the only solution is to increase the size of the cable.

The use and application of the correction factors will require you to increase the physical cross sectional area of the cable, which in turn increases the cable's capacity to carry current and give off heat.

When the right correction factors have been applied, we then know the effective current carrying capacity of the conductor (I_z).

We then need to apply the result from the formula to the correct cable table to find the correct cable size.

This value of current tabulated in Appendix 4 of BS 7671 is given the symbol I_t which must be greater than I_z. Hence:

$$I_t \geq I_z \geq \frac{I_n}{C_a \times C_g \times C_i \times C_c}$$

(if they apply)

Now you can use the value of I_z, together with the correct cable table from BS 7671 Table 4D1A to 4J4B, to find the size of cable that you need to use, given these correction factors.

There are some examples at the end for you to practise working with rating factors. However, this is not the end of cable selection and you must check for voltage drop, shock protection and thermal constraints. All these subjects will be covered in the next section.

Example 1

A single-phase circuit has a design current of 30 A. It is to be wired in flat two core 70°C PVC insulated and sheathed cable with cpc to BS 6004. It will have copper conductors and the cable is enclosed in trunking with four other similar cables. If the ambient temperature is 30°C and the circuit is to be protected by a BS 3036 fuse, what should be the nominal current rating of the fuse and the minimum csa of the cable conductor?

In answering these types of questions, it is a good idea to construct a table of information. This will help you understand and retain the process of cable selection. It also makes cable selection easier to understand.

Installation Method number will be 6/7 (installed in trunking), therefore Reference Method = B

As ambient temperature equals 30°C, then from Table 4B1 for 70° Celsius thermoplastic, $C_a = 1.0$.

As grouped, then from Table 4C1, $C_g = 0.6$ (there are five circuits and Reference Methods A to F apply).

Design current (I_b) was given as 30 A.

As the protective device rating (I_n) must be ≥ the rating of the design current (I_b) we can say that:

In = 30A.

But as the protective device is a BS 3036 fuse, a factor C_c equal to 0.725 must be applied

$$I_z = \frac{I_n}{C_a \times C_g \times C_c} = \frac{30}{1.0 \times 0.6 \times 0.725} = 69\text{ A}$$

We said that I_t must be ≥ I_z, therefore using table 4D2A (column 4), we see that a 16 mm^2 conductor size has a tabulated current rating of 69 A.

Therefore **the minimum conductor csa that can be used is 16 mm^2.**

Voltage drop

Regulation 525.3 of BS 7671 states that 'the voltage drop between the origin of the installation (usually the supply terminals) and a socket outlet (or the terminals of fixed current using equipment) doesn't exceed that stated in Appendix 12.

Appendix 12 goes on to state that for low voltage installations supplied from a public distribution system, then voltage drop should be no greater than 3 per cent of the supply voltage for lighting circuits and no greater than 5 per cent for other circuits.

This means that where the supply voltage is 230 V, a maximum voltage drop of 6.9 V for lighting or 11.5 V for other circuits is allowed.

In unusual circumstances where the final circuit has a length in excess of 100 m, BS 7671 now allows an increase in voltage drop of 0.005 per cent per metre of that circuit above 100 m, up to a maximum value of 0.5 per cent.

The voltage drops either because the resistance of the conductor becomes greater as the length of the cable increases or the cross-sectional area of the cable is reduced. This means that on long cable runs, the cable cross-sectional area may have to be increased, reducing the resistance allowing current to 'flow' more easily and reducing the voltage drop across the circuit.

We can calculate the maximum allowed voltage drop like this:

1. For single phase 230 volt lighting circuits:

   $$3\% = 230 \times \frac{3}{100} = 6.9 \text{ volts}$$

2. For three phase 400 volt systems:

   $$5\% = 400 \times \frac{5}{100} = 20 \text{ volts}$$

Therefore, when carrying out circuit calculations it is worth remembering we are allowed to have a minimum system voltage as low as 223.1 V on lighting circuits and 218.5 V on other circuits. It is better to keep the voltage drops as low as possible, because low voltages can reduce the efficiency of the equipment being supplied, thus lamps will not be as bright and heaters may not give off full heat. The values for cable voltage drop are given in the accompanying tables of current carrying capacity in Appendix 4 of the *IEE Wiring Regulations*. The values are given in milli volts per ampere per metre(mV/A/m).

You should use the formula below to calculate the actual voltage drop.

$$\text{Voltage drop (VD)} = \text{mV/A/m}\,\frac{I_b \times L}{1000}$$

Where mV/A/m is the value given in the Regulation Tables.

I_b is the circuit's design current.

L is the length of cable in the circuit measured in metres.

NB: As mV/A/m uses the prefix milli (thousandths) then you are going to have to convert I_b and L into the common unit and you do this by dividing them by 1000.

Note: Alternatively in some areas of the country people prefer the following equation for working out voltage drop.

$$\text{Voltage drop (VD)} = \text{mV/A/m} \times I_b \times L \times 10^{-3}$$

Example 2

To reinforce understanding, this is the same question that was used in Example 1, but this time you are also required to work out the voltage drop for the cable and consider its effect.

A single phase circuit has a design current of 30 A. It is to be wired in flat two core 70°C PVC insulated and sheathed cables with cpc to BS 6004. It will have copper conductors and the cable is enclosed in trunking with four other similar cables. If the ambient temperature is 30°C, the circuit is to be protected by a BS 3036 fuse and the circuit length is 27 m, what should be the nominal current rating of the fuse (I_n) and the minimum csa of the cable conductor?

Installation Method number will be 6/7 (installed in trunking) therefore Reference Method = B

As ambient temperature = 30°C then, from Table 4B1, for 70°C thermoplastic: $C_a = 1.0$

As grouped, then from Table 4C1, $C_g = 0.6$ (there are five circuits and Reference Methods A to F applies)

Design current (I_b) was given as 30 A.

As the protective device rating (I_n) must be ≥ the rating of the design current (I_b) we can say that **I_n = 30 A.**

But as the protective device is a BS 3036 fuse, the factor $C_c = 0.725$ must be applied

$$I_z = \frac{I_n}{C_a \times C_g \times C_c} = \frac{30}{1.0 \times 0.6 \times 0.725} = 69\text{ A}$$

We said that I_t must be ≥ I_z, therefore using Table 4D2A (column 4), we see that a 16 mm^2 conductor size has a tabulated current rating of 69 A and is therefore acceptable in this respect. We must now check the voltage drop for this cable.

From Table 4D2B (column 3) we see that the voltage drop is given as 2.8 mV/A/m. We need to work in common units, so we must divide I_b and L by 1000.

$$\text{Voltage Drop} = \text{mV/A/m} \times \frac{I_z \times L}{1000} = 2.8 \times \frac{(30 \times 27)}{1000} = 2.27\text{ V}$$

BS 7671 states that for other than a lighting circuit, voltage drop must not exceed 5 per cent of the supply voltage, which in this case would be 5 per cent of 230 V or 11.5 V. As the actual voltage drop of our proposed cable was calculated to be 2.27 V, this means that **the minimum csa of the cable conductor will be 16 mm^2.**

Shock protection

On page 53 of this book we looked at the protection measure Automatic Disconnection of Supply (ADS). With this measure, all metalwork including extraneous-conductive parts and exposed-conductive parts are connected to earth. Therefore, should an earth fault occur, we need the protective device to work quickly. The current must then be large enough to operate the device within given times and the earth fault loop impedance must be low enough to allow this to happen.

BS 7671 Regulation 411.3.2.1 states that a protective device shall automatically interrupt the supply to the line conductor (of a circuit or of equipment) in the event of a fault of negligible impedance between the line conductor and an exposed-conductive-part or protective conductor in the circuit or equipment, within the times given in Table 41.1 of BS 7671 (see table 2.13).

From this table we can see that for a 230 V final circuit (e.g. lighting, socket outlets, cooker circuits etc.) on a TN system the disconnection time is 0.4 seconds, but on a TT system is 0.2 seconds. Only for distribution circuits (i.e. a circuit supplying a DB or switchgear) is a time on TN of 5.0 seconds acceptable.

To achieve these times, the value of earth fault loop impedance (Z_s) should not be larger than the values given in Tables 41.2, 41.3 and 41.4 of BS 7671.

System	50 V < U_o < 120 V seconds		120 V < U_o < 230 V seconds		230 V < U_o < 400 V seconds		U_o < 400 V seconds	
	a.c.	d.c.	a.c.	d.c.	a.c.	d.c.	a.c.	d.c.
TN	0.8	NOTE 1	0.4	5	0.2	0.4	0.1	0.1
TT	0.3	NOTE 1	0.2	0.4	0.07	0.2	0.04	0.1

Table 2.13 Maximum disconnection times (Table 41.1 BS 7671)

You need to do a calculation to check that the circuit protective device will operate within the required time.

You do this by checking that the actual Z_s is lower than the maximum Z_s given in the relevant table, for the protective device you have chosen.

Maximum Z_s can also be found from *IEE Wiring Regulations* or manufacturers' data. Use the following formula to calculate the actual Z_s:

$$\text{Actual } Z_s = Z_e + (\{R_1 + R_2\} \times (\text{mf}) \times \frac{\text{length}}{1000})$$

Where:

- Z_e = the external impedance on the supply authority's side of the earth fault loop. You can get this value from the supply authority. Typical maximum values are: TN-C-S (PME) system 0.35 ohms, TN-S (cable sheath) 0.8 ohms, TT system 21 ohms.

- $R_1 + R_2$ = the resistance of the phase conductor plus the cpc resistance. You can find values of resistance/metre for $R_1 + R_2$, for various combinations of phase and cpc conductors up to and including 50 mm^2 in Table G1 of the *Unite Guide to Good Electrical Practice.*
- The following table gives you multipliers (mf) to apply to the values given in Table G1 to calculate the resistance under fault conditions. If the conductor temperature rises, resistance in the conductors will increase. This table is based on the type of insulation used and whether the cpc isn't incorporated in a cable or bunched with cables (54.2), or is incorporated as a core in a cable or bunched with cables (54.3).

	Insulation material		
Conductor installed as	**70° thermoplastic (pvc)**	**85° thermosetting (rubber)**	**90° thermosetting**
54.2 (not incorporated)	1.04	1.04	1.04
54.3 (incorporated)	1.20	1.26	1.28

Table 2.14 Multipliers to be applied to Table G1

This information, together with the length of the circuit, can now be applied to the formula. Then you can check to see that the actual Z_s is less than the maximum Z_s given in the appropriate tables.

When you have done this, you can prove that the protective device will disconnect the circuit in the time that is specified in the appropriate table. In other words, if the actual Z_s is less than the maximum Z_s then we have compliance for shock protection.

Example 3

To reinforce understanding, this is the same question that was used in Examples 1 and 2, but this time you are also required to work out the requirement for shock protection and consider its effect.

A single phase circuit supplying a domestic cooker has a design current of 30 A and is to be wired in flat two core 70°C PVC insulated and sheathed cables with cpc to
BS 6004. It will have copper conductors and the cable is enclosed in trunking with four other similar cables. If the ambient temperature is 30°C, the circuit is to be protected by a BS 3036 fuse and the circuit length is 27 m, what should be the nominal current rating of the fuse (I_n) and the minimum csa of the cable conductor? You are informed that Z_e is 0.8 Ω.

Installation Method number will be 6/7 (installed in trunking) therefore Reference Method = B.

As ambient temperature = 30°C then, from Table 4B1, for 70°C thermoplastic: $C_a = 1.0$

As grouped, then from Table 4C1, $C_g = 0.6$ (there are five circuits and Reference Methods A to F applies)

Design current (I_b) was given as 30 A.

As the protective device rating (I_n) must be ≥ the rating of the design current (I_b) we can say that **$I_n = 30$ A.**

But as the protective device is a BS 3036 fuse, the factor $C_c = 0.725$ must be applied.

$$I_z = \frac{I_n}{C_a \times C_g \times C_c} = \frac{30}{1.0 \times 0.6 \times 0.725} = 69\text{ A}$$

We said that I_t must be ≥ I_z, therefore using Table 4D2A (column 4), we see that a 16 mm² conductor size has a tabulated current rating of 69 A and is therefore acceptable in this respect. We must now check the voltage drop for this cable.

From Table 4D2B (column 3) we see that the voltage drop is given as 2.8 mV/A/m. We need to work in common units, so we must divide I_b and L by 1000.

$$\text{Voltage Drop} = \text{mV/A/m} \times \frac{I_z \times L}{1000} = 2.8 \times \frac{(30 \times 27)}{1000} = 2.27\text{ V}$$

BS 7671 states that for other than a lighting circuit, voltage drop must not exceed 5 per cent of the supply voltage, which in this case would be 5 per cent of 230 V or 11.5 V. As the actual voltage drop of our proposed cable was calculated to be 2.27 V, this is acceptable. We must now check for shock protection.

From Table 41.2, the maximum permitted Z_s for our 30 A BS 3036 fuse is 1.09 Ω and from Table G1 of the Unite guide, for a 16 mm² with 6 mm² cpc, the Resistance/m ($R_1 + R_2$/m) is given as 4.23 mΩ/m.

Additionally we need the multiplier for a grouped cable, which from our table on page 81 is 1.2. We also need to convert length from m to mm by dividing by 1000.Therefore:

$$Z_s = Z_e + ([R_1 + R_2] \times \text{mf} \times \frac{\text{length}}{1000}) = 0.8 + ([4.23] \times 1.2 \times \frac{27}{1000}) = 0.937\ \Omega$$

As actual Z_s (0.937 Ω) is less than the permitted value (1.09 Ω), this means that **16 mm² cable with 6 mm² cpc is the minimum csa of the cable conductor**.

Thermal constraints

Now that you have chosen the type and size of cable to suit the conditions of the installation, we must look at 'thermal constraints'. This is a check to make sure that the size of the cpc, 'the earth conductor', complies with the *IEE Wiring Regulations*.

If there is a fault on the circuit, which could be a short circuit, or earth fault, a fault current of hundreds or thousands of amperes could flow. Imagine that this is a 1 mm^2 or 2.5 mm^2 cable; if this large amount of current was allowed to flow for a short period of time, i.e. a few seconds, the cable would melt and a fire would start.

We need to check that the cpc will be large enough to be able to carry this fault current without causing any heat/fire damage. The formula that is used to check this situation is the **adiabatic equation**. The cpc will only need to carry the fault current for a short period of time, until the protective device operates.

Regulation 543.1.1 states that 'The cross-sectional area of every protective conductor shall be calculated in accordance with Regulation 543.1.3 (adiabatic equation) or selected in accordance with Regulation 543.1.4 (Table 54.7)'. You will see that Regulation 543.1.4 asks that reference should be made to Table 54.7. This table shows that, for cables 16 mm^2 and below with the cpc made from the same material as the line conductor, the cpc should be the same size as the line conductor. A line conductor between 16 mm^2 and 35 mm^2 requires a cpc to be 16 mm^2. A line conductor above 35 mm^2 requires a cpc to be at least half the cross-sectional area.

Multicore cables have cpcs smaller than their respective line conductors, except for 1 mm^2, which has the same-sized cpc. Regulation 543.1.2 of BS 7671, gives two options, calculation or selection. If we were to apply option (ii) of this Regulation then selection would make these cables contravene the Regulations. Clearly, it is not intended that composite cables should have their cpcs increased in accordance with the table and therefore calculation by the adiabatic equation required in option (i) of the same Regulation should be applied.

On the job: Intermittent fault

You have been asked by your employer to investigate an 'earthing' problem that has been reported by a customer. The customer concerned has just moved into a large 80-year-old house that will allow him to work from home as a computer software designer. The house has large gardens, a separate garage, three outside security lights, a gas-fired central-heating system and the customer will be working from one of the bedrooms on his computers.

The customer has reported that he is getting intermittent RCD trips on the main consumer unit. The only things the customer has noticed is that the trips seem to occur only at night, and only when it has been raining. Rain during the day does not cause the same problems. There is also a circuit out to the garage that has its own RCD and this does not trip.

1 What do you think could be causing the problem?

(A suggested answer can be found on page 332)

The adiabatic equation referred to in the introduction enables a designer to check the suitability of the cpc in a composite cable. If the cable does not incorporate a cpc, a cpc installed as a separate conductor may also be checked.

The equation is as follows:

$$S = \frac{\sqrt{I_f^2 \times t}}{k}$$

Where:

S = the cross-sectional area of the cpc in mm^2.

I_f = the value of the fault current in amperes.

t = the operating time of the disconnecting device in seconds.

k = a factor that takes into account resistivity, temperature coefficient and heat capacity.

In order to apply the adiabatic equation, we first need to calculate the value of I (fault current) from the following equation:

$$I_f = \frac{U_o}{Z_s}$$

Where:

U_o is the nominal supply line voltage to earth.

Z_s is the earth fault loop impedance.

If you are using method (i) from Reg 543.1.2 and applying the adiabatic equation, you must find out the time/current characteristics of the protective device. A selection of time/current characteristics for standard overcurrent protective devices is given in Appendix 3 of *IEE Wiring Regulations*. You can get the time (t) for disconnection to the corresponding earth fault current from these graphs.

If you look at the time/current curve you will find that the scales on both the time (seconds) scale and the prospective current (amperes) scale are logarithmic and the value of each subdivision depends on the major division boundaries into which it falls.

For example, on the current scale, all the subdivisions between 10 and 100 are in quantities of 10, while the subdivisions between 100 and 1000 are in quantities of 100 and so on. This also occurs with the time scale, subdivisions between 0.01 and 0.1 being in hundredths and the subdivisions between 0.1 and 1 being in tenths, etc.

As an example, if you look at the graph shown in Figure 2.20 you will see that for a BS 88 fuse with a rating of 32 A, a fault current of 200 A will cause the fuse to clear the fault in 0.6 seconds.

In addition to this graph, the IEE has produced a small table at the top right-hand corner showing some of the more common sizes of protective devices and the fault currents for a given disconnection time.

Next, you need to select the k factor using Tables 54.2 to 54.6. The values in these tables are based on the initial and final temperatures shown in each table. This is where you may need to refer to the cable's operating temperature shown in the cable tables.

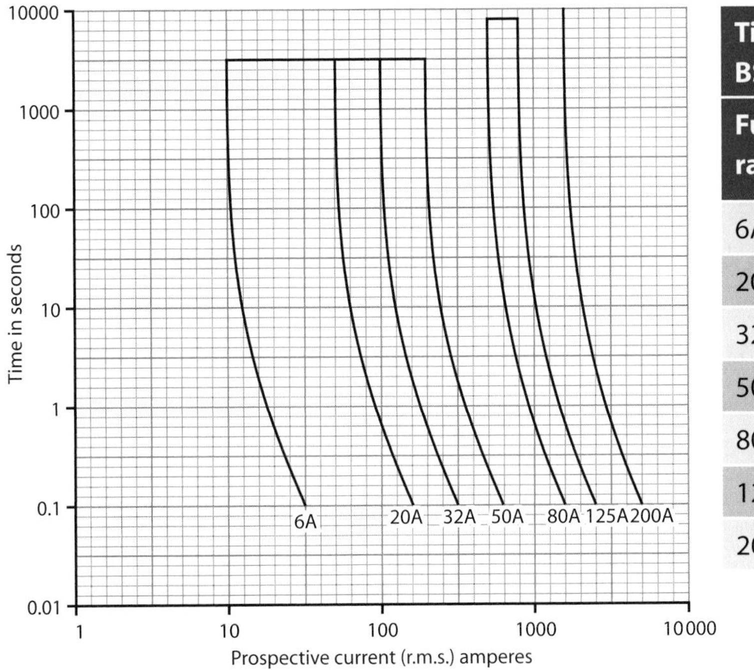

Time/Current characteristics for fuses to BS 88.2.2 and BS 88.6					
Fuse rating	**Current for time**				
	0.1 sec	**0.2 sec**	**0.4 sec**	**1 sec**	**5 secs**
6A	36A	31A	27A	23A	17A
20A	175A	150A	130A	110A	79A
32A	320A	260A	220A	170A	125A
50A	540A	450A	380A	310A	220A
80A	1100A	890A	740A	580A	400A
125A	1800A	1500A	1300A	1050A	690A
200A	3000A	2500A	2200A	1700A	1200A

Figure 2.20 Time/current characteristics graph

Now you must substitute the values for I, t and k into the adiabatic equation. This will give you the minimum cross-sectional area for the cpc. If your calculation produces a non-standard size, you must use the next largest standard size.

From a designer's point of view, it is advantageous to use the calculation method as this may lead to savings in the size of cpc.

Now try the following example to give you a complete understanding of cable selection. The example is similar to those used in Cable Selection Examples 1, 2 and 3, but this time you also need to complete calculations for thermal constraints.

Example 4

To reinforce understanding, this is the same question that was used in Examples 1, 2 and 3, but this time we will only concern ourselves with the calculations relevant to thermal constraints. You are also required to work out the requirement for shock protection and consider its effect.

A single phase circuit supplying a domestic cooker has a design current of 30 A and is to be wired in flat two core 70°C PVC insulated and sheathed cables with cpc to BS 6004. It will have copper conductors and the cable is enclosed in trunking with four other similar cables. If the ambient temperature is 30°C, the circuit is to be protected by a BS 3036 fuse and the circuit length is 27 m, what should be the nominal current rating of the fuse (I_n) and the minimum csa of the cable conductor?

You are informed that Z_e is 0.8 Ω.

From our previous calculations in Example 3, we know that **I_n = 30 A** and that **16 mm^2 cable** with 6 mm^2 cpc **is the minimum csa of the cable conductor.**

For thermal constraints, we need to know the fault current before we can apply the adiabatic equation. We find this by:

$$I_f = \frac{U_o}{Z_s} = \frac{230}{0.937} = 245.5\ \text{A}$$

We also need to know the value of t and the value of k. Therefore from Table 3.2A within Appendix 3 of BS 7671 we can see that our fault current of 245.5 A will operate the device in a time of 0.3 second. We therefore state t as being 0.3 second.

From Table 54.3 we can see that the value of k will be 115, as the cable is under 300 mm^2.

Now using the adiabatic equation:

$$S = \frac{\sqrt{I_f^2 \times t}}{K} = \frac{\sqrt{245.5^2 \times 0.3}}{115} = \frac{\sqrt{18081.08}}{115} = \frac{134.5}{115} = 1.17\ mm^2$$

As 1.17 mm^2 is less than our 6 mm^2 cpc, our choice of cable remains as a cable with 16 mm^2 conductors and a 6 mm^2 cpc to ensure compliance with thermal constraints.

Diversity

In this section we will look at Maximum Demand and Diversity as considered in BS7671 Chapter 31. The current demand for a final circuit is determined by adding up the current demands of all points of utilisation and equipment in the circuit and, where appropriate, making an allowance for diversity. Diversity makes allowances on the basis that not all of the load or connected items will be in use at the same time.

In most cases main or sub-main cables will supply a number and/or variety of final circuits. Use of the various loads must now be considered, otherwise if all the loads are totalled, a larger cable than necessary will be selected at considerable extra cost. Therefore a method of assessing the load must be used.

Section 3 of the *Unite Guide to Good Electrical Practice* contains such a method, which allows diversity to be applied depending upon the type of load and installation premises. The individual circuit/load figures are added together to determine the total 'Assumed Current Demand' for the installation. This value can then be used as the starting point to determine the rating of a suitable protective device and the size of cable, considering any influencing factors in a similar manner to that applied to final circuits.

Please also remember that the calculation of maximum demand is not an exact science and a suitably qualified electrical engineer may use other methods of calculating maximum demand. This subject is also discussed under 'Load balancing' in Book 1.

Example

A 230 volt domestic installation consists of the following loads:

- 15 × filament lighting points
- 6 × fluorescent lighting points, each rated at 40 watts
- 4 × fluorescent lighting points each rated at 85 watts
- 3 × ring final circuits supplying 13 A socket outlets
- 1 × radial circuit protected by a 20 A device supplying 13 A sockets for the adjoining garage
- 1 × 3 kW immersion heater with thermostatic control
- 1 × 13.6 kW cooker with a 13 A socket outlet incorporated in the control unit.

Determine the maximum current demand for determining the size of the sub-main cable required to feed this domestic installation? The circuit protection is by the use of BS 1361 fuses.

Answer

Lighting

Tungsten light points (See page 40 of the *Unite Guide to Good Electrical Practice*)

15 × 100W minimum 1500

Fluorescent light points (See page 40 and note of the *Unite Guide to Good Electrical Practice*)

6 × 40W with multiplier of 1.8 (40 x 1.8 = 72) 432

4 × 85W with multiplier of 1.8 (85 x 1.8 = 153) 612

Total 2544 W

Using Item 1 of the table on page 41 of the *Unite Guide to Good Electrical Practice*, we can apply diversity as being 66% of the total current demand. Therefore:

66% of 2544 = 1679 W and since $I = \frac{P}{V}$ this gives us $\frac{1679}{230} =$ 7.3 A

Power (Item 9 on page 41 of the *Unite Guide to Good Electrical Practice*)

3 Ring final circuits

1 × ring at 100% rating (30 A) 30 A

2 × ring at 40% rating (40% of 30 A) 24 A

20 A radial circuit (see item 9 on page 41 of the *Unite Guide to Good Electrical Practice*)

1 × radial at 40% rating (40% of 20 A) 8 A

3 kW Immersion Heater (See Item 6 on page 41 of the *Unite Guide to Good Electrical Practice*)

3 kW heater with no diversity $I = \frac{P}{V} = \frac{3000}{230}$ which gives us 13 A

13.6 kW Cooker with socket outlet (See item 3 on page 41 of the *Unite Guide to Good Electrical Practice*)

The first 10 A, plus 30% of the remainder of the overall rated current, plus 5 A for the socket.

$I = \frac{P}{V}$ giving $\frac{136000}{230}$ which gives us a total rated current of 59 A

59 A – 10 A = 49 A and therefore 30% of 49 A = 14.7 A

The allowable total cooker rating is therefore 10 A + 14.7 A + 5 A = 29.7 A

Our total assumed current demand is therefore :

7.3 + 30 + 24 + 8 + 13 + 29.7 giving a total of **112 A**

Knowledge check

1. What could be the result of a bad earth in the event of an earth fault?
2. In a TT system what does the first letter T represent?
3. In a TN-S system what does the third letter S represent?
4. With which type of system is the PME associated?
5. List the disconnection times for lighting and socket outlet circuits in domestic premises.
6. What reference methods should be used for cables in thermal insulation?
7. Explain the formula for calculating voltage drop.
8. Describe the earth-fault loop impedance.
9. Explain the term overload.
10. Why must you always have a fuse in a circuit?
11. Why does a circuit breaker have both thermal and magnetic devices?
12. What is the type of MCB required for protecting a domestic installation and why?
13. What value of impedance is likely to be recorded with a short circuit?
14. Explain the effects of a short circuit current.
15. What is meant by the term 'discrimination between protective devices'?
16. What type of protective device would normally have a fusing factor of around 1.8–2.0?

Chapter 3

Lighting

OVERVIEW

Illumination by means of electricity has been available for over 100 years. In that time it has changed in many ways, though many of the same ideas are still in use. The first type of electric lamp was the 'arc lamp', which used electrodes to draw an electrode through the air, this is now known as discharge lighting. This was quite an unsophisticated use of electricity, and many accidents and fires resulted from it. Regulations had to be developed to control discharge lighting installations, and it is interesting that the first edition of the Regulations, introduced in 1882, was entitled 'Rules and Regulations for the Prevention of Fire Risks Arising from Electric Lighting'.

This chapter will cover:

- Lamp caps
- Incandescent lamps
- Regulations concerning lighting circuits
- Discharge lighting
- Calculating lighting requirement

Lamp caps

Before we look at the types of lamp available, we should look at lamp caps. The cap is that part of the lamp that allows an electrical connection to be made with the supply. There are many different types, some of which are shown below.

The Bayonet Cap

The Bayonet Cap (BC) is probably the lamp you have come across most often. It is 22 mm diameter and has two locating lugs, the electrical contact made over two pins on the base of the cap. Two popular variations of this are the SBC cap which is 15 mm in diameter and the SCC cap. The SCC cap only has one contact on the base, the other contact being the cap itself.

Bayonet caps

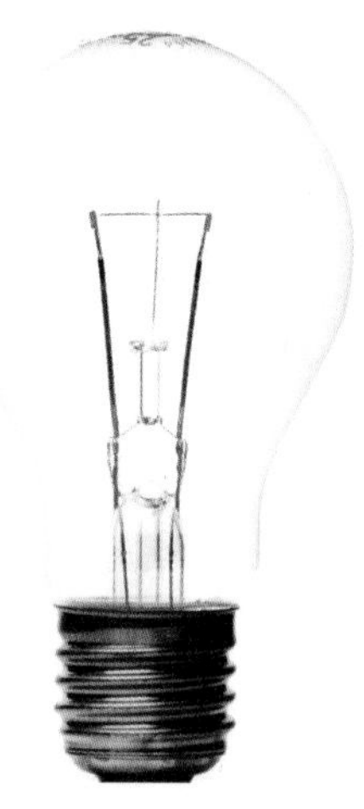

The Edison Screw cap

Most Edison Screw (ES) lamps are represented as the letter 'E' followed by a number. This number denotes the diameter of the cap. The most popular types for domestic use in the UK are E14 (14 mm and also known as SES) and E27. There is also a version used in street lighting and industrial situations with mercury fluorescent lamps, the E40 or GES (Giant Edison Screw).

Edison Screw caps

Halogen capsule lamp

Halogen lamp caps

There are three common types of halogen lamp camp:

- **Halogen capsule lamps** are generally designated by the distance in millimetres between the pins of the lamp. The most common of these, G4 (2 pins – 4 mm apart) is used in low voltage applications such as desk lamps.

- **Linear tungsten halogen lamps** are mains voltage and normally seen in security lights, floodlights and some up-lighters. They have what is known as an R7 cap at each end of a thin gas-filled quartz tube.

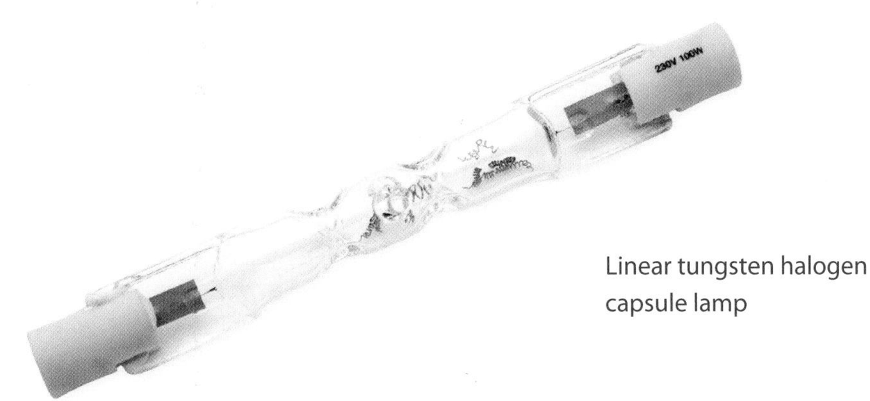

Linear tungsten halogen capsule lamp

- **Halogen spotlights** have become ever more popular and are seen in many domestic applications such as bathrooms, dining rooms and kitchens or commercially in display applications. The most common is the GU10 version shown opposite.

Halogen spotlights

Low pressure mercury (fluorescent) caps)

Using their common name, fluorescent tubes have a bi-pin cap at both ends of the gas filled tube. Diameters of the tube range from T5 (16 mm) through T8 (25 mm) to T12 (38 mm). T8 and T12 tubes normally have pins that are 13 mm apart, whereas the T5 tube has pins that are 5 mm apart.

Low pressure mercury (fluorescent) caps

Incandescent lamps

Remember

The filament wire in an incandescent lamp reaches a temperature of about 2500–2900°C

In this method of creating light a fine filament of wire is connected across an electrical supply, which makes the filament wire heat up until it is white-hot and gives out light. The filament wire reaches a temperature of 2500–2900°C. These lamps are very inefficient and only a small proportion of the available electricity is converted into light; most of the electricity is converted into heat as infrared energy. The light output of this type of lamp is mainly found at the red end of the visual spectrum, which gives an overall warm appearance.

Operation of GLS lamps

The General Lighting Service (GLS) lamp is one type of incandescent lamp and is commonly referred to as the 'light bulb'. It has at its 'core' a very thin tungsten wire that is formed into a small coil and then coiled again.

A current is passed through the tungsten filament, which causes it to reach a temperature of 2500°C or more so that it glows brightly. At these temperatures, the oxygen in the atmosphere would combine with the filament to cause failure, so all the air is removed from the glass bulb and replaced by gases such as nitrogen and argon. Nitrogen is used to minimise the risk of arcing and argon is used to reduce the evaporation process. On low-power lamps such as 15 and 25 watt, the area inside the bulb remains a vacuum. The efficiency of a lamp is known as the efficacy. It is expressed in lumen per watt, lm/w. For this type of lamp the efficacy is between 10 and 18 lumens per watt. This is low compared with other types of lamp, and its use is limited. However, it is the most familiar type of light source used and has many advantages including:

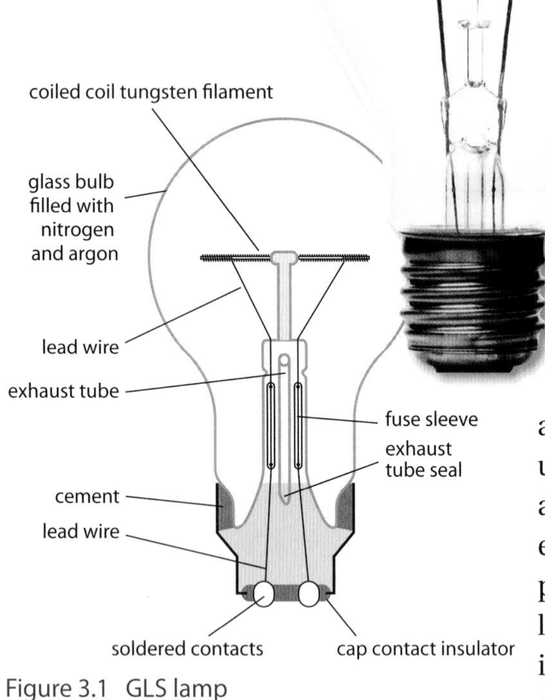

Figure 3.1 GLS lamp

- comparatively low initial costs
- immediate light when switched on
- no control gear
- it can easily be dimmed.

When a bulb filament finally fails it can cause a very high current to flow for a fraction of a second – often sufficient enough to operate a 5 or 6 amp miniature circuit breaker which protects the lighting circuit. High-wattage lamps, however, are provided with a tiny integral fuse within the body of the lamp to prevent damage occurring when the filament fails.

Did you know?

The first lamp that was developed for indoor use was the carbon-filament lamp. Although this was a dim lamp by modern standards it was cleaner and far less dangerous than the exposed 'arc lamp'

If the lamp is run at a lower voltage than that of its rating this results in the light output of the lamp being reduced at a greater rate than the electricity used by the lamp, and the lamp's efficacy is poor. This reduction in voltage, however, increases the lifespan and can be useful where lamps are difficult to replace or light output is not the main consideration.

It has been calculated that an increase in 5 per cent of the supply voltage can reduce the lamp life by half. However, if the input voltage is increased by just 1 per cent this will produce an increase of 3.5 per cent in lamp output (lumens). When you consider that the Electricity Supply Authority is allowed to vary its voltage up to and including 6 per cent it is easy to see that if this was carried on for any length of time the lamps would not last very long .

Did you know?

The average life of this type of lamp is 1,000 hours, after which the filament will rupture

Tungsten halogen lamps

These types of lamps were introduced in the 1950s. For their operation the tungsten filament is enclosed in a gas-filled quartz tube together with a carefully controlled amount of halogen such as iodine. Figure 3.2 illustrates the linear tungsten halogen lamp.

Did you know?

Running the lamp at a higher voltage than it was designed to do results in a shorter lamp life

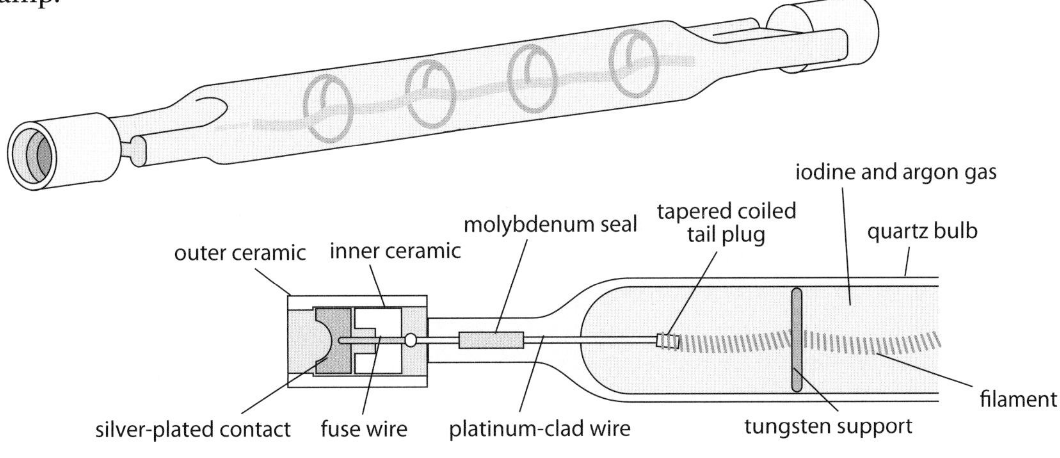

Figure 3.2 Linear tungsten lamp

Operation of tungsten halogen lamps

The inclusion of argon and iodine in the quartz tube allows the filament to burn at a much higher temperature than the incandescent lamp. The inclusion of the halogen gas produces a regeneration effect which prolongs the life of the lamp.

As small particles of tungsten fall away from the filament, they combine with the iodine passing over the face of the quartz tube, forming a new compound. Convection currents in the tube cause this new compound to rise, passing over the filament. The intense heat of the filament causes the compound to separate into its component parts, and the tungsten is deposited back on the filament.

The lamp should not be touched with bare fingers as this would deposit grease on the quartz glass tube; this would lead to small cracks and fissures in the tube when the lamp heats rapidly, causing the lamp to fail. If accidentally touched on installation, the lamp should be cleaned with methylated spirit before being used.

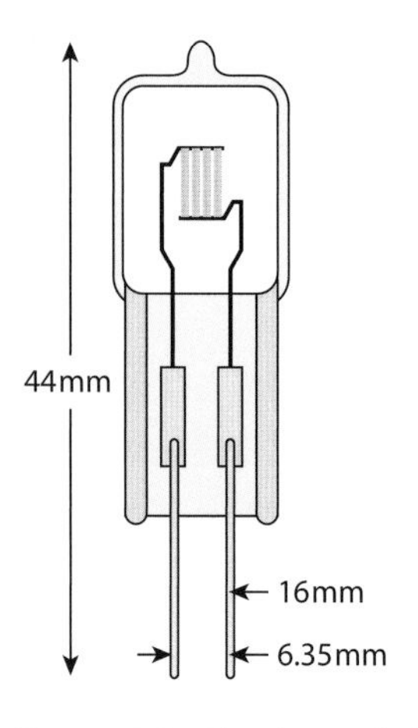

Figure 3.3 Single-ended filament lamp

The linear type of lamp must be installed within 4° of the horizontal to prevent the halogen vapour migrating to one end of the tube, causing early failure. These types of lamps have many advantages, which include:

- increased lamp life (up to 2000 hours)
- increase in efficacy (up to 23 lumens per watt)
- reduction in lamp size.

There are two basic designs that have been produced: the double-ended linear lamp and the single-ended lamp, with both contacts embedded in the seal at one end (see Figure 3.3). This type of lamp has been produced to work on extra-low voltages: they are used extensively in the automobile industry for vehicle headlamps. They may also be used for display spotlights where extra-low voltage is desirable. They may be supplied from an in-built 230 volt/12 volt transformer.

FAQ

Q Why does a MCB trip when a GLS lamp stops working?

A In the cheaper lamps there are no internal fuses; therefore, what sometimes happens is that when the filament eventually burns out, some of the filament falls way, the remaining ends touch together, completing the circuit, and current flows. The filament now has a much lower resistance, which allows more current to flow, and the lamp often appears to be much brighter. The raised level of current causes the MCB to trip.

Q Where would I use a tungsten halogen luminaire?

A These lamps are much brighter than a GLS lamp and are therefore used where a cheap source of area floodlighting is required – frequently in conjunction with a PIR motion detector. Modern downlighters, either extra low voltage or low voltage, are now commonplace in both commercial and domestic installations.

Regulations concerning lighting circuits

The 17th Edition of the IEE Regulations (BS 7671) saw the introduction of a new section (559) within Chapter 55 that applies to:

- the selection and erection of luminaires and lighting installations that are intended to be part of the fixed installation (559.1)
- highway power supplies and street furniture (559.1).

Particular requirements are specified for the following:

- Fixed outdoor lighting installations.
- Extra low voltage installations supplied at up to 50 V a.c. or 120 V d.c.
- Lighting for display stands.

The definitions within Regulation 559.3, mean that lighting installations for roads, parks, car parks, gardens, places open to the public, sporting areas, illuminating monuments, floodlighting, telephone kiosks, bus shelters, advertising panels, road signs and road traffic signalling systems are now all covered by BS 7671.

Road signs and traffic signals are now covered by BS 7671

However, BS 7671 does not cover the following:

- High voltage signs supplied at low voltage (e.g. neon tubes)
- Signs and luminous discharge tube installations in excess of 1 kV.

Although Section 559 sees the introduction of some 44 new regulations, we're only going to look at the most appropriate ones in the following text.

Regulation 559.4 – requires that all luminaires and track systems comply with the relevant standards for manufacture.

Regulation 559.5 – requires that the selection and erection process considers the thermal effect of radiant and convected energy on to surroundings. This means considering the power of lamps and the fire resistance of nearby materials.

Regulation 559.6 requires that at each fixed lighting point, one of the following shall be used:

- ceiling rose
- luminaire
- luminaire support coupler
- batten lampholder or pendant set
- suitable socket outlet/connection unit
- suitable lighting distribution unit
- device for connecting a luminaire (DCL).

Regulation 559.6 also states that:

- a batten lampholder/ceiling rose for a filament lamp shall not be used for a circuit in excess of 250 V
- a ceiling rose can't be used for attaching more than one pendant unless it has been designed for that purpose
- luminaire support couplers provide for connection to, and support of, the luminaire. They must not be used to connect any other item of equipment
- adequate means of fixing the luminaire must be provided
- lighting circuits that include B15, B22, E14, E27 or E40 lampholders shall have a protective device for the circuit of no greater than 16 A
- except for E14 and E27 lampholders to BS EN 60238, in circuits on a TN or TT system, the outer contact of every ES lampholder or single-centre bayonet cap lampholder shall be connected to the neutral conductor
- the installation of through wiring in a luminaire is only permitted if the luminaire has been designed accordingly
- if through wiring is permitted then any cabling must be selected in accordance with the temperature information on that luminaire. In the absence of such information then heat resistant cables and/or conductors must be used
- groups of luminaires divided between the three line conductors of a three phase system but having a common neutral, must have a linked circuit breaker that simultaneously disconnects all the line conductors.

Did you know?

E14 (SES) and E27 (ES) holders are in common use. The reason for this exemption regarding polarity is that the structure of the lampholder to BS EN 60238 ensures that with the bulb in place it is impossible to touch any live metal, and on removing the lamp, before the metal threaded part begins to show, the centre contact has already been broken.

It is still good practice to connect the outer threaded contact to Neutral, however for this type of lampholder it isn't mandatory. This is now reflected in inspection and testing as well.

Regulation 559.9 recognises the stroboscopic effect mentioned later in this chapter.

Regulations 559.10.1 and 559.10.2 generally state that for outdoor lighting installations, highway power supplies and street furniture, the protective measures 'placing out of reach', 'obstacles' and 'earth-free-equipotential bonding' shall not be used.

Regulation 559.10.3 states that:

- where the protective measure ADOS is used, then all live parts of equipment shall be protected by insulation or barriers/enclosures providing basic protection. However, a door in street furniture used for access to electrical equipment cannot be classed as a barrier/enclosure
- for every accessible enclosure, all live parts must only be accessible by using a key/tool unless the enclosure is located where it can only be accessed by a skilled person
- for a luminaire mounted less than 2.8 m above ground level, access shall only be possible after removing a barrier/enclosure that requires the use of a tool
- for an outdoor lighting installation, any nearby metallic structure such as a fence that is not part of the installation, need not be connected to the main earthing terminal

- it is recommended that equipment within the lighting arrangements of places such as telephone kiosks, bus shelters etc., be provided with additional protection in the form of an RCD
- a maximum disconnection time of 5 seconds shall apply to all circuits feeding fixed equipment in highway power supplies.

Regulation 559.10.6.1 states that as long as certain measures are taken, a suitable rated fuse carrier may be used as the means of isolation. However, **559.10.6.2** states that if the distributor's cut out is used as the means of isolation, then distributor permission must be obtained.

Regulation 559.11.5 states that the protective measure FELV cannot be used. The introduction of this new section in the Regulations also sees the introduction of new symbols and these are contained in Table 55.2 as detailed in Figure 3.4.

Description	Symbol
Luminaire with limited surface temperature (BS EN 60598 series)	D
Luminaire suitable for direct mounting on normally flammable surfaces	F
Luminaire suitable for direct mounting on non-combustible surfaces only	F
Luminaire suitable for mounting in/on normally flammable surfaces where a thermal insulating material may cover the luminaire **Note**: The marking of the symbols corresponding to IP numbers is optional.	F
Use of heat-resistant supply cables, interconnecting cables or external wiring. (The number of cores shown is optional)	t......°C
Luminaire designed for use with bowl mirror lamps	
Rated maximum ambient temperature	t_a .....°C
Warning against the use of cool beam lamps	COOL BEAM
Minimum distance from lighted objects (m)	---m
Rough service luminaires	
Luminaire for use with high pressure sodium lamps that require an external ignitor (to the lamp)	E
Luminaire for use with high pressure sodium lamp having an internal starting device	I
Replace any cracked protective shield	
Luminaire designed for use with self shielded tungsten halogen lamps only	
Transformer – short-circuit proof (both inherently and non-inherently)	
Electronic convertor for an extra-low voltage lighting installation	110
A "class P" thermally protected ballast(s)/transformer(s)	P
A temperature declared thermally protected ballast(s)/transformers(s) with a marked value equal to or below 130°C	• • •
The generally recognised symbol is of an independent ballast of EN60417	

Note: These symbols are referenced within BS EN 60598-1:2004. However, some of these symbols at the time of going to press, are the subject of change; the reader is advised to consult the latest edition of BS EN 60598 for current luminaire marking requirements

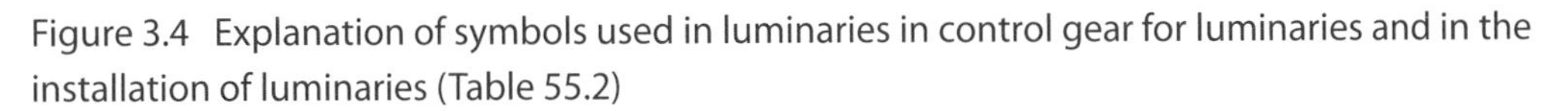

Figure 3.4 Explanation of symbols used in luminaries in control gear for luminaries and in the installation of luminaries (Table 55.2)

Discharge lighting

This is a term that refers to illumination derived from the ionisation of gas. This section looks at:

- low-pressure mercury vapour lamps
- the glow-type starter circuit
- semi-resonant starting
- high frequency
- stroboscopic effect
- other methods of starting the fluorescent tube
- other discharge lamps.

Low-pressure mercury vapour lamps

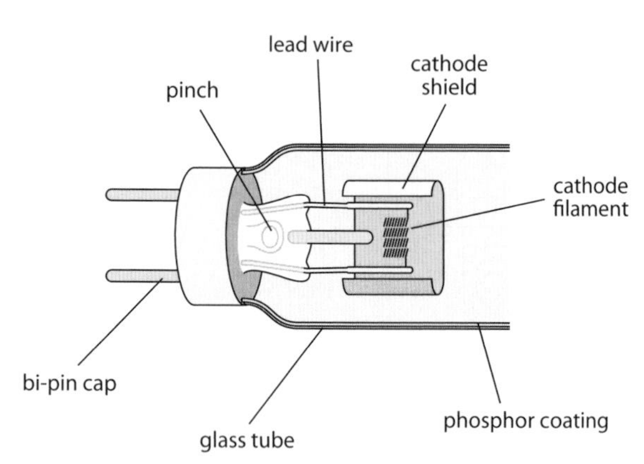

Figure 3.5 Detail of one end of a fluorescent tube

The fluorescent lamp, or, more correctly, the low-pressure mercury vapour lamp, consists of a glass tube filled with a gas such as krypton or argon and a measured amount of mercury vapour. The inside of the glass tube has a phosphor coating, and at each end there is a sealed set of oxide-coated electrodes, known as cathodes.

When a voltage is applied across the ends of a fluorescent tube the cathodes heat up, and this forms a cloud of electrons, which ionise the gas around them. The voltage to carry out this ionisation must be much higher than the voltage required to maintain the actual discharge across the lamp. Manufacturers use several methods to achieve this high voltage, usually based on a **transformer** or **choke**. This ionisation is then extended to the whole length of the tube so that the arc strikes and is then maintained in the mercury, which evaporates and takes over the discharge. The mercury arc, being at low pressure, emits little visible light but a great deal of ultraviolet, which is absorbed by the phosphor coating and transformed into visible light.

The cathodes are sealed into each end of the tube and consist of tungsten filaments coated with an electron-emitting material. Larger tubes incorporate cathode shields – iron strips bent into an oval shape to surround the cathode. The shield traps material given off by the cathodes during the tube life and thereby prevents the lamp-ends blackening.

The type of phosphorus used on the lamp's inner surface will determine the colour of light given out by the lamp

The gas in standard tubes is a mercury and argon mix, although some lamps (the smaller ones and the new slim energy-saving lamps) have krypton gas in them. The phosphor coating is a very important factor affecting the quantity and quality of light output. When choosing different lamps there are three main areas to be considered:

- lamp efficacy
- colour rendering
- colour appearance.

Lamp efficacy

This refers to the lumen output for a given wattage. For fluorescent lamps this varies between 40 and 90 lumens per watt.

Colour rendering

This describes a lamp's ability to show colours as they truly are. This can be important depending on the building usage. For example, it would be important in a paint shop but less so in a corridor of a building. The rendering of colour can affect people's attitude to work etc. – quite apart from the fact that in some jobs true colour may be essential. By restoring or providing a full colour range the light may also appear to be better or brighter than it really is.

Colour appearance

This is the actual look of the lamp, and the two ends of the scale are warm and cold. These extremes are related to temperatures: the higher the temperature, the cooler the lamp. This is important for the overall effect, and generally warm lamps are used to give a relaxed atmosphere while cold lamps are used where efficiency and businesslike attitudes are the priorities. The subject of lamp choice has become very complicated, and a programme of lamp rationalisation has begun. The intention is that the whole range currently available will be reduced. Also, new work has resulted in lamps with high lumen outputs and good colour rendering possibilities.

The glow-type starter circuit

In the starter, a set of normally open contacts is mounted on bi-metal strips and enclosed in an atmosphere of helium gas. When switched on, a glow discharge takes place around the open contacts in the starter, which heats up the bi-metal strips, causing them to bend and touch each other. This puts the electrodes at either end of the fluorescent tube in circuit and they warm up, giving off a cloud of electrons; simultaneously an intense magnetic field is building up in the choke, which is also in circuit.

The glow in the starter ceases once the contacts are touching so that the bi-metal strips now cool down and they spring apart again. This momentarily breaks the circuit, causing the magnetic field in the choke to rapidly collapse. The high back-e.m.f. produced provides the high voltage required for ionisation of the gas and

Did you know?

The most economic tube life is limited to around 5,000 or 6,000 hours. In industry, tubes are changed at set time intervals and all the tubes, whether still working or not, are replaced. This saves money on maintenance, stoppage of machinery and scaffolding erection etc.

Remember

The capacitor connected across the supply terminals is to correct the poor power factor that has been created by this type of light

enables the main discharge across the lamp to take place. The voltage across the tube under running conditions is not sufficient to operate the starter and so the contacts remain open.

The resistance of the ionised gas gets lower and lower as it warms up and conducts more current. This could lead to disintegration of the tube. However, the choke has a secondary function as a current-limiting device: the impedance of the choke limits the current through the lamp, keeping it in balance. This is one reason why it is often referred to as ballast.

This type of starting may not succeed first time and can result in the characteristic flashing on/off, when initially switching it on.

Remember

Once the tube strikes, current no longer passes through the starter; therefore the starter takes no further part in the circuit

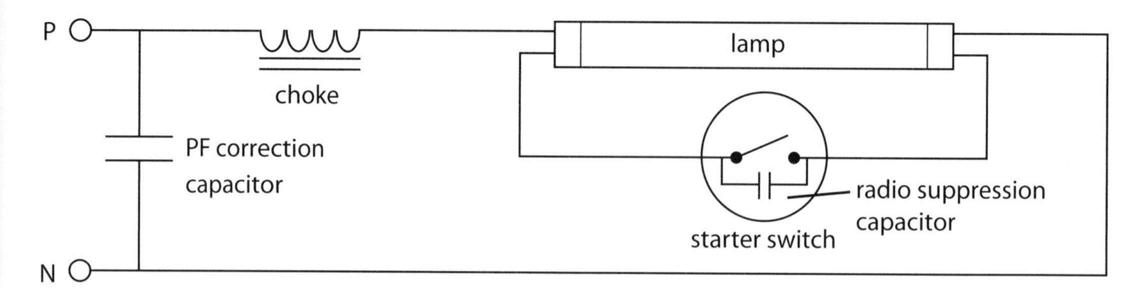

Figure 3.6 Glow-type starter circuit

Semi-resonant starting

In this circuit the place of the **choke** is taken by a specially wound transformer. Current flows through the primary coils to one cathode of the lamp and then through the secondary coil, which is wound in opposition to the primary coil. A fairly large **capacitor** is connected between the secondary coil and the second cathode of the lamp (the other end of which is connected to the neutral).

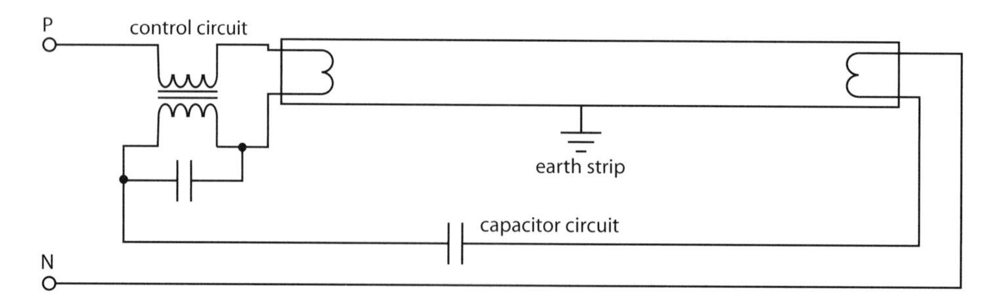

Figure 3.7 Semi-resonant circuit

The current that flows through this circuit heats the cathodes and, as the circuit is predominantly capacitive, the pre-start current leads the main voltage. Owing to the primary and secondary windings being in opposition, the voltages developed across them are 180° out of phase, so that the voltage across the tube is increased, causing the arc to strike. The primary windings then behave as a choke, thus stabilising the current in the arc. The circuit has the advantage of high power factor and easy starting at low temperatures.

High frequency

Standard fluorescent circuits operate on a mains supply frequency of 50 Hz; however, high-frequency circuits operate on about 30,000 Hz. There are a number of advantages to high frequency:

- higher lamp efficacy
- first-time starting
- noise-free
- the ballast shuts down automatically on lamp failure.

The higher efficacy for this type of circuit can lead to savings of at least 10 per cent, and in some large installations they may be as high as 30 per cent. Many high-frequency electronic ballasts will operate on a wide range of standard fluorescent lamps. The high-frequency circuit will switch on the lamp within 0.4 of a second and there should be no flicker (unlike glow-type starter circuits). Stroboscopic effect is not a problem as this light does not flicker on/off due to its high frequency.

However, a disadvantage of this type of circuit is that supply cables installed within the luminaire must not run adjacent to the leads connected to the ballast output terminals as interference may occur. Also, the initial cost of these luminaires is greater than traditional glow-type switch starts.

Stroboscopic effect

A simple example of this effect is that, when watching the wheels rotate on a horse-drawn cart on television, it appears that the wheels are stationary or even going backwards. This phenomenon is brought about by the fact that the spokes on the wheels are being rotated at about the same number of revolutions per second as the frames per second of the film being shot. This effect is known as the 'stroboscopic effect' and can also be produced by fluorescent lighting.

The discharge across the electrodes is extinguished 100 times per second, producing a flicker effect. This flicker is not normally observable but it can cause the stroboscopic effect, which can be dangerous. For example, if rotating machinery is illuminated from a single source it will appear to have slowed down, changed direction of rotation or even stopped. This is a potentially dangerous situation to any operator of rotating machines in an engineering workshop.

However, this stroboscopic effect can be harnessed – for example, to check the speed of a CD player or the speed of a motor vehicle for calibration purposes.

By using one of the following methods, the stroboscopic effect can be overcome or reduced. The first three maintain light falling on the rotating machine. The fourth makes the effective flicker at a different frequency from the operating frequency of the machines.

Did you know?

Certain frequencies of stroboscopic flash can induce degrees of drowsiness, headaches, eye fatigue and, in extreme cases, disorientation

(i) Tungsten filament lamps can be fitted locally to lathes, pillar-drilling machines etc. This will lessen the effect but will not eliminate it completely.

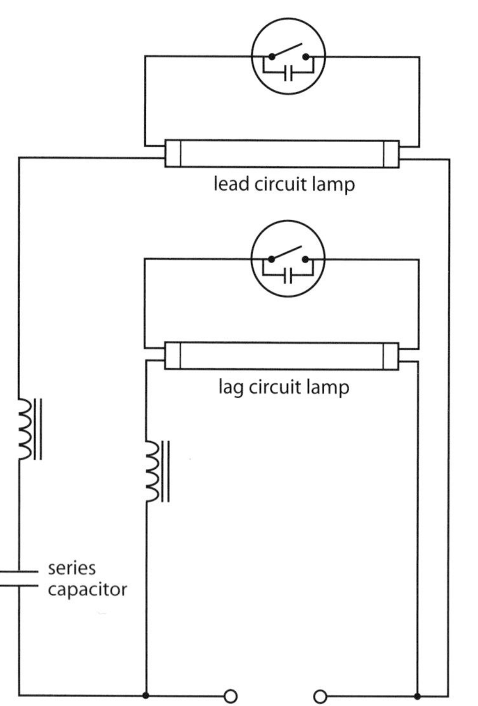

Figure 3.8 Lead lag circuit

(ii) Adjacent fluorescent fittings can be connected to different phases of the supply. Because in a three-phase supply the phases are 120° out of phase with each other, the light falling on the machine will arrive from two different sources. Each of these will be flickering at a different time and will interfere with each other, reducing the stroboscopic effect.

(iii) Twin lamps can be wired on lead-lag circuits, thus counteracting each other. The lead-lag circuit, as the name implies, is a circuit that contains one lamp in which the **power factor** leads the other – hence the other lags. Using the leading current effect of a capacitor and the lagging current effect of an inductor produces the lead-lag effect. The lagging effect is produced naturally when an inductor is used in the circuit as shown in Figure 3.8. The leading effect uses a series capacitor, which has a greater effect than the inductor in the circuit. When these two circuits are combined as shown there is no need for further power factor correction as one circuit will correct the other.

(iv) The use of high-frequency fluorescent lighting reduces the effect by about 60 per cent.

Other methods of starting the fluorescent tube

Quick start

The electrodes of this type of circuit are rapidly pre-heated by the end windings of an autotransformer so that a quick start is possible. The method of ionisation of the gas is the same as in the semi-resonant circuit. Difficulties may occur in starting if the voltage is low.

Thermal starter circuit

This type of circuit has waned in popularity over the years. However, there are still thousands of these fittings in service, so it is worth describing them. In this starter, the normally closed contacts are mounted on a bi-metal strip. A small heater coil heats one of these when the supply is switched on. This causes the strip to bend and the contacts to open, creating the momentary high voltage and starting the circuit discharge. The starter is easily recognised as it has four pins instead of the usual two, the extra pins being for the heater connection.

Other discharge lamps

- high-pressure mercury vapour
- low-pressure sodium
- high-pressure sodium.

Typically these consist of a glass tube known as an arc tube, in which an electrode has been fitted at either end. Rare metals such as lanthanum, lutetium or mercury may be added to the tube. The tube is pressurised with gases such as argon and neon before sealing. Lamps such as these will not normally start using a raw mains voltage but require an external igniter to provide a high-voltage injection in order to set up an electromagnetic field. This excites ions present in the gas, causing them to collide with molecules of gas, resulting in the emission of visible light. This may take some time to occur and is a characteristic of this type of lamp.

Calculating lighting requirements

Before we can perform any calculations, it's probably a good idea to know what we are trying to calculate and what we can use to achieve it. In plain English, two early things to consider are:

- How powerful is something at source? (How bright is the light?)
- How much light has landed on an object a certain distance away from the light?

Measuring light

We refer to the 'brightness' of a source (the power of the light) as **luminous intensity** and this is given the symbol *I* and is measured in candelas (cd).

Luminous flux is the measure of the flow (or amount) of light being emitted from that source and one of the factors used when designing lighting systems is **Illuminance**.

Formerly called illumination, illuminance is our measure of the amount of light falling on a surface. This is defined as 'the density of the luminous flux striking a surface'.

Using the symbol *E*, illuminance has the unit of measurement (Lux), with one Lux being the illuminance at a point on a surface that is one metre from, and perpendicular to, a uniform point source of one candela.

Let's explore that a bit more. Take a ruler 1 metre long and place it flat on the floor with one end touching a 1 m^2 wall. Fix a candle to the other end of it and then light the candle. If we assume that the candle has a luminous intensity of one candela, then the amount of light hitting the wall is one Lux. In other words, one lumen uniformly distributed over one square metre of wall surface provides an illuminance of 1 Lux (1 lux = 1 lumen/square metre).

Remember

A luminarie is more commonly known as a lighting fitting and refers to the fully assembled enclosure, lamp, control gear and reflector etc.

If we were now to move the ruler and candle further away from the wall, then the wall will appear less brightly lit. However, the amount coming from the candle has remained the same. This is inversely proportional.

As a concept what we are saying is the closer you are to a luminaire, the brighter that luminaire is. Or, put another way, if we can't change how much light comes out of the luminaire, to make more light land on an object we either have to move the luminaire closer, or add more luminaires.

Other factors that affect illuminance

Whatever type of luminaires we eventually decide to install will be affected by age, collection dust etc. All of these factors will effect our level of illuminance and are grouped under an overall title of **Maintenance Factor**.

Maintenance Factor (MF) is the ratio of the illuminance provided by an installation after a period of use against its initial illuminance when it first started use. This is expressed as a number or percentage and has no unit.

As a simple example, let's say that a shift manager's office in a garage workshop has a ceiling luminaire with one 65 W fluoresent lamp inside it installed from new. When first installed the lamp had a lumen output of 1000 lm, but when measured again after six months in operation the output had fallen to 850 lm.

The output has decreased by a ratio of: $\frac{850}{1000} = 0.85$

We therefore have a Maintenance Factor of 0.85.

The Maintenance Factor is based on how often the lights are cleaned and replaced. It takes into account such factors as decreased efficiency with age, accumulation of dust within the fitting itself and the depreciation of reflectance as walls and ceiling age. It is fully represented by the following formula:

$MF = LLMF \times LSF \times LMF \times RSMF$

Where:

- LLMF (Lamp Lumen Maintenance Factor) – the reduction in lumen output after specific burning hours.
- LSF (Lamp Survival Factor) – the percentage of lamp failures after specific burning hours.
- LMF (Luminaire Maintenance Factor) – the reduction in light output due to dirt deposited on or in the luminaire.
- RSMF (Room Surface Maintenance Factor) – the reduction in reflectance due to dirt deposition in the room surfaces.

As a rough guide, for convenience, MF is usually taken as being around the following values:

- Good = 0.70
- Medium = 0.65
- Poor = 0.55

One other consideration is the **Utilisation Factor** (UF), once referred to as the coefficient of Utilisation (CU). Using tables available from manufacturers, it is possible to determine the Utilisation Factor for different lighting fittings if the reflectance of both the walls and ceiling is known, the room index has been determined and the type of luminaire is known.

In other words, UF is a number used to represent the amount of luminous flux emitted by a lamp that reaches a working surface.

Factors that make up the UF include:

- the light output of luminaire
- the flux distribution of the luminaire
- Room Index (room dimensions and spacing and mounting height of luminaires)
- room reflectances.

When checking existing light levels in a building, you should be aware that as the light output of lamps varies depending on their operating temperature, it is essential that the luminaires have been operating under normal thermal conditions before checking. This may require, for example, both lighting and heating or air conditioning systems to be switched on for long enough to achieve steady conditions.

Where lamps are known to be new, they should be run for about 100 hours under normal operating conditions. Also, where possible the line voltage supply to the lighting circuit should also be monitored, as fluctuations in lumen output are caused by variations in supply voltage.

Lets now look at the calculations.

The Lumen Method

This method of calculation is only applicable in square or rectangular rooms with a uniform array of luminaires as shown in Figure 3.9.

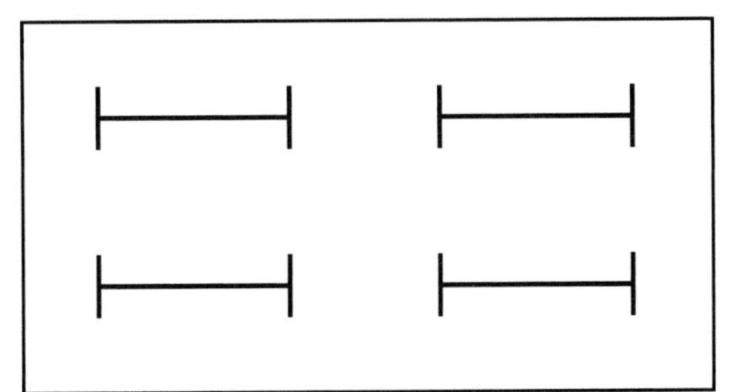

Figure 3.9 A uniform array of luminaires in a room

Using the factors mentioned previously, it is the simplest method of calculating the overall illumination level for such areas. It is accurate enough for the majority of purposes, and is the calculation most used by lighting engineers when determining the number of luminaires for a given lighting level. The formula is as follows:

$$E = \frac{F \times n \times N \times MF \times UF}{A}$$

Where:

E = average illuminance

F = initial lamp lumens

N = number of luminaires

n = number of lamps in each luminaire

MF = maintenance factor

UF = utilisation factor

A = area

Example 1

You have been given the following information and asked to calculate how many luminaires we need to give 300 lux at desk level within a room in a primary school.

UF = 0.44

MF = 0.85 (as the building is clean and without any air conditioning system)

n = 4 lamps per luminaire

F = 2350 lumens for the fluorescent tube

E = 300 lux at the level of the table (good for such a school)

$A = 9\text{ m} \times 4\text{ m} = 36\text{ m}^2$

By transposition of the formula:

$$N = \frac{A \times E}{F \times n \times MF \times UF}$$

Therefore:

$$N = \frac{300 \times 36}{2350 \times 4 \times 0.85 \times 0.44} = \frac{10800}{3515} = 3.07$$

Therefore 3 × 4 tube luminaires are required.

The Inverse Square Law

In physics, an inverse-square law is any physical law stating that some physical quantity or strength is inversely proportional to the square of the distance from the source of that physical quantity.

In English, what this means is that an object that is twice the distance from a point source of light will receive a quarter of the illumination. Or put another way, if you moved an object from 3 m to 6 m (**twice** the distance) away from a light source, you would need four times (**2^2**) the amount of light to maintain the same level of illumination.

We can see this in real life very easily. Consider a campfire at night – a pool of light surrounded by darkness. Or a torch being shone into the night sky – a bright beam of light that rapidly fades to nothing. You might think that when you double the distance from a light source you are now getting half as much light, but it doesn't work like that – you actually get just a quarter as much light. Figure 3.10 shows how the inverse square law works.

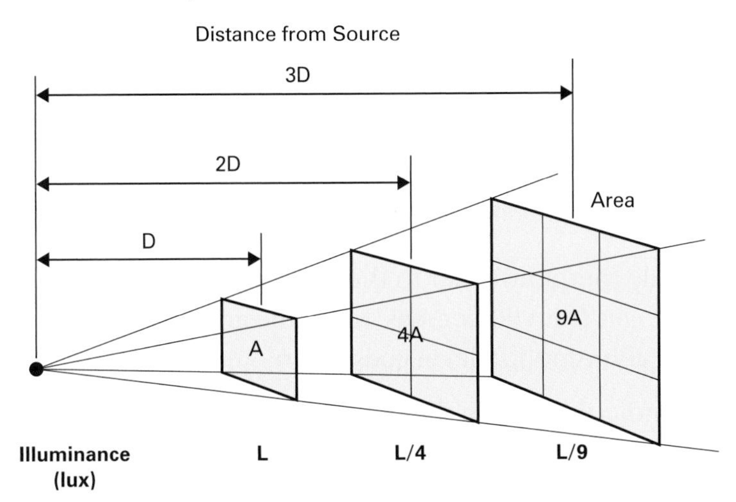

Figure 3.10 The inverse square law

Notice in the diagram how as the distance from the source increases to three times the original distance from the light source (3D), that the intensity of illumination at the new distance is nine times less ($3D^2$). This is because the amount of illumination is inversely proportional to the distance from the source.

We express this phenomenon with the following formula and apply it when the light source is directly above a surface:

$$\text{Illuminance in lux } E = \frac{I \text{ (luminous intensity in candela)}}{d^2 \text{ (Distance between source and point of measurement in m}^2\text{)}}$$

Example 2

A luminaire producing a luminous intensity of 1000 cd is installed 3 m above a surface, what is the illuminance on that surface directly beneath the luminaire?

$$E = \frac{I}{d^2} = \frac{1000}{3^2} = \frac{1000}{9} = 111.1 \text{ lux}$$

If the luminaire was now installed 1m higher, what would be the new level of illuminance on the surface?

$$E = \frac{I}{d^2} = \frac{1000}{4^2} = \frac{1000}{16} = 62.5 \text{ lux}$$

Lambert's cosine law

Inverse Square law applies when the light source is directly above the work surface and the measurement of illumination applied to a straight line beneath that luminaire. Lambert's Cosine Law (commonly referred to as the Cosine Rule) allows us to measure the illumination on the work surface but at an angle to the light source.

The law states that the illuminance on any surface varies as the cosine of the angle of incidence (the angle of incidence is the angle between the normal to the surface and the direction of the incident light).

When light strikes a surface normally (perpendicular to the surface), it gives a certain illumination level. As the angle changes from 90 degrees, the same amount of light is spread out over a larger area, so the illumination level goes down.

If we call this angle from the perpendicular x, then the illumination level is proportional to cos x. This is demonstrated in Figure 3.11.

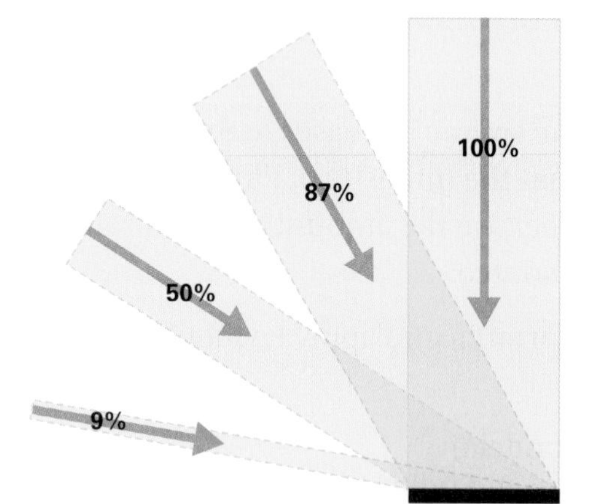

Figure 3.11 Lambert's cosine law

In environmental language, this is why it's cold in winter, and warm during the summer. This is because the sun's rays hit the Earth at a steep angle. The light does not spread out as much, thus increasing the amount of energy hitting any given spot.

But during the winter, the sun's rays hit the Earth at a shallow angle. The rays are therefore more spread out, which minimises the amount of energy that hits any given spot.

From our lighting perspective, if we use the pendant light in Figure 3.12 below, then we can see that the level of illuminance at point A must be higher than it is at point B, and that this reduced level at point B depends on the cosine of the angle.

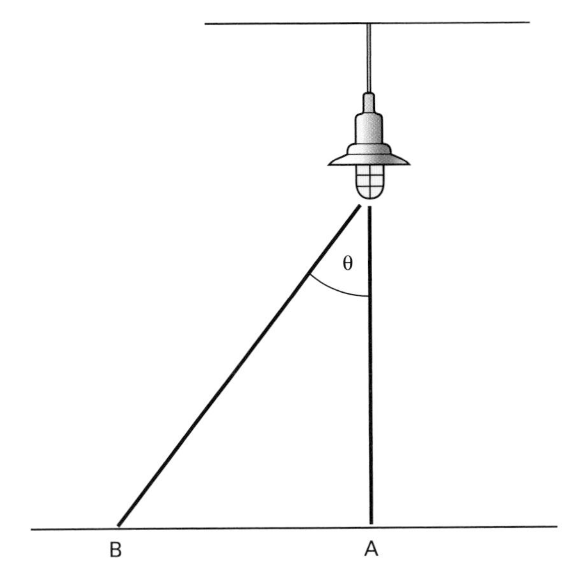

Figure 3.12 Illuminance at a series of points

We express this with the following formula, which is a composite of this and the inverse square law, namely:

$$E = \frac{I \times \cos\theta}{d^2}$$

Example 3

If the above pendant light produces 2000 candela and is suspended 2 m above a horizontal surface, calculate the illuminance on the surface both directly beneath the lamp (point A) and 3 m away from the lamp (point B).

Point A: $E = \frac{I}{d^2} = \frac{2000}{2^2} = \frac{2000}{4} = 500$ lux

Point B: $E = \frac{I \times \cos\theta}{d^2}$ but we don't know $\cos\theta$

We therefore need to use some logic and trigonometry.

Effectively, we have been given the dimensions of two sides of a right-angled triangle:

- The distance from the pendant to point A (2 m)
- the distance between points A and B (3 m).

The right angle 'points' at the hypotenuse (side H) and this is the measurement that we are trying to find as it is the distance from the pendant to point B.

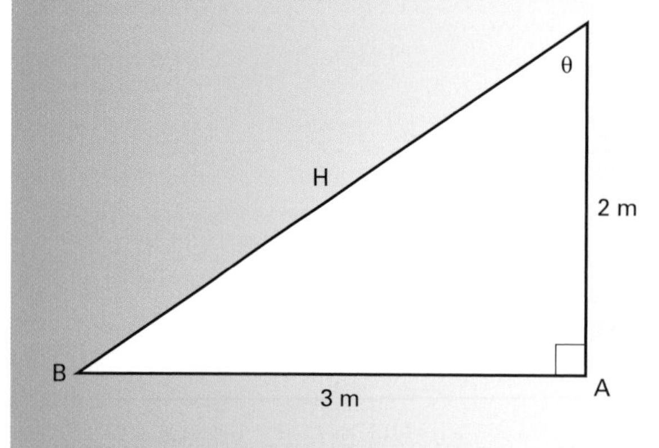

Therefore using Pythagoras' theorem:

$H^2 = A^2 + B^2$

$H^2 = 2^2 + 3^2$

$H^2 = 4 + 9$

Therefore $H = \sqrt{4+9} = 3.6$

In other words, the distance from the pendant to Point B (side H) is 3.6 m

For our illuminance formula we are trying to find cos θ, and a cosine is the ratio between the adjacent side and the hypotenuse. We now know the hypotenuse, and the adjacent side is the one 'next to' the angle we are working with (θ), which in this case has a measurement of 2 m.

The cosine formula is:

$$\cos\theta = \frac{\text{Adjacent}}{\text{Hypotenuse}} = \frac{2}{3.6} = 0.56$$

We can now use this value in our formula

$$E = \frac{I \times \cos\theta}{d^2} = \frac{2000 \times 0.56}{12.96} = \frac{1120}{12.96} = 86.4$$

Therefore the illuminance at point B = 86.4 lux.

On the job: Fluorescent fault

Your employer asks you to visit the owners of an engineering manufacturing factory to discuss a problem with them. Staff have reported that some rotating machines in the main factory area seem to be slowing down. The manager has investigated this and in actual fact they are not. The main factory area is lit by single tube fluorescent lighting fittings hung from chains.

1 How would you go about investigating the problem?

2 What do you think the problem might be?'

(A suggested answer can be found on page 332)

FAQ

Q Where would I use a high-pressure mercury vapour fitting?

A High-pressure mercury vapour gives a high-quality light output, with high levels of colour rendering. Therefore it is used for lighting sports events, such as evening football and cricket matches.

Q What does it mean if, when turned on, the ends of a glow-start fluorescent tube glow but it does not strike?

A The contacts of the glow-type starter have shorted together and are not opening – if they did, the tube would strike. Replace the starter.

Q How can I tell the difference between a high-pressure sodium lamp and a high-pressure mercury vapour lamp when they are illuminated?

A By the colour: the high pressure sodium lamp gives of a warm golden-coloured light, while the high-pressure mercury lamp gives a blue/green light.

Knowledge check

1. List four advantages and one disadvantage of a GLS lamp.
2. What is the typical temperature at which GLS lighting will eventually settle after being switched on?
3. Which inert gases are used in GLS lamps?
4. State the efficacy and average life of a linear tungsten halogen lamp.
5. State three methods of overcoming stroboscopic effects of discharge lighting.
6. Give one reason why a fluorescent type luminaire is slow to start.

Chapter 4

Circuits and systems

OVERVIEW

This chapter will look at various circuits that are common in electrical installation work. It will look at how electricity circulates around domestic premises and how it is used in heating, security and control circuits.

This chapter will cover the following areas:

- **Timers and programmers**
- **Emergency lighting, fire alarms and standby power supplies**
- **Water heating**
- **Space heating**
- **Cooker thermostats and controllers**
- **Closed-circuit television and camera systems**
- **Intruder alarm systems**

Timers and programmers

Did you know?

Manufacturer Danfoss Randall is one of many that supply programmers

Timers and programmers come in a wide variety of shapes and sizes and can control simple and/or complex switching operations in all aspects of modern living. Simple timers include:

- plug-in clock timers that switch lights or heaters on and off
- timers that delay the operation of contactors in motor control circuits
- timers that switch Christmas lights on and off rapidly or in a pattern.

More complex timers can be programmed to carry out multiple functions when external inputs are received or a certain time is reached. One of the most common programmers that most of you will see each day is the one used to switch the heating and hot water on at home. It is mains operated, and in the event of a power failure a rechargeable battery takes over and maintains the settings for up to two days.

These programmers can be set to switch on/off either heating or hot water or both at several different times each day, seven days per week. The component parts of a typical central heating/hot-water system consist of a multi-way connection box, circulating pump, boiler, room thermostat, hot-water tank thermostat, changeover/diverter valve and, of course, the programmer itself.

The control circuits for these are many and varied depending on the type of heating system installed. For typical control circuits the use of Danfoss Randall Control Packs with WB12 wiring centres is recommended. The WB12 includes terminal-to-terminal wiring details for Heatshare (HSP) and Heatplan (HPP) Packs.

The supply to the programmer, and hence the rest of the circuit, is usually via a fused spur with a 3 A fuse fitted. The various cables are installed to the respective component parts via the central connecting block and the programmer itself.

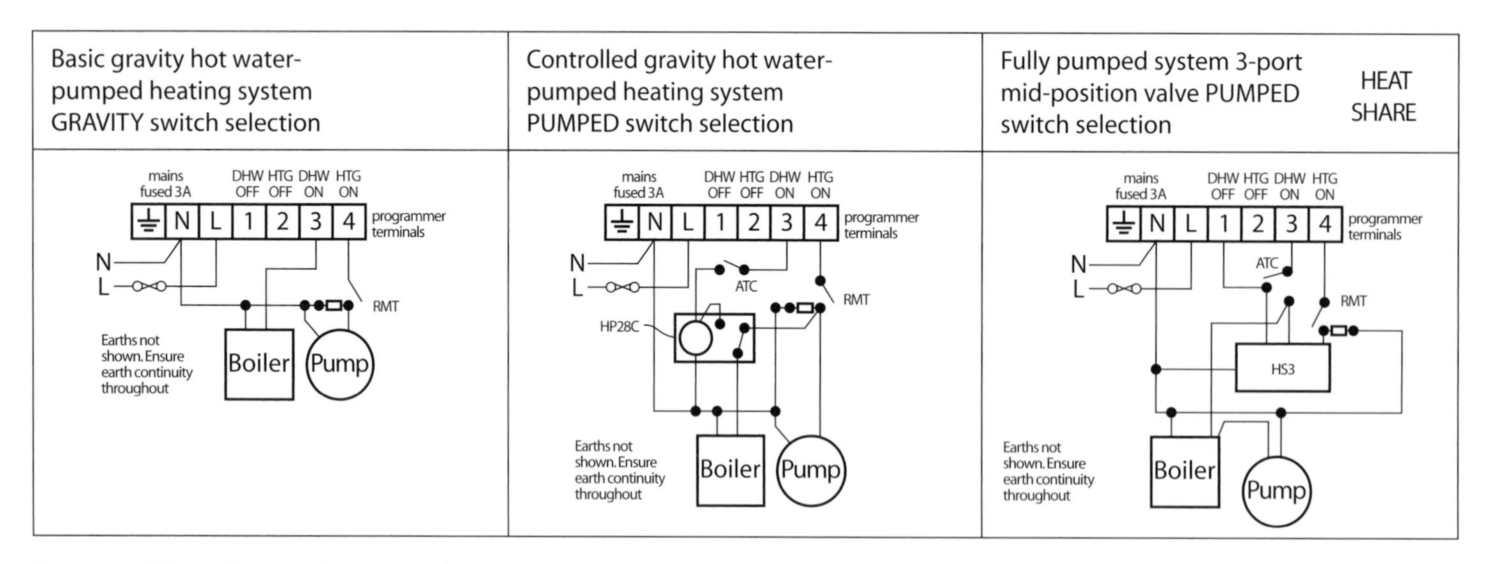

Figure 4.1 Wiring diagram for timer and programming domestic system (continued overleaf)

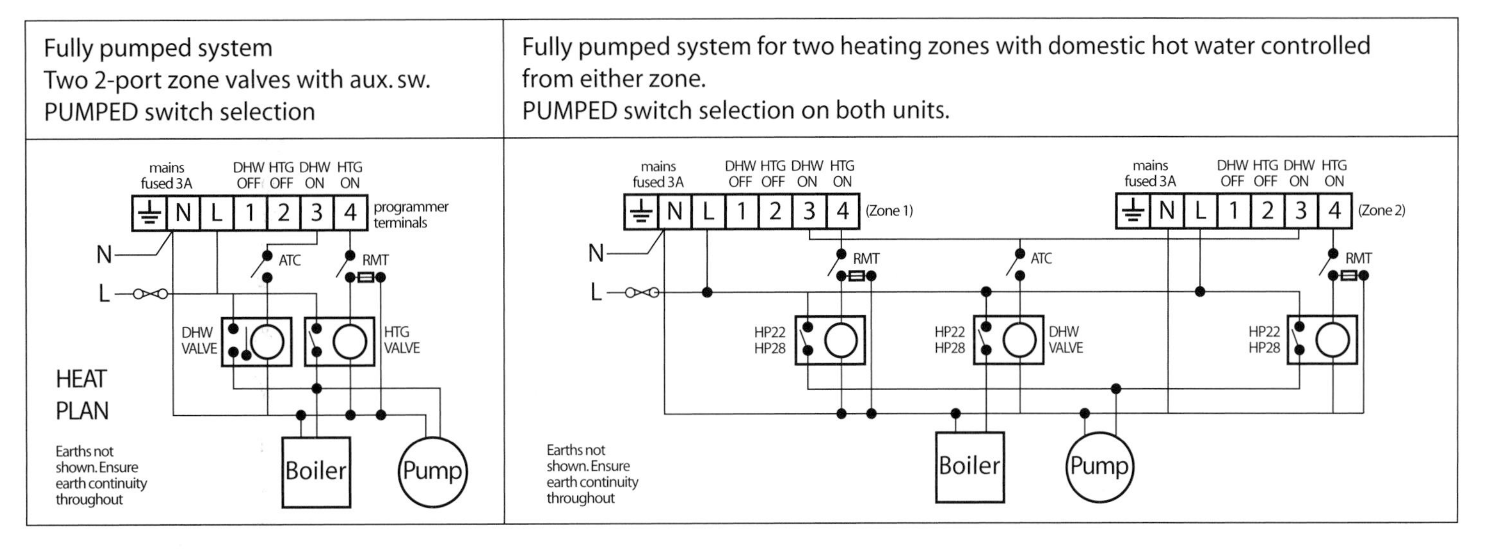

Figure 4.1 Wiring diagram for timer and programming domestic system (continued)

When the clock part of the programmer switches on to, say, central heating, then the supply is fed via the room thermostat to the boiler and the pump until the thermostat switches itself – and thus the heating – off. The same principle applies when hot water is selected, only this time the thermostat on the hot-water cylinder cuts out.

If both heating and hot water are selected, then a diverter valve switches the flow of water via a motorised valve (one part of the system is closed off) within the system to prioritise the heating first; when that reaches temperature the diverter valve changes over and prioritises the hot-water side of the system.

On the job: Customer technical query

You have been asked to install an additional light fitting in the kitchen of a house. The central heating in the house is controlled through a central programmer located in the kitchen and an electronic digital thermostat mounted on a wall in the entrance hall. The owners ask how the electronic thermostat works with the programmer.

1 What could you tell them?

(A suggested answer can be found on page 333)

Emergency lighting, fire alarms and standby power supplies

This section looks at the following emergency lighting systems, fire alarms and standby power supplies:

- emergency lighting (maintained, non-maintained and sustained)
- fire-alarm systems
- standby power supplies.

Remember

Emergency lighting helps to prevent the panic that often occurs in emergency situations

Emergency lighting

Emergency lighting is not required in private homes because the occupants are familiar with their surroundings. However, in public buildings, people are in unfamiliar surroundings and in an emergency they will require a well-illuminated and easily identified exit route.

Emergency lighting should be planned, installed and maintained to the highest standards of reliability and integrity, so that it will operate satisfactorily when called into action. It must be installed in accordance with the British Standard Specification BS 5266: Part 1: 1999 – Code of Practice for Emergency Lighting.

Emergency-lighting terminology

For the purposes of the European Standard EN 1838, emergency lighting is regarded as a general term. There are actually several types, as shown in the diagram below:

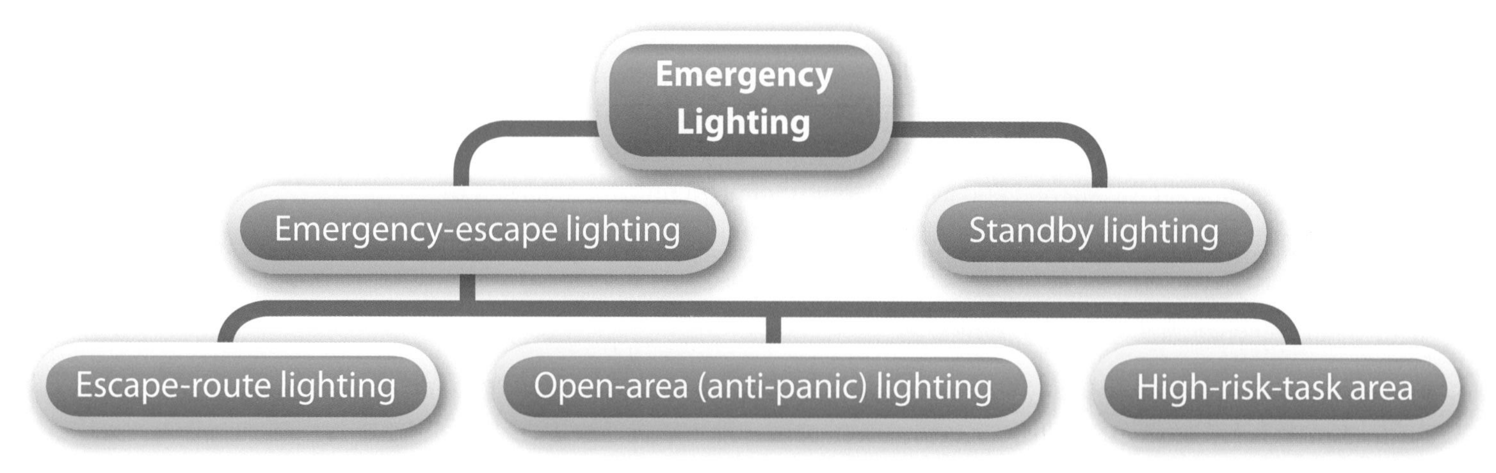

Figure 4.2 Specific forms of emergency lighting

Emergency-escape lighting: provided to enable safe exit in the event of failure of the normal supply.

Standby-lighting: provided to enable normal activities to continue in the event of failure of the normal mains supply.

Escape-route lighting: provided to enable safe exit for occupants by providing appropriate visual conditions and direction-finding on escape routes and in special areas/locations, and to ensure that fire-fighting and safety equipment can be readily located and used.

Open area (or anti-panic area) lighting: provided to reduce the likelihood of panic and to enable safe movement of occupants towards escape routes by providing appropriate visual conditions and direction-finding.

High-risk-task area lighting: provided to ensure the safety of people involved in a potentially dangerous process or situation and to enable proper shut-down procedures to be carried out for the safety of other occupants of the premises.

Types

Emergency lighting comes in two main formats: individual, self-contained systems with their own emergency battery power source, and centralised battery-backup systems. In using both these formats there are then three types available:

Maintained: The same lamp is used by both the mains and the emergency backup system and therefore operates continuously. The lamp is supplied by an alternative supply when the mains supply fails.

The advantage of this system is that the lamp is continuously lit and therefore we can see whether a lamp needs replacing. The disadvantage is that, although the lamp is lit, we do not know whether it is being powered by the mains supply or the batteries. It is therefore common to find a buzzer and indicator lamp that show which supply is being used. Emergency lighting should be of the maintained type in areas in which the normal lighting can be dimmed, e.g. theatres or cinemas, or where alcohol is served.

Non-maintained: The emergency lighting lamp only operates when the normal mains lighting fails. Failure of the mains supply connects the emergency lamps to the battery supply. The disadvantage of this system is that a broken lamp will not be detected until it is required to operate. It is therefore common to find an emergency-lighting test switch available that disconnects the mains supply for test purposes.

Sustained: An additional lamp housed in the mains luminaire is used only when the mains fails.

The duration of the emergency lighting is normally three hours in places of entertainment and for sleeping risk, or where evacuation is not immediate, but one hour's duration may be acceptable in some premises if evacuation is immediate and re-occupation is delayed until the system has recharged.

Siting of luminaires

BS 5266 and IS 3217 provide detailed guidance on where luminaires should be installed and what minimum levels of illuminance should be achieved on escape routes and in open areas. It also specifies what minimum period of duration should be achieved after failure of the normal mains lighting.

Local and national statutory authorities, using legislative powers, usually require escape lighting. Escape-lighting schemes should be planned so that identifiable features and obstructions are visible in the lower levels of illumination that will occur during an emergency.

Current UK regulations require the provision of a horizontal illuminance at floor level on the centre line of a defined escape route, of not less than 0.2 lux (similar to the brightness of a full moon). In addition, for escape routes of up to 2 m wide, 50 per cent of the route width should be lit to a minimum of 0.1 lux. Wider escape routes can be treated as a number of 2 m wide bands.

Emergency-escape lighting should:

- indicate the escape routes clearly, allowing for changes of direction or of level
- provide illumination along escape routes to allow safe movement towards the final exits
- ensure that fire-alarm call points and fire-fighting equipment can be readily located.

Did you know?

Cashpoints in commercial buildings need to be illuminated at all times to discourage acts of theft occurring during a mains failure

Standby lighting is required in, for example, hospital operating theatres and in industry, where an operation or process, once started, must continue even if the mains lighting fails.

Additional emergency lighting should also be provided in:

- lift cars – potential for the public to be trapped
- toilet facilities – particularly disabled toilets – and open tiled areas over 8 m^2
- escalators – to enable users to get off them safely
- motor generator, control or plant rooms. These require battery-supplied emergency lighting to help any maintenance or operating personnel
- covered car parks along the normal pedestrian routes.

Illuminance levels for open areas

Emergency lighting is required for areas larger than 60 m^2 or open areas with an escape route passing through. Illuminance BS 5266 requires 1 lux average over the floor area. The European standard EN 1838 requires 0.5 lux minimum anywhere on the floor level excluding the shadowing effects of contents. The core area excludes the 0.5 m to the perimeter of the area.

High-risk-task area lighting

BS 5266 requires that higher levels of emergency lighting are provided in areas of particular risk, although no values are defined. The European standard EN 1838 says that the average horizontal illuminance on the reference plane (note that this is not necessarily the floor) should be as high as the task demands in areas of high risk. It should not be less than 10 per cent of the normal illuminance, or 15 lux, whichever is the greater. It should be provided within 0.5 seconds and continue for as long as the hazard exists. This can normally only be achieved by a tungsten or permanently illuminated maintained fluorescent lamp source. The required illuminance can often be achieved by careful location of emergency luminaires at the hazard, and may not require additional fittings.

Maintenance

Essential servicing should be defined to ensure that the system remains at full operational status. This would normally be performed as part of the testing routine, but for consumable items, such as replacement lamps, spares should be provided for immediate use.

Fire-alarm systems

A correctly installed fire-alarm system installation is of paramount importance and can be compared to any other electrical undertaking, as life could be lost and property damaged as a result of carelessly or incorrectly connected fire-detection and alarm equipment. The subject is detailed, and therefore this section sets out only to give an overview of requirements.

Did you know?

The Building Regulations require the installation of mains-fed smoke detectors in new-build domestic installations

BS 5839 Part 1 classifies fire-alarm systems, perhaps better described as fire-detection and alarm systems, into the following general types:

- Type M: Break-glass contacts operating sounders for protection of life. No automatic detection
- Type L: Automatic detection systems for the protection of life
- Type P: Automatic detection systems for the protection of property.

It is essential that the installation of fire-alarm systems is carried out in compliance with the requirements of BS 5839 Part 1, BS 7671 and manufacturers' instructions, but remember: local government can enforce even stricter requirements in the interests of public safety. BS 5839 and BS 7671 (528-01-04) state that fire-alarm circuits must be segregated from other circuits, and in order to comply with BS 7671 a dedicated circuit must be installed to supply mains power to the fire-alarm control panel.

Fire-alarm systems can be designed and installed for one of two reasons:

- property protection
- life protection.

Property protection

A satisfactory fire-alarm system for the protection of property will automatically detect a fire at an early stage, indicate its location and raise an effective alarm in time to summon fire-fighting forces (both the resident staff and the fire service). The general attendance time of the fire service should be less than 10 minutes. Therefore an automatic direct link to the fire service is a normal part of such a system.

A fire-alarm system might have prevented this

Protection for property is classed as:

- P1: All areas of the building must be covered with detectors with the exception of lavatories, water closets and voids less than 800 mm in height, such that spread of fire cannot take place in them prior to detection by detectors outside the void.
- P2: Only defined areas of high risk are covered by detectors. A fire-resisting construction should separate unprotected areas.

Life protection

A satisfactory fire-alarm system for the protection of life can be relied upon to sound a fire alarm in sufficient time to enable the occupants to escape. Life protection is classed as:

- M: The most basic and minimum requirement for life protection. It relies upon manual operation of call points and therefore requires people to activate the system. Such a system can be enhanced to provide greater cover by integrating any, or a combination, of the following:
- L1: Same as P1 (as previously mentioned)
- L2: Only provides detection in specified areas where a fire could lead to a high risk to life, e.g. sleeping areas, kitchens, day accommodation etc., and places where the occupants are especially vulnerable owing to age or illness or are unfamiliar with the building. An L2 system always includes L3 coverage.
- L3: Protection of escape routes. The following areas should therefore be included:
  (i) corridors, passageways and circulation areas
  (ii) all rooms opening onto escape routes
  (iii) stairwells
  (iv) landing ceilings
  (v) the top of vertical risers, e.g. lift shafts
  (vi) at each level within 1.5 m of access to lift shafts or other vertical risers.

Types of fire-alarm system

All fire-alarm systems operate on the same general principle, i.e. if a detector detects smoke or heat or if a person operates a break-glass contact, then the alarm will sound. We will look at the devices that may be incorporated into the system later. That said, most fire-alarm systems belong to one of the following categories:

Conventional

In this type of system, a number of devices (break-glass contacts/detectors) are wired as a radial circuit from the control panel to form a zone (e.g. one floor of a building). The control panel would have lamps on the front to indicate each zone, and if a device operates then the relevant zone lamp would light up on the control panel. However, the actual device that has operated is not indicated.

Identifying accurately where the fire has started would therefore depend on having a number of zones and knowing where in the building each zone is. Such systems are therefore normally found in smaller buildings or where a cheap, simple system is required.

Addressable

The basic principle here is the same as for a conventional system, the difference being that by using modern technology the control panel can identify exactly which device initiated the alarm.

These systems have their detection circuits wired as loops, with each device then having an 'address' built in. Such systems therefore help fire location by identifying the precise location of an initiation, and thus allow the fire services to get to the source of a fire more quickly.

Radio addressable

These are the same as addressable systems, but have the advantage of being wireless and can thus reduce installation time.

Analogue

Sometimes known as intelligent systems, analogue systems incorporate more features than either conventional or addressable systems. The detectors may include their own mini-computer, and this evaluates the environment around the detector and is therefore able to let the control panel know whether there is a fire, a change in circumstance likely to lead to a fire, a fault, or even if the detector head needs cleaning. Consequently these systems are useful in preventing the occurrence of false alarms.

Fire-prevention systems

Although still incorporating fire-detection systems, one recent innovation has been the introduction of the fire-reduction system. This type of system is still under development, but works by reducing levels of one of the main components in the fire triangle – oxygen – and thus seeks to create a 'fire-free' area. Although not without problems, usage of these systems could be appropriate in critical areas such as historical archives or identified unmanned areas such as chemical storage.

Zones

To ensure a fast and unambiguous identification of the source of fire, the protected area should be divided into zones. Although less essential in analogue addressable systems, the following guidelines relate to zones:

- If the floor area of each building is not greater than 300 m^2 then the building only needs one zone, no matter how many floors it has. This covers most domestic installations.
- The total floor area for one zone should not exceed 2,000 m^2.

Remember

A fire compartment is an area bordered by a fire-resisting structure that usually offers at least 30 minutes' resistance

- The search distance should not exceed 30 m. This means the distance that has to be travelled by a searcher inside a zone to determine visually the position of a fire should not be more than 30 m. The use of remote indicator lamps outside of doors may reduce the number of zones required.
- Where stairwells or similar structures extend beyond one floor but are in one fire compartment, the stairwell should be a separate zone.
- If the zone covers more than one fire compartment then the zone boundaries should follow compartment boundaries.
- If the building is split into several occupancies, no zone should be split between two occupancies.

System devices

Remember

The control panel may contain standby batteries and must always be located where the fire services can easily find and see it, e.g. near to the front entrance of a building

The control panel

This is the heart of any system, as it monitors the detection devices and their wiring for faults and operation. If a device operates, the panel operates the sounders as well as any other related equipment and gives an indication of the area in which the alarm originated.

Break-glass contacts (manual call points)

The break-glass call point is a device to enable personnel to raise the alarm in the event of a fire, by simply breaking a fragile glass cover (housed in a thin plastic membrane to protect the operative from injury sustained by broken or splintered glass). A sturdy thumb pressure is all that is required to rupture the glass and activate the alarm. The following guidance relates to the correct siting and positioning of break-glass call points:

Break-glass call point

- They should be located on exit routes and in particular on the floor landings of staircases and at all exits to the open air.
- They should be located so that no person need travel more than 30 m from any position within the premises in order to raise the alarm.
- Generally, call points should be fixed at a height of 1.4 m above the floor, at easily accessible, well-illuminated and conspicuous positions free from obstruction.
- The method of operation of all manual call points in an installation should be identical unless there is a special reason for differentiation.
- Manual and automatic devices may be installed on the same system, although it may be advisable to install the manual call points on separate zones for speed of identification.

Automatic detectors

When choosing the type of detector to be used in a particular area it is important to remember that the detector has to discriminate between fire and the normal environment existing within the building, for example smoking in hotel bedrooms, fumes from fork-lift trucks in warehouses, or steam from kitchens and bathrooms. There are several automatic detectors available, as described below.

Remember

It is wise not to install a rate-of-rise heat detector unit in a boiler-room or kitchen, where fluctuations in ambient temperature occur regularly. This will help to avoid nuisance alarms

Heat detectors (fixed-temperature type)

The fixed-temperature heat detector is a simple device designed to activate the alarm circuit once a predetermined temperature is reached. Usually a choice of two operational temperatures is available: either 60°C or 90°C. This type of detector is suitable for monitoring boiler-rooms or kitchens where fluctuations in ambient temperature are commonplace.

Heat detector (rate-of-rise type)

This type of detector responds to rapid rises of temperature by sampling the temperature difference between two heat-sensitive thermocouples or thermistors mounted in a single housing (a thermistor is a device whose resistance quickly decreases with an increase in temperature).

Smoke detectors

May be either of the 'ionisation' or 'optical' type. Smoke detectors are not normally installed in kitchens, as burning toast and so on could activate the alarm.

Smoke detector

The ionisation detector is very sensitive to smoke with fine particles such as that from burning paper or spirit, whereas the optical detector is sensitive to 'optically dense' smoke with large particles such as that from burning plastics.

The optical smoke detector, sometimes known as the photoelectric smoke detector, operates by means of the light-scattering principle. A pulsed infrared light is targeted at a photo-receiver but separated by an angled non-reflective baffle positioned across the inner chamber. When smoke and combustion particles enter the chamber, light is scattered and reflected on to the sensitive photo receiver, triggering the alarm.

Detector heads for fire-alarm systems should only be fitted after all trades have completed work, as their work could create dust, which impairs the detector operation. Strict rules exist regarding the location of smoke detectors.

Alarm sounders

These are normally either a bell or an electronic sounder, which must be audible throughout the building in order to alert (and/or evacuate) the occupants of the building. The following gives guidance for the correct use of alarm sounders:

- A minimum level of either 65 dBA, or 5 dBA above any background noise likely to persist for a period longer that 30 seconds, should be produced by the sounders at any occupiable point in the building.
- If the alarm system is to be used in premises such as hotels, boarding houses etc. where it is required to wake sleeping persons, then the sound level should be a minimum of 75 dBA at the bedhead.

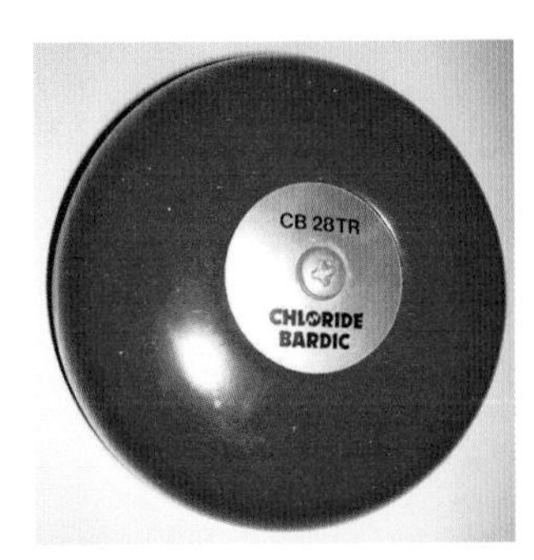

Alarm bell

- All audible warning devices used in the same system should have a similar sound.
- A large number of quieter sounders rather than a few very loud sounders may be preferable. At least one sounder will be required per fire compartment.
- The level of sound should not be so high as to cause permanent damage to hearing.

Wiring systems for fire alarms

BS 5839 Part 1 recommends 11 types of cable that may be used where prolonged operation of the system in a fire is not required. However, only two types of cable may be used where prolonged operation in a fire is required.

It is obvious that the cabling for sounders and any other device intended to operate once a fire has been detected must be fireproof. However, detection wiring can be treated differently as it can be argued that such wiring is only necessary to detect the fire and sound the alarm.

Find out

What other fire-resistant cabling is available?

In reality fire-resistant cabling tends to be used throughout a fire-alarm installation for both detection and alarm wiring. Consequently, as an example, MICC cable used throughout the system is considered by many as the most appropriate form of wiring, but there are alternatives, such as Fire-tuf.

Irrespective of the cable type and the circuit arrangements of the system, all wiring must be installed in accordance with BS 7671. Where possible, cables should be routed through areas of low fire risk, and where there is risk of mechanical damage they should be protected accordingly.

Because of the importance of the fire-alarm system, it is wise to leave the wiring of the system until most of the constructional work has been completed. This will help prevent accidental damage occurring to the cables. Similarly, keep the control panel and activation devices in their packing cartons, and only remove them when building work has been completed in the area where they are to be mounted, thus preventing possible damage to the units.

Standby back-up for fire-alarm systems

The standby supply, which is usually a battery, must be capable of powering the system in full normal operation for at least 24 hours, and, at the end of that time period, must still have sufficient capacity to sound the alarm sounders in all zones for a further 30 minutes.

Typical maintenance checks for a fire-alarm system

BS 5839 Part 1 makes the following recommendations:

Daily inspection	Annual test
• Check that the control panel indicates normal operation. Report any fault indicators or sounders not operating to the designated responsible person.	• Repeat the quarterly test • Check all call points and detectors for correct operation • Enter details of test in logbook.
Weekly test	**Every two to three years**
• Check panel key operation and reset button • Test fire alarm from a call point (different one each week) and check sounders • Reset fire-alarm panel • Check all call points and detectors for obstruction • Enter details of test in logbook.	• Clean smoke detectors using specialist equipment • Enter details of maintenance in logbook.
Quarterly test	**Every five years**
• Check all logbook entries and make sure any remedial actions have been carried out. • Examine battery and battery connections • Operate a call point and detector in each zone • Check that all sounders are operating • Check that all functions of the control panel are operating by simulating a fault • Check sounders operate on battery only • Enter details of test in logbook.	• Replace battery (see manufacturer's information).

Table 4.1 BS 5839 recommendations

Remember

Large installations need a standby generating system, whereby a large combustion engine cuts in automatically and drives a generator capable of supplying the load needed to continue working safely

Standby power supplies

A new section within BS 7671, Chapter 56 now covers general requirements for the selection and erection of supply systems for safety services such as emergency lighting, fire detection and alarm systems as well as essential medical and industrial systems (e.g. air traffic control).

An electrical safety service supply is classed as either being:

- **Non-automatic** – its operation is started by an operator.
- **Automatic** – its operation is automatic and independent of an operator.

Automatic systems are then classified according to the their changeover time, some examples being:

- **No break** – automatic supply with continuous supply within specified conditions.
- **Very short break** – automatic supply available within 0.15 seconds.
- **Short break** – automatic supply available between 0.15 and 0.5 seconds.
- **Medium break** – automatic supply available between 5 and 15 seconds.

There are essentially two sources – batteries (normally for fire alarms and emergency lighting) and diesel generators (supplies for essential services) with a third technology (rotary systems) providing an 'overlap' between them. You should also be aware of some important terminology, such as auxiliary supplies, back-up supplies, standby supplies, emergency supplies and uninterruptible power supplies (UPS).

Hospital facilities provide a good example of these. Hospital operations are typically protected by local (i.e. within the hospital grounds) standby generators should a mains failure occur. These will provide power to allow essential services to continue as normal. Equally, if a mains failure occurs, we would wish to see fire alarms and emergency lighting to continue operation in both essential and 'non-essential' areas and this could be provided by a central battery system. This would be considered an auxiliary supply (or back-up, standby or emergency supply).

An Uninterruptible Power Supply (UPS), maintains a continuous supply to connected equipment by supplying power from a separate source (e.g. batteries). It differs from an auxiliary power supply or standby generator, as these don't provide instant protection from a momentary power interruption. Integrated systems that have both UPS and standby generators are often referred to as emergency power systems.

We might also see some official systems within the hospital having UPS protection, where there is an issue over quality of supply, or the need to ensure power whilst a generator starts or mains failure, where the UPS battery pack should be sized to allow the powering up of on-site generators and/or sufficient time to allow any medical procedure to be safely finished.

There are two main types of UPS – off-line and on-line.

- An **off-line UPS** remains idle until a power failure occurs, and then switches to its own power source, almost instantaneously.
- An **on-line UPS** (sometimes called double conversion) continuously powers the protected load from its reserves (usually lead-acid batteries), while at the same time refilling the reserves from the main supply. It also provides protection against power fluctuations and for this reason is used with IT equipment where a 'steady' supply is required.

Bearing in mind the slight lag between loss of supply and the standby supply 'kicking in', some installations use a secondary 'rotary' system in conjunction with the diesel generator.

The rotary system uses the inertia of a continuously spinning flywheel to provide short-term provision (e.g. 120 kW for 20 seconds) in the event of power loss. The flywheel can also act as a buffer against power fluctuations as these are not really capable of affecting the speed of the flywheel. However, as it is only capable of providing reserve power for a few seconds, it is traditionally used in conjunction with standby diesel generators, providing backup power only for the brief period of time the engine needs to start running.

Regulation 560.6 requires electrical sources for safety services to be installed as fixed equipment in such a manner that they cannot be affected by a failure of the normal supply. They must consequently also be placed in a suitable, well ventilated location to allow exhaust gas to escape and only be accessible to skilled or instructed persons.

In the case of the central battery system, there is no upper limit of the supply capacity. However, Regulation 560.6.9 requires the battery design to be in accordance with BS EN 50171 with a declared life of 10 years. Regulation 560.7 also requires that circuits of safety services must:

- be independent of other circuits
- not pass through zones exposed to risk of explosion
- not pass through zones exposed to fire risk unless they are fire resistant (last resort)
- not be installed in shafts unless for a rescue service lift.

Regulation 560.7.10 additionally requires that a drawing showing the electrical safety service provision must be provided and displayed at the origin of the installation and must show the exact location of all connected equipment, circuit designation and switching arrangements.

Water heating

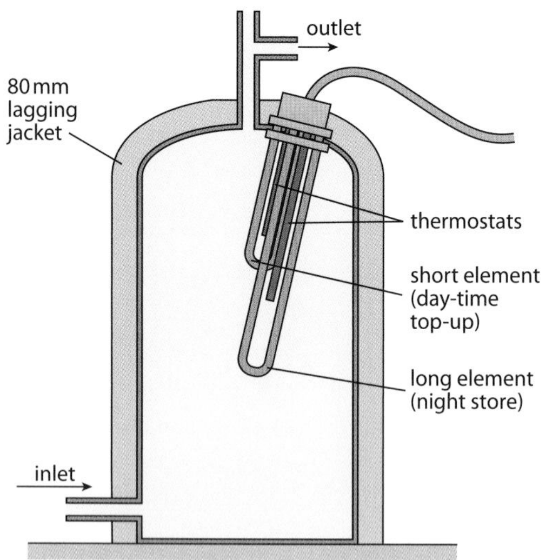

Figure 4.3 Dual-element immersion heater, hot water

There are two main methods of heating water electrically: either heating a large quantity stored in a tank or heating only what is required when it is needed. With both of these types of heater it is important to ensure that the exposed and extraneous conductive parts (refer to Chapter 2 Earthing and protection) are adequately bonded to earth: water and electricity do not mix together well! It is also important to ensure that the cables selected are of the correct size for full load current, since no diversity is allowed for water heaters (refer to the Cable Selection section in Chapter 2).

Heating large tanks of stored water (typically 137+ litres) is done using an immersion heater (see Figure 4.3) fitted into a large water tank and then controlling when it is on or off via either a timer switch or an on/off switch.

The temperature of the water is controlled by a stem-type thermostat (similar to the oven type described later in the Cooker section), which is incorporated within the housing of the heating element. This type of heater is used in domestic situations, although larger multiple immersion heaters can be used in industrial situations. The heater in a domestic situation must be fed from its own fuse/MCB in the consumer unit and have a double pole isolator fitted next to the storage tank. The final connection to the heating element must be with heatproof flexible cable due to the high ambient temperatures where the water tank is normally located.

This type of system sometimes has two elements, one of which is controlled via a separate supply which only operates at night time (Economy 7 or white-meter supply) when cheap electricity is available, thus heating a full tank of water ready for use the next day; the other is switched on as and when needed during the day to boost the amount of hot water available.

The rest of this section will describe the different types of water system:

- cistern type
- non-pressure
- instantaneous.

Cistern-type

Where larger volumes of hot water are needed, for example in a large guest house, then a cistern-type water heater (9 kW+) is used which is capable of supplying enough hot water to several outlets at the same time.

Figure 4.4 Cistern-type water heater

Non-pressure

Non-pressure water heaters, which are typically rated at less than 3 kW and contain less than 15 litres of water, heat the water ready for use and are usually situated directly over the sink, such as in a small shop or hairdresser's salon.

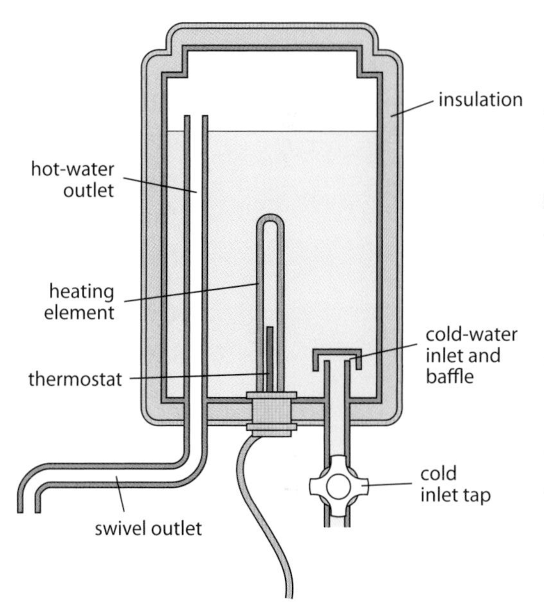

Figure 4.5 Non-pressure water heater

Instantaneous

Instantaneous water heaters heat only the water that is needed. This is done by controlling the flow of water through a small internal water tank which has heating elements inside it; the more restricted the flow of water then the hotter the water becomes.

The temperature of the water can therefore be continuously altered or stabilised locally at whatever temperature is selected. This is how an electric shower works, and showers in excess of 10 kW are currently available. The shower-type water heater must be supplied via its own fuse/MCB in the consumer unit and have a double pole isolator located near the shower.

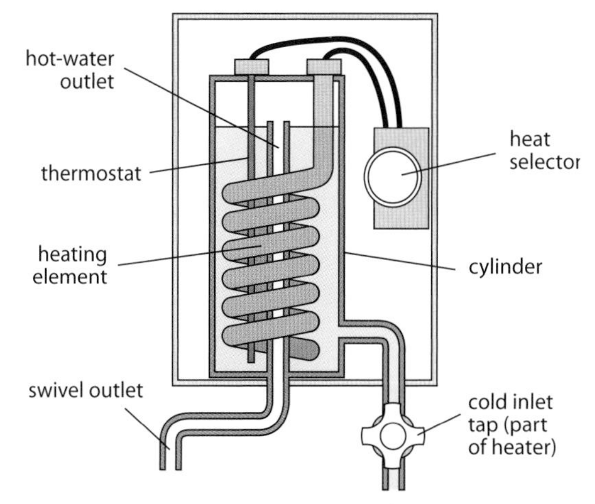

Figure 4.6 Instantaneous water heater

Space heating

The type of electric heating available falls into two main categories: direct acting heaters and thermal storage devices.

Direct acting heaters

Direct acting heaters are usually just switched on and off when needed; some of them can be thermostatically controlled. Direct heaters fall into two categories: radiant and convection.

Radiant heaters

The radiant-type heaters reflect heat and come in a variety of shapes, sizes and construction as follows.

Traditional electric fire: has a heating element supported on insulated blocks with a highly polished reflective surface behind it; these range in size from about 750W to 3kW.

Infrared heater: consists of an iconel-sheathed element or a nickel-chrome spiral element housed in a glass silica tube which is mounted in front of a highly polished surface. Sizes vary from about 500 W to 3kW; the smaller versions are usually suitable for use in bathrooms and may be incorporated with a bulb to form a combined heating and lighting unit.

Oil-filled radiator: consists of a pressed steel casing in which are housed heating elements; the whole unit is filled with oil. Oil is used because it has a lower specific gravity than water and so heats up and cools down more quickly. Surface temperature reaches about 70°C, and power sizes range from about 500 W to 2kW.

Tubular heater: low-temperature unit designed to supplement the main heating in the building. Consists of a mild steel or aluminium tube of about 50mm diameter in which is mounted a heater element. The elements themselves are rated at 200 W to 260 W per metre length and can range in length from about 300mm up to 4.5m. The surface temperature is approximately 88°C.

Under-floor heater: consists of heating elements embedded under the floor which heat up the tiles attached to the floor surface. The floor then becomes a large low-temperature radiant heater, and a room thermostat controls the temperature within the room. The floor temperature does not normally exceed 24°C. The elements have conductors made from a variety of materials such as chromium, copper, aluminium, silicon or manganese alloys. The insulating materials used are also made from a variety of materials such as asbestos, PVC, silicon rubber and nylon.

Convection heaters

Convection heaters consist of a heating element housed inside a metal cabinet that is insulated both thermally and electrically from the case so that the heat produced warms the surrounding air inside the cabinet. Cool air enters the bottom of the cabinet and warm air is passed out at the top of the unit at a temperature of between 80 and 90°C. A thermostatic control is usually fitted to this type of heater.

Fan heater: operates in the same way as a convector heater but uses a fan for expelling the warm air into the room. Fan heaters usually have a two-speed fan incorporated into the casing and up to 3 kW of heating elements.

Thermal storage devices

Thermal storage heaters heat up thermal blocks within the unit during off-peak times to enable use of cheap-rate electricity. The heat stored is then released during the day when it is needed.

A thermal storage unit consists of several heating elements mounted inside firebricks, which in turn are surrounded by thermal insulation such as fibreglass, all housed inside a metal cabinet. The firebricks are made from clay, olivine, chrome and magnesite, which have very good heat-retaining properties. The bricks are heated up during off-peak hours (usually less than half the normal price per electrical unit) and the heat is stored within the bricks until the outlet vent is opened the following day and allows the warm air to escape and hence heat up the room.

Cooker thermostats and controllers

This section covers:

- simmerstat
- oven thermostats.

Simmerstat (energy regulator)

This device is used to control the temperature of electric cooking plates. It uses a bi-metal strip as its main principle of operation; it is not controlled by the temperature of the hotplates. Operation is by the opening and closing of a switch at short definite time intervals by the heating up (via an internal heating coil) of a bi-metal strip. The length of time that the switch is opened or closed is determined by the control knob mounted on the front of the device, and hence the length of time that the hotplate has power is varied. The control knob is normally calibrated from either 0 to 10 or 0 to 5, with the highest number being the hottest temperature that the hotplate will reach.

There are two basic ways that these devices are arranged. One is with a shunt (parallel)-connected heating coil, and the other is with a series-connected heater coil. The heater coil in each case responds to the current flowing through it and hence determines the control of the bi-metal strip.

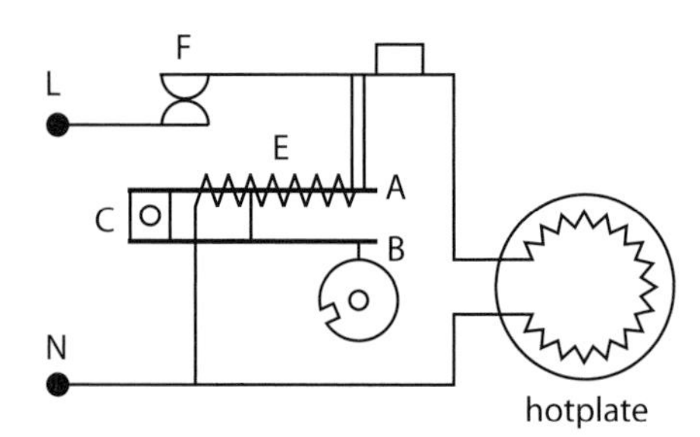

Figure 4.7 Shunt-wired regulator

The shunt-connected thermal regulator (Figure 4.7) consists of a two-part bi-metal strip block where one strip 'A' has a small-gauge heater wire wrapped around it; this is in turn connected in parallel with the hotplate element. The second part of the bi-metal strip block 'B' is in mechanical contact with the cam of the control knob. Both of these strips are connected together at one end and pivoted on a fulcrum point 'C'. When the control knob is in the 'off' position, the cam is pushing against the bi-metal strip and hence keeps the contacts at 'F' open, so that no current flows and therefore the hotplate does not heat up.

When the control knob is moved to one of the 'on' positions, the cam moves and pressure on the bi-metal strip is reduced, thus allowing the contacts at 'F' to close and start to heat up the hotplate. As this happens current flows through the heater coil as well as the hotplate and causes the bi-metal strip to bend, causing the contacts at 'F' to open. When the heater coil and bi-metal strip cool down the bi-metal strip bends and allows the contacts at 'F' to close again, thus repeating the cycle.

The hotplate therefore has power switched 'on' and 'off' rapidly and hence stays at a constant temperature. This technique is known as 'simmering', hence the name of the device 'simmerstat'. At low-temperature settings the contacts will be open for longer, and in the fully 'on' position the contacts at 'F' will be closed and no regulation occurs.

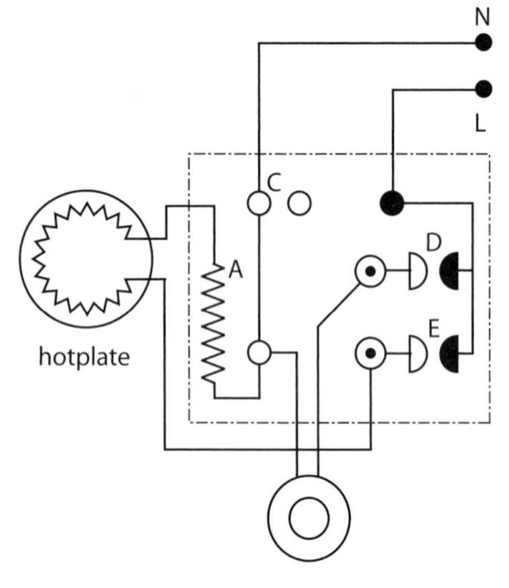

Figure 4.8 Series-wired regulator

In a series type of simmerstat as shown in Figure 4.8, 'A' is the bi-metallic heater coil, which is connected in series with the hotplate. On the right of the diagram, there are two sets of contacts, 'D' and 'E'. When the control knob is operated as if to turn the hotplate 'on', both contacts will close. One of them 'D' will bring a pilot lamp into the circuit to show that the hotplate is being heated, the other 'E' is the main contact that will energise the hotplate.

As the heater coil 'A' transmits heat it causes the bi-metal strip to bend, and eventually this will cause the contacts at 'C' to open, thus breaking the neutral to the coil and the pilot lamp. Consequently the pilot lamp will go out and the hotplate is switched 'off'.

When this happens the heater coil begins to cool and the bi-metal strip returns to its original position, allowing the contacts at 'C' to close again, and thus the cycle is repeated. Each time the contacts open, the pilot light goes off and gives a visual indication that the hotplate is up to temperature.

Oven thermostats

There are two basic types of oven thermostat, both of which work in similar ways. A capillary type has a capillary tube (typically about 800 mm long) filled with liquid which, when the phial containing it heats up, expands. The liquid then pushes against the capsule in the control housing, which in turn pushes against the plunger in the pressure block and causes the contacts to open and hence switch off the oven-heating elements.

The other type of oven housing is the stem type. These come in various lengths to suit different types of oven. The operation of the stem type relies on two dissimilar metals

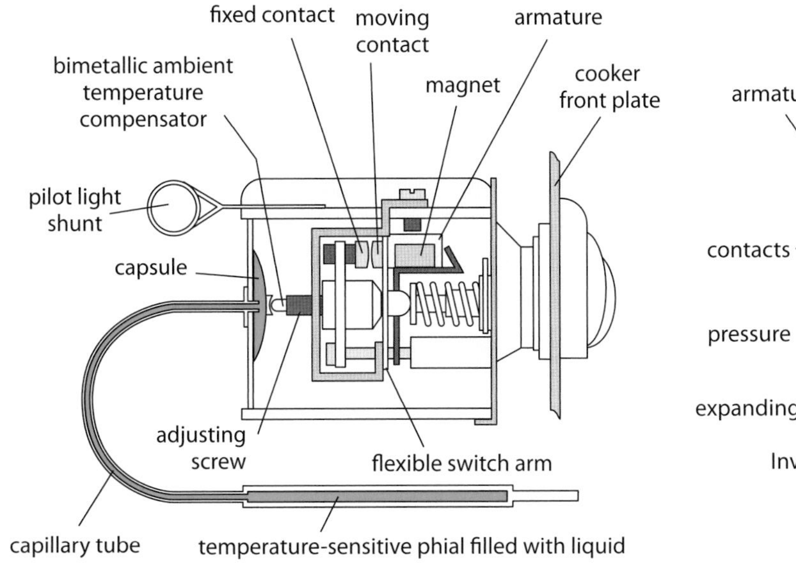

Figure 4.9 Capillary type oven thermostat

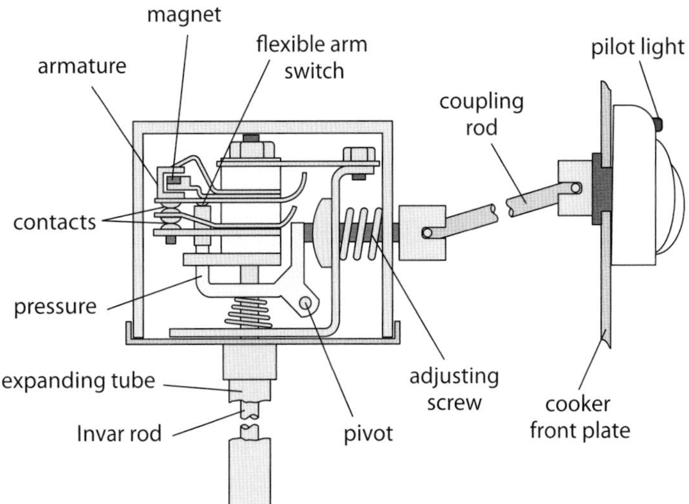

Figure 4.10 Stem type oven thermostat

expanding at different rates. It consists of a copper tube with an **Invar** rod inside. When the temperature increases, the copper tube expands faster than the invar rod and pushes against the pressure rod; this opens the contacts and the oven element is switched off. When it cools down, the copper tube 'shrinks' and allows the invar plunger rod to move down and close the contact for the oven element. This type of thermostat is also used in water-heating systems.

Closed-circuit television and camera systems (CCTV)

There are many different types of CCTV systems in use today, ranging from those suitable for domestic properties through to sophisticated multi-camera/multi-screen monitoring for large commercial and industrial premises.

Typical systems

In this section we will look at the component parts that make up a typical system. The following topics will be looked at:

- wireless CCTV
- wired CCTV
- cameras
- light levels
- monitoring and recording
- other systems.

Wireless CCTV

These systems do not require cabling back to a monitor or video recorder; they have an in-built transmitter which transmits the image seen back to these pieces of equipment. Typically they can transmit 100 m outdoors and 30 m indoors. They do, however, still require a power supply (usually 9–12 V d.c.), which is usually obtained via a small power supply connected to the mains. These systems are useful where it is difficult to install video cable back to the monitor or video recorder but they can suffer from interference problems.

Wired CCTV

These systems do require cabling back to the monitor or video recorder but can be positioned many hundreds of metres away from them. Usually the same cable will provide power and the video signal back to the recording device, so all the power supplies for many cameras can be located at one central control point.

Cameras

There are many different types of camera available, ranging from very cheap (less than £100) to those costing many thousands of pounds. There are two common types: CMOS and CCD. The CMOS type is the cheapest but the images produced are not very clear or sharp. The CCD camera, on the other hand, produces very clear and sharp images from which people are easily identifiable.

Most cameras are installed outside and therefore virtually all cameras available are weatherproof; if they are not they will need to be fitted into a weatherproof housing. Virtually all cameras have the lens integrated into the camera and are sealed to prevent moisture getting in; thus they do not need a heater built in to keep the lens dry. With the lens being sealed into the camera, the former cannot be adjusted, so only one field of view is possible.

Colour and monochrome types of camera are available, with colour cameras being the more expensive. Colour cameras can only transmit colour if the light level is high, so generally speaking they will not transmit colour images at night. Monochrome cameras, on the other hand, can incorporate infrared (IR) sensitivity, allowing for clearer images where discreet IR illumination is available.

Light levels

Light levels available where the camera is to be used are an important consideration. Table 4.2 shows some typical light levels.

When choosing a suitable camera for a particular environment, it is best to select one that is specified at approximately 10 times the minimum light level for the environment. One that is specified at the same level of light will not produce the clear images needed, because the camera will not have enough light to 'see'.

Environment	Typical light level
Summer sunlight	50,000 lux
Dull daylight	10,000 lux
Shop/office	500 lux
Main street lighting	30 lux
Dawn/dusk	1–10 lux
Side street lighting	3 lux

Table 4.2 Typical light levels

Monitoring and recording

Most CCTV systems use several cameras, each relaying images back to a central control where they are either viewed or recorded. Three methods for recording or viewing these images are by using a video switcher, a quad processor or a multiplexer.

Switcher

A CCTV switcher, as the name suggests, is a device that switches between camera images one at a time. The image can either be viewed or recorded onto a video recorder; only one image at a time can be accessed.

Quad processor

This device enables four camera images to be viewed on one screen at the same time, or one image or all four to be recorded at the same time. The quality of the image when recording all four is not as good.

Multiplexer

This device allows simultaneous recording of multiple full-sized images onto one VCR, or can allow more than one camera image to be displayed at the same time without losing picture quality.

For recording purposes, a slower-moving tape can record the images for long periods of time. The time lapse can be set for either 24, 240 or 960 hours of recording on standard tapes.

Other systems

PC-based systems

By adding a video capture card and surveillance software to a PC, a powerful digital system can be created. Some of the advantages are:

- it is easy to expand the system
- it is easy to record (via hard drive)
- images can be emailed
- text alerting is possible
- software allows many configurations for monitoring
- remote viewing is possible.

Motion detectors

The camera and recording facilities are only activated when movement is detected within the camera's range. Typically this is activated by the use of passive infrared sensors (PIRs) similar to those used on security lighting and alarm systems.

Intruder alarm systems

Alarm systems are increasingly seen as standard equipment in a house or office. They act as a deterrent to some intruders but will never stop the more determined ones. People feel more secure when they have an alarm installed, and in most cases it will reduce their insurance premiums. There are basically two ways to protect a property: one is called perimeter protection and the other is space detection.

Perimeter protection detects a potential intruder before they gain entry to the premises, whereas space detection only detects when the intruder is already on the premises. Sometimes both types are used together for extra security.

Typical systems

In this section we look at some of the component parts of an alarm system and some of the more common types of detection devices available and what they do:

- proximity switches
- inertia switches
- passive infrared
- ultrasonic devices
- control panels
- audible and visual warning devices.

Proximity switch

This is a two-part device: one part is a magnet and the other contains a reed switch. The two parts are fixed side by side (usually less than 6 mm apart) on a door or window, and when the door or window is opened the reed switch opens (because the magnet no longer holds it closed) and activates the alarm panel. The switch can be surface-mounted or can be recessed into the door or window frame. This device is generally used for perimeter protection and does not rely on a power supply to operate.

Inertia switch

This type of switch detects the vibration created when a door or window is forced open. This then sends a signal to the alarm panel and activates the sounder. The sensitivity of these devices can be adjusted, and they are used for perimeter protection. These need a 12 volt d.c. supply to operate.

Passive infrared

These devices are used to protect large areas of space and are only activated when the intruder has already gained entry. The device monitors infrared which detects the movement of body heat across its viewing range; this in turn sends a signal to the panel and activates the sounder. These can be adjusted for range and, by fitting different lenses, the angle of detection can also be adjusted. These need a 12 volt d.c. supply to operate.

Ultrasonic devices

These devices send out sound waves and receive back the same waves when no-one is in the building. However, when an intruder enters the detection range, the sound waves change (because of deflection) and trigger the alarm panel. These devices also require a 12 volt d.c. supply for operation and are used for space-detection systems.

Control panels

Control panels are the 'brains' of the system to which all the parts of the system are connected. They used to be key-operated but nowadays they virtually all use a digital keypad, either on the panel itself or mounted remotely elsewhere in the building, for switching the system on or off.

The panels are all programmable whereby entry- and exit-route zone delays can be adjusted, new codes selected for switching on/off, automatic telephone diallers set to ring any phone selected etc.

Control panels have a mains supply installed which is reduced (via a transformer) down to 12 volts d.c. for operation of all the component parts that need it. A rechargeable battery backup is provided in case of mains failure.

Audible and visual warning devices

When an alarm condition occurs a means of attracting attention is obviously needed, either audibly or visually or sometimes both. The most common audible sounder is the electronic horn (I'm sure you've all heard them before!), which will sound for 20 minutes (the maximum allowed by law) before being switched off by the panel automatically. The panel then re-arms itself and monitors the system again.

To help identify which alarm has sounded (especially when there are several in the same area) a visual warning is usually fitted to the sounder box, which activates at the same time. This is a xenon light (strobe light) and can be obtained in a variety of colours. This light usually remains on after the alarm has automatically been reset to warn the occupant upon their return that an alarm condition has occurred. It is only reset when the control panel itself is reset by the occupant.

FAQ

Q Can I use PVC/PVC cable to wire in sounders?

A No, these need to be wired in a flameproof cable such as MICC, FP200 or Fire-tuf.

Q Why is it necessary to put emergency lighting outside buildings?

A To avoid panic and accidents if there is a need to evacuate a building during the hours of darkness. In the event of a power failure, people would be moving out of an illuminated area into the darkness, possibly tripping and blocking the exit for others. Putting an emergency light outside exits prevents this.

Knowledge check

1. What are the two reasons why emergency lighting is provided?
2. Describe the main difference between maintained and non-maintained emergency lighting.
3. List the segregation requirements of fire alarms to BS 5839.
4. Name three situations requiring emergency lighting.
5. List four types of automatic fire detectors.
6. State the primary reasons for zoning a fire-alarm system within a building.
7. Why is it necessary to use a fire-resistant cable on a fire-alarm sounder circuit?
8. Typical maintenance checks are carried out on fire-alarm systems. What is checked annually?
9. State three types of water heater.
10. State two types of device used in a cooker to control temperature and regulate energy.
11. Why should a CCTV camera have a minimum lux level of ten times the lux level of the location it is monitoring?
12. List the typical six component parts of an intruder alarm-system.

Chapter 5

Inspection, testing and commissioning

OVERVIEW

Electrical installations should be designed into separate circuits, e.g. power, lighting, cooker circuits, to avoid danger in the event of faults and to facilitate the safe operation of the inspection and testing process. It is a requirement of the *Electricity at Work Regulations 1989* that this information will be available as an on-site record of the design. This chapter examines the requirements for inspecting and testing an installation. It will cover the following:

- **General requirements**
- **Initial verification, certification and commissioning**
- **The sequence of initial tests**
- **Certificates**
- **Initial inspection**
- **Periodic inspection and testing**

General requirements

When to inspect and test

Initial inspection and testing is necessary on all newly completed installations. In addition, because all electrical installations deteriorate due to a number of factors such as damage, wear and tear, corrosion, excessive electrical loading, ageing and environmental influences, periodic inspection and testing must be carried out at regular intervals determined by the following:

- Legislation requires that all installations must be maintained in a safe condition and therefore must be periodically inspected and tested. Table 5.1 details the maximum period between inspections of various types of installation.
- Licensing authorities, public bodies, insurance companies and other authorities may require public inspection and testing of electrical installations.
- The installation must be checked to ensure that it complies with BS 7671.
- It is also recommended that inspection and testing of installations should occur when there is a change of use of the premises, any alterations or additions to the original installation, any significant change in the electrical loading of the installation, and where there is reason to believe that damage may have been caused to the installation.

Type of installation	Frequency
Domestic Or if change of occupancy	10 years
Commercial Or if change of occupancy	5 years
Educational premises	5 years
Industrial	3 years
Hospitals	5 years
Residential accommodation Or if change of occupancy	1 year
Offices	5 years
Shops	5 years
Laboratories	5 years
Cinemas	3 years
Churches	5 years
Leisure complexes	3 years
Places of public entertainment	3 years
Restaurants	5 years
Hotels	5 years
Theatres	3 years
Public houses	5 years
Village halls/community centres	5 years
Agricultural/horticultural	3 years
Caravans	3 years
Caravan parks	1 year
Highway power supplies	6 years
Marinas	1 year
Fish farms	1 year
Swimming pools	1 year
Emergency lighting	3 years
Fire alarms	1 year
Launderettes	1 year
Petrol filling stations	1 year
Construction sites	3 months

Table 5.1 Frequency of inspection

The inspection process

In new installations inspection should be carried out progressively as the installation is installed, and must be done before it is energised. As far as reasonably practicable an initial inspection should be carried out to verify:

- all equipment and material is of the correct type and complies with applicable British Standards or acceptable equivalents
- all parts of the fixed installation are correctly selected and erected
- no part of the fixed installation is visibly damaged or otherwise defective
- the equipment and material used are suitable for the installation relative to the environmental conditions.

Regulation 611.3 states that as a minimum, the inspection shall include the following items, where relevant to the installation, and if necessary during erection

i. Connection of conductors

ii. Identification of conductors

iii. Routing of cables in safe zones or protection against mechanical damage

iv. Selection of conductors for current carrying capacity and voltage drop

v. Connection of single pole devices in line conductors only

vi. Correct connection of accessories and equipment

vii. Presence of fire barriers and protection against thermal effects

viii. Methods of protection against electric shock:

a) both basic and fault protection

- SELV
- PELV
- double insulation
- reinforced insulation

b) basic protection (including measurement of distances)

- protection by insulation of live parts
- protection by a barrier or enclosure
- protection by obstacles
- protection by placing out of reach

c) fault protection

i) automatic disconnection of supply

- presence of earthing conductor
- presence of cpcs
- presence of protective bonding conductors
- presence of supplementary bonding conductors
- presence of earthing arrangements for protective and functional purposes
- presence of arrangements for alternative sources
- FELV
- choice and setting of protective devices

ii) non-conducting location

- absence of protective conductors

iii) earth-free local equipotential-bonding

- presence of earth-free protective bonding conductors

iv) electrical separation

d) additional protection

ix. Prevention of mutual detrimental influence

x. Presence of appropriate devices for isolation and switching

xi. Presence of undervoltage protective devices

xii. Labelling of protective devices

xiii. Selection of equipment/protective devices appropriate to external influences

xiv. Adequacy of access to switchgear and equipment

xv. Presence of danger notices and warning signs

xvi. Presence of diagrams, instructions and similar information

xvii. Erection methods.

Periodic inspection

As previously mentioned, all installations should be regularly inspected and, as necessary, tested. In some cases there is a statutory requirement for inspections at specified intervals, but in others the period is left to the discretion of the designer, installer or the inspector.

Where a diagram, chart or tables are not available a degree of exploratory work may be necessary so that inspection and testing can be carried out safely and effectively. Note should be made of any known changes in environmental conditions, building structure and alterations, which may have affected the suitability of the wiring for its present load and method of installation.

A careful check should be made of the type of equipment on site so that the necessary precautions can be taken, where conditions permit, to disconnect or short out electronic and other equipment, which may be damaged by subsequent testing. Special care must be taken where control and protective devices contain electronic components. It is essential to determine the degree of these disconnections before planning the detailed inspection and testing. If inspection and testing cannot be carried out safely without diagrams or equivalent information, they must be prepared for compliance with Section 6 of the Health and Safety at Work Act 1974.

For safety, it is necessary to carry out a visual inspection of the installation before beginning any tests or opening enclosures, removing covers, etc. So far as is reasonably practicable, the visual inspection must verify that the safety of persons, livestock and property is not endangered. A thorough visual inspection should be made of all electrical equipment that is not concealed and should include the accessible internal condition of a sample of the equipment. External conditions should be noted and damage identified or, if the degree of protection has been impaired, the matter should be recorded on the schedule of the report. This inspection should be carried out without power supplies to the installation, wherever possible, in accordance with the *Electricity at Work Regulations 1989*.

The inspection should include a check on the condition of all electrical equipment and materials, taking into account any available manufacturer's information with regard to the following:

- safety
- wear and tear
- corrosion
- damage
- excessive loading (overloading)
- age
- external influences
- suitability.

Remember

With periodic inspection and testing, inspection is the vital initial operation and testing is subsequently carried out in support of that inspection

The assessment of condition should take account of known changes in conditions influencing and affecting electrical safety, e.g. extraneous conductive parts, plumbing, structural changes, etc. It would not be practicable to inspect all parts of an installation; thus a random sample should be inspected. This should include:

- Check that joints and connections are properly secured and that there is no sign of overheating.
- Check switches for satisfactory electrical and mechanical conditions.
- Check that protective devices are of the correct rating and type; check for accessibility and damage.
- Check that conductors have suffered no mechanical damage and have no signs of overheating.
- Check that the condition of enclosures remains satisfactory for the type of protection required.

Initial verification, certification and commissioning

Inspection and testing of electrical installations

This section is intended to act as a guide for those engaged in the inspection and testing of electrical installations both at the initial construction stage or when carrying out a periodic inspection and test as required by the *IEE Wiring Regulations* (BS 7671).

To ensure that this most important aspect of an electrician's work is carried out satisfactorily, the inspection and test procedure must be carefully planned and carried out and the results correctly documented. It is hoped that the principles described here will enable that goal to be achieved with the minimum of errors.

The process of commissioning is one that combines a full inspection of the installation and all tests required to be carried out before the installation is energised (referred to as pre-commissioning tests), followed by those tests which require power to be available i.e. the measurement of earth-fault loop impedance and the functional testing of residual current devices.

Commissioning also involves the initial energising of an installation, which has to be carried out in a controlled manner and with the knowledge of everyone involved. Other persons who may be working on the site at the time of energising must be informed of the likelihood of power being applied to the installation in order that all precautions can be taken to prevent danger.

The final act of the commissioning process is to ensure the safe and correct operation of all circuits and equipment which have been installed, and that the customer's requirements have been met. All of these points will be discussed in detail later

Initial verification – general requirements

BS 7671 Part 6 states that every electrical installation shall, either during construction and/or on completion, but certainly before being put into service, be inspected and tested to verify, so far as is reasonably practicable, that the requirements of the Regulations have been met.

When carrying out such inspection and test procedures, precautions must be taken to ensure no danger is caused to any person or livestock and to avoid damage to property and installed equipment.

Periodic inspection and test procedures must also be carried out on existing installations to ensure that the installation has not deteriorated and still meets all regulatory requirements. Details of the recommended periods between such tests are given later in this chapter.

Inspection and testing is also required where minor alterations or additions have been carried out to existing installations. In such cases care has to be taken to ensure that the original design has not been changed in such a way that a dangerous situation might occur e.g. overloading of existing cables or switchgear.

Did you know?

The recording of inspection and test results is also a recommendation contained in *The Memorandum of Guidance on the Electricity at Work Regulations (EAWR)*, which states that records of all maintenance including test results should be kept throughout the life of an installation. This can enable the condition of equipment and the effectiveness of maintenance to be monitored

Initial verification is intended to confirm that the installation complies with the designer's requirements and has been constructed, inspected and tested in accordance with BS 7671. Inspection is a very important part of this procedure and should be carried out prior to electrical tests being applied and normally with that part of the installation under inspection disconnected from the supply.

The results of all such inspections and tests must be recorded and compared with the relevant design criteria. The relevant criteria will for the most part be the requirements of the Regulations, although there may be some instances where the designer has specified requirements that are particular to the installation concerned. In these cases the person carrying out the inspection and test should be provided with the necessary data for comparison purposes. In the absence of such data the inspector should apply the requirements set out in BS 7671.

Certification

Although the different forms of certification available will be discussed in some detail later in this chapter it is worth pointing out at this stage that BS 7671 provides examples of three different types of certificate:

- The Electrical Installation Certificate is designed for new installations or major alterations and is available in two different forms. One is for use where the responsibility for design, construction and inspection and test is the responsibility of one person and the second is for use where design, construction and inspection and test have been carried out by, and require the signatures of, three different parties.
- Minor Works Certificates should only be used for small works that do not include the provision of a new circuit, such as an additional socket outlet or lighting point to an existing circuit.
- Periodic Inspection and Test Certificates are for use where periodic inspections and tests have been carried out on existing installations and would normally have both an Inspection Schedule and a Test Results Schedule attached.

(In addition to the above, a standard form of Inspection Schedule and a Test Results Schedule are provided for use as attachments to the appropriate certificate.)

On confirmation that the results of the inspection and test comply with the relevant criteria the competent person, who may be the person carrying out the inspection and testing, should sign the inspection and test section of the appropriate certificate. If that person has also been responsible for the design and construction of the installation then they should make use of the combined certificate.

Alterations and additions

Where an installation has been the subject of an alteration or addition, then the existing installation must be inspected and tested so far as is necessary to ensure the safety of the alteration and addition. This should include as a minimum the continuity of the circuit protective conductor and earth fault loop impedance.

Safety

The testing of electrical installations can cause some degree of danger and it is the responsibility of the person carrying out the tests to ensure the safety of themselves and others. *Health and Safety Executive Guidance Note GS38* (*Electrical test equipment for use by electricians*) details relevant safety procedures and should be observed in full.

When using test instruments the following points will help to achieve a safe working environment:

- The person carrying out the tests must have a thorough understanding of the equipment being used and its rating.
- The person carrying out the tests must ensure that all safety procedures are being followed e.g. erection of warning notices and barriers where appropriate.
- The person carrying out the tests should ensure that all instruments being used conform to the appropriate British Standard i.e. BS EN 61010 or older instruments manufactured to BS 5458 provided these are in good condition and have been recently calibrated.
- The person carrying out the tests should check that test leads including probes and clips are in good condition, are clean and have no cracked or broken insulation. Where appropriate the requirements of GS38 should be observed, including the use of fused test leads.

A further consideration in relation to safety when carrying out inspection and test procedures is the competence of the inspector. All persons called upon to carry out an inspection and test of an installation must be skilled and experienced and have sufficient knowledge of the type of installation to be inspected and tested to ensure no risk of injury to persons or livestock or damage to property.

Safety tip

Particular attention should be paid when using instruments capable of generating a test voltage in excess of 50 volts e.g. Insulation Resistance Testers. If the live terminals of such an instrument are touched a shock will be received and, although this may not be harmful in itself, it may cause a loss of concentration that could be dangerous, especially if working at heights. Care must also be taken when working with instruments that use the supply voltage for the purpose of the test, such as earth loop impedance testing or when testing a residual current device (RCD). Either of these tests can impose a voltage on associated-earthed metalwork, and precautions must be taken to avoid the risk of shock to other persons

Information

BS 7671 requires that the following information be provided to the person carrying out the inspection and test of an installation:

- the maximum demand of the installation expressed in amperes per phase
- the number and type of live conductors at the point of supply
- the type of earthing arrangements used by the installation, including details of equipotential bonding arrangements
- the type and composition of circuits, including points of utilisation, number and size of conductors and types of cable installed (this should also include details of the 'reference installation method' used)
- the location and description of protective devices (fuses/circuit breakers etc.)
- details of the method selected to prevent danger from shock in the event of an earth fault e.g. earthed equipotential bonding and automatic disconnection of supply
- the presence of any sensitive electronic devices.

It is a requirement of the Regulations that this information shall be available as an on-site record of the design. Such details would normally be included in the project Health and Safety File required by the Construction (Design and Management) Regulations.

The customer

Following initial verification, an Electrical Installation Certificate together with a schedule of test results shall be given to the person ordering the work. The requirements of the Regulations have not been met until this has been done. Sometimes the person ordering the work is not the end-user, e.g. the builder of a new housing estate sells the individual houses to various occupiers. In these cases it is recommended that copies of the Inspection and Test Certificates together with a Test Results Schedule be passed on to the new owners.

Statutory and non-statutory requirements

For domestic electrical installations compliance with BS 7671, which is a non-statutory document, is the only requirement. For commercial or industrial installations the requirements of the *Electricity Supply Regulations 1988* and the *Electricity at Work Regulations 1989*, both of which are statutory instruments, should also be taken into account.

Compliance with BS 7671 will in most cases satisfy the requirements of statutory Regulations such as the *Electricity at Work Regulations* but this cannot be guaranteed. It is essential to establish which statutory and other Regulations apply, and to carry out the design, the construction and the inspection and testing accordingly.

Did you know?

For certain installations where there are increased risks or occupation by members of the public, such as cinemas, public houses, restaurants and hotels etc, the local licensing authority may impose additional requirements, especially in the area of regular inspection and testing

Commissioning

Prior to commencing work on a new electrical installation it is essential that a full specification is prepared. The specification should set out the detailed design and provide sufficient information to enable competent persons to construct the installation and to commission it.

Precise details of all equipment installed should be obtained from the manufacturers or suppliers in order to check that the required standards have been met, to ensure satisfactory methods of installation have been used and to provide the information necessary to confirm its correct operation.

All the above information must be included in the Operation and Maintenance Manual prepared for the project, which should also contain a description of how the installation is to operate and technical data for all items installed such as switchgear, luminaires and any special control systems that may have been incorporated.

The Health and Safety at Work Act 1974 and the *Construction (Design and Management) Regulations 1994* both state the requirements for the provision of such information. The purpose of commissioning is to ensure the safe energising and operation of all aspects of the installation and consists of the following activities:

- inspection of the installation for compliance with BS 7671, the project specification and other relevant standards
- pre-commissioning tests such as continuity, polarity and insulation resistance (including that of any fixed equipment)
- energising of individual circuits within the installation
- measurement of earth loop impedance and functional testing of any residual current devices
- checking the correct operation of all current using equipment and associated control systems
- handover of the installation to the client.

Inspection and testing procedures are fully covered in other sections. However, the requirements for energising circuits and commissioning of equipment are discussed below.

Before energising any circuit it must be assured that it is safe to do so. All required inspection and testing procedures must have been followed and the results confirmed as satisfactory. All other persons in the building e.g. other contractors, users of the building, members of the public etc. must be advised of any potential danger, and warning notices or barriers must be erected where required. All devices used to control current using equipment must be in the 'off' position.

After energising the installation, measurement of earth loop impedance and the functional testing of any RCDs must be completed and the results confirmed as satisfactory before any further commissioning is carried out.

Commissioning of individual items of equipment may now take place. This includes ensuring the correct settings of overload devices, thermostats, time controllers etc. and finally energising the equipment being commissioned in order to check its correct operation. It may be necessary where a piece of equipment requires more than one service to combine the electrical commissioning with that of other contractors e.g. a gas fired central heating system requires both gas and water as well as an electricity supply, and commissioning of the system should preferably take place in conjunction with the heating engineer.

Remember

Many industrial processes have very complicated control systems and may require the supplier of the control panel as well as the client to be present when commissioning is carried out

Handover to client

Handover of the installation to the client is the final activity and should include a tour of the installation, an explanation of any specific controls or settings and, where necessary, a demonstration of any particularly complicated control systems. The Operation and Maintenance Manuals produced for the project should be formally handed to the client at this stage.

The sequence of initial tests

BS 7671 Regulation 610.1 states 'Every installation shall, during erection and/or on completion before being put into service, be inspected and tested to verify, so far as is reasonably practicable, that the requirements of the Regulations have been met. Precautions shall be taken to avoid danger to persons, livestock, and to avoid damage to property and installed equipment during inspection and testing'.

Regulation 612 lists the sequence in which tests should be carried out. If any test indicates a failure to comply, that test and any preceding test, the results of which may have been influenced by the fault indicated, must be repeated after the fault has being rectified.

Remember

The tests in this section are not prescriptive; you may use your own testing methods provided they give valid results

The sequence of tests

Initial tests should be carried out in the following sequence where applicable, before the supply is connected or with the supply disconnected as appropriate:

1. Continuity of protective conductors including main and supplementary bonding (612.1)
2. Continuity of ring final circuit conductors (612.2)
3. Insulation resistance (612.3)
4. Protection by SELV, PELV or electrical separation (612.4)
5. Insulation resistance/impedance of floors and walls (612.5)
6. Polarity (612.6)
7. Earth electrode resistance (612.7)
8. Protection by Automatic Disconnection of Supply (ADOS) (612.8)
9. Earth fault loop impedance (612.9)
10. Additional protection (612.10)
11. Prospective fault current (612.11)
12. Phase sequence (612.12)
13. Functional testing (612.13)
14. Verification of voltage drop (not normally part of initial verification) (612.14).#

The test results should be recorded on an installation schedule similar to the diagram in Figure 5.1.

Forms of completion or periodic inspection, inspection, test and an installation schedule (including test results) should be provided to the person ordering the work.

These tests are fully described in *Guidance Note 3 to BS 7671*.

Contractor	Address/location of dist board	Type of supply	Instruments
Test date		Ze at origin	RCD tester
	Equipment vulnerable to testing		Continuity
Signature		PSSC	Insulation
			Others

		overcurrent device			circuit conductors		Test results											
Circuit	No. of points	type	rating	Short Circuit Cap	live	CPC	Continuity		Insulation resistance				Earth	Functional testing			polarity	remarks
							R1+R2	R2	Phase/ Phase	Phase/ neutral	Phase/ earth	Neutral/ earth	loop impedance	RCD	Time ms	other		
1	2	3	4	5	6	7	8	9	10	11	12	13	14	15	16	17	18	19
1																		
2																		
3																		
4																		
5																		
6																		
7																		
8																		
9																		
10																		
11																		
12																		

Main bonding check	size
Gas	
Water	
Other	

Gas
Earth electrode resistance

Figure 5.1 Installation schedule

Continuity of protective conductors including main and supplementary bonding

Regulations state that every protective conductor, including each bonding conductor, shall be tested to verify that it is electrically sound and correctly connected. The test described below as well as checking the continuity of the protective conductor will also measure $R_1 + R_2$ which, when corrected for temperature, will enable the designer to verify the calculated earth fault loop impedance Z_s. For this test you need a low reading ohmmeter.

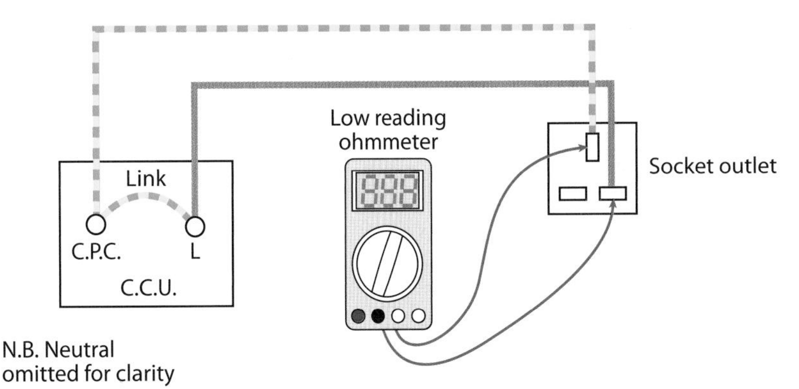

Figure 5.2 Test method 1

Test method 1. Before carrying out this test (shown in Figure 5.2) the leads should be 'nulled out'. If the test instrument does not have this facility, the resistance of the leads should be measured and deducted from the readings. The line conductor and the protective conductor are linked together at the consumer unit or distribution board. The ohmmeter is used to test between the line and earth terminals at each outlet in the circuit. The measurement at the circuit's extremity should be recorded and is the value of $R_1 + R_2$ for the circuit under test. On a lighting circuit the value of R_1 should include the switch wire at the luminaires. This method should be carried out before any supplementary bonds are made.

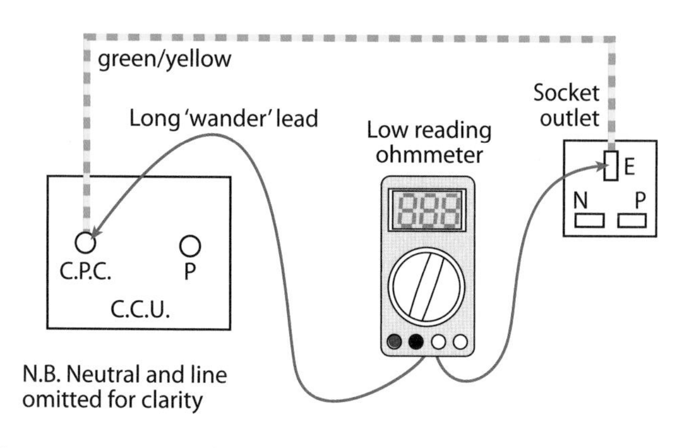

Figure 5.3 Test method 2

Test method 2. One lead of the continuity tester is connected to the consumer's main earth terminals (see Figure 5.3). The other lead is connected to a trailing lead, which is used to make contact with protective conductors at light fittings, switches, spur outlets etc. The resistance of the test leads will be included in the result; therefore the resistance of the test leads must be measured and subtracted from the reading obtained if the instrument does not have a nulling facility. In this method the protective conductor only is tested and this reading R_2 is recorded on the installation schedule.

Test of the continuity of supplementary bonding conductors

Test method 2 as described above is used for this purpose. The ohmmeter leads are connected between the points being tested, between simultaneously accessible extraneous conductive parts i.e. pipe work, sinks etc. or between simultaneously accessible extraneous conductive parts and exposed conductive parts (metal parts of the installation). This test will verify that the conductor is sound. To check this, move the probe to the metalwork to be protected as shown in Figure 5.4. This method is also used to test the main equipotential bonding conductors.

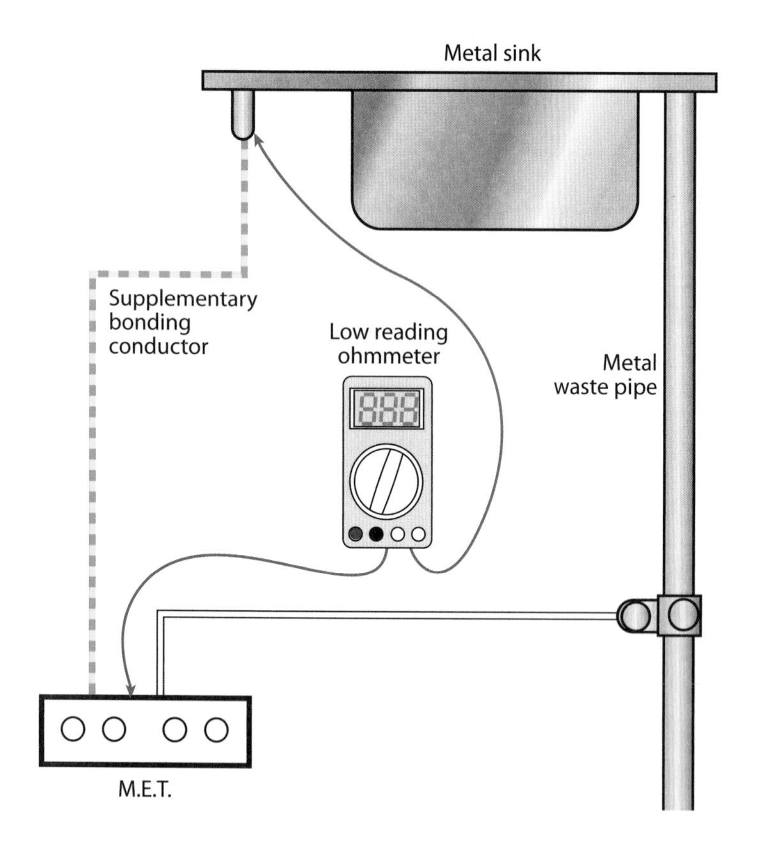

Figure 5.4 Test of the continuity of supplementary bonding conductors

Where ferrous enclosures have been used as the protective conductors, e.g. conduit, trunking, steel-wire armouring etc., the following special precautions should be followed:

- Perform the standard ohmmeter test using the appropriate test method described above. Use a low resistance ohmmeter for this test.
- Inspect the enclosure along its length to verify its integrity.
- If there is any doubt as to the soundness of this conductor a further test should be performed using a phase-earth loop impedance tester after the connection of the supply.
- If there is still doubt, a further test may be carried out using a high-current, low-impedance ohmmeter. The high-current, low-impedance ohmmeter has a test voltage not exceeding 50 volts and can provide a current approaching 1.5 times the design current of the circuit, but the current need not exceed 25 A.

Continuity of ring final circuit conductors

A test is required to verify the continuity of each conductor including the circuit protective conductor (cpc) of every ring final circuit. The test results should establish that the ring is complete and has no interconnections. The test will also establish that the ring is not broken. Figure 5.5 shows a ring circuit illustrating these faults.

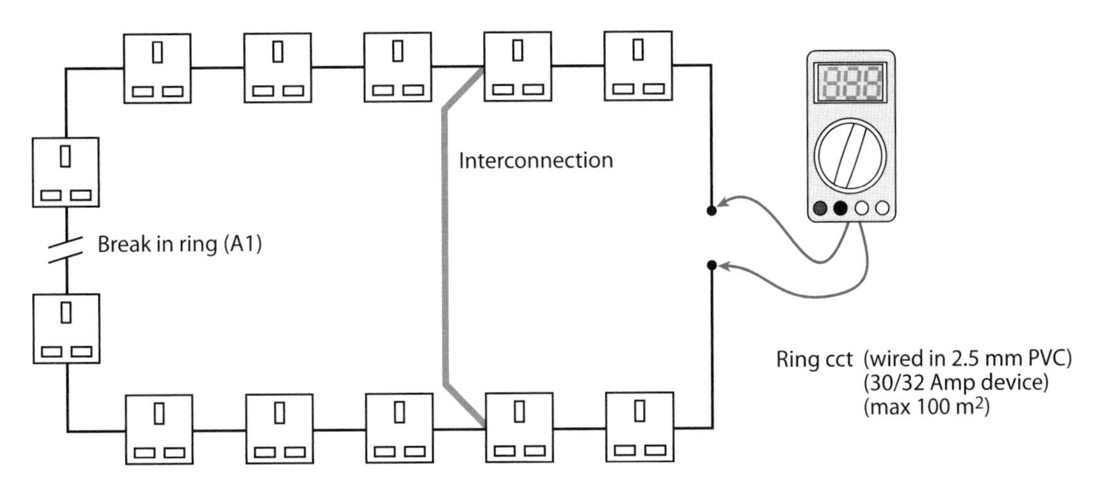

Figure 5.5 Test of continuity of ring final circuit conductors

It may be possible, as an alternative and in order to establish that no interconnected multiple loops have been made in the ring circuit, for the inspector to check visually each conductor throughout its entire length. However, in most circumstances this will not be practicable and the following test method for checking ring circuit continuity is recommended.

The line, neutral and protective conductors are identified and their resistances are measured separately (see Figure 5.6). The end-to-end resistance of the protective conductor is divided by four to give R_2 and is recorded on the installation schedule form. A finite reading confirms that there is no open circuit on the ring conductors under test.

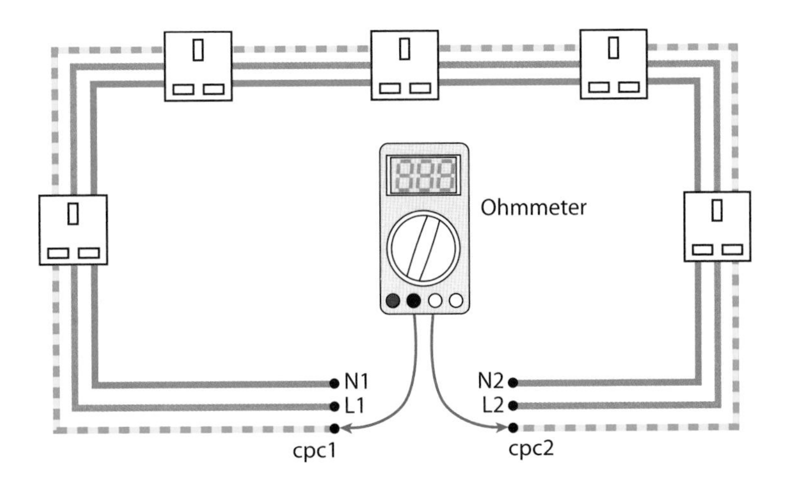

Figure 5.6 Measurement of line, neutral and protective conductors

The resistance values obtained should be the same (within 0.05 ohms) if the conductors are the same size. If the protective conductor has a reduced csa, the resistance of the protective loop will be proportionally higher than that of the line or neutral loop. If these relationships are not achieved then either the conductors are incorrectly identified or there is a loose connection at one of the accessories.

The line and neutral conductors are then connected together so that the outgoing line conductor is connected to the returning neutral conductor and vice versa as in Figure 5.7. The resistance between line and neutral conductors is then measured at each socket outlet. The readings obtained from those sockets wired into the ring will be substantially the same and the value will be approximately half the resistance of the line or the neutral loop resistance. Any sockets wired as spurs will have a proportionally higher resistance value corresponding to the length of the spur cable.

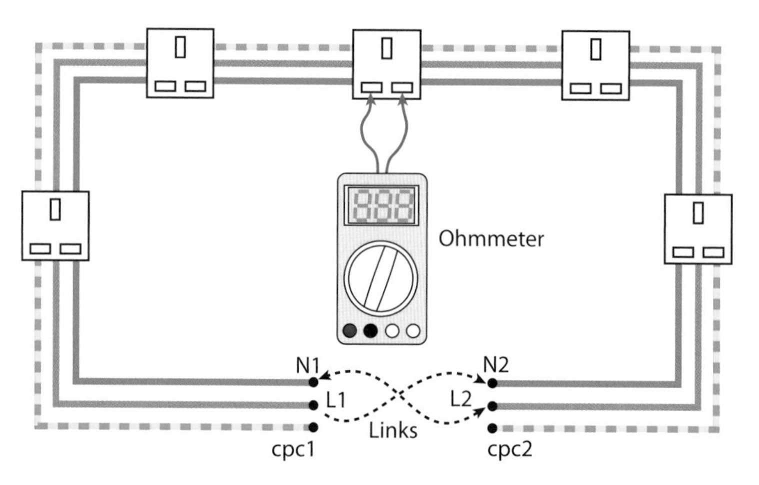

Figure 5.7 Line and neutral conductors connected together

The latter exercise is then repeated but with the line and cpc cross-connected as in Figure 5.8. The resistance between line and earth is then measured at each socket. The highest value recorded represents the maximum $R_1 + R_2$ of the circuit and is recorded on the test schedule. It can be used to determine the earth loop impedance (Z_s) of the circuit to verify compliance with the loop impedance requirements of the Regulations.

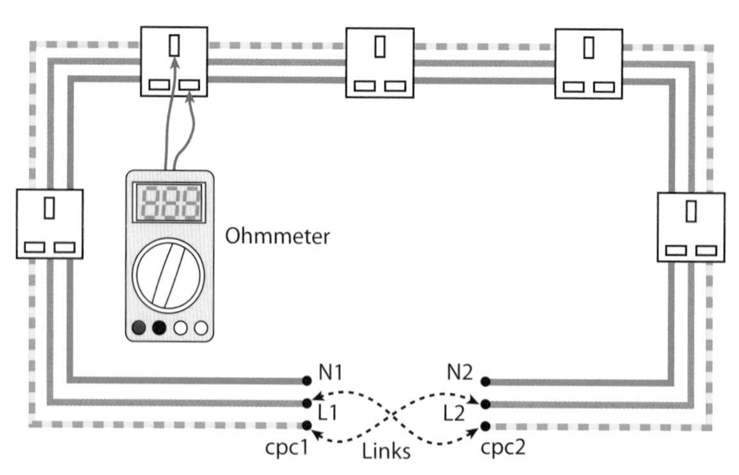

Figure 5.8 Line and cpc cross-connected

Insulation resistance

Insulation resistance tests are to verify, for compliance with BS 7671, that the insulation of conductors, electrical accessories and equipment is satisfactory and that electrical conductors and protective conductors are not short-circuited, or do not show a low insulation resistance (which would indicate defective insulation). In other words, we are testing to see whether the insulation of a conductor is so poor as to allow any conductor to 'leak' to earth or to another conductor. Before testing ensure that:

- Pilot or indicator lamps and capacitors are disconnected from circuits to avoid an inaccurate test value being obtained.
- Voltage sensitive electronic equipment such as dimmer switches, delay timers, power controllers, electronic starters for fluorescent lamps, emergency lighting, Residual Current Devices etc. are disconnected so that they are not subjected to the test voltage.
- There is no electrical connection between any line and neutral conductor (e.g. lamps left in).

To illustrate why we remove lamps, consider Figure 5.9. Should we leave the lamp in? The coil that is the lamp filament is effectively creating a short circuit between the line and neutral conductors.

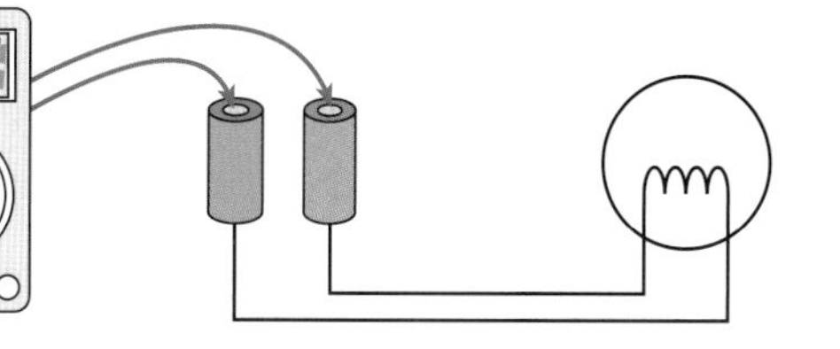

Figure 5.9 Insulation resistance test

The test equipment to be used would be an insulation resistance tester meeting the criteria as laid down in BS 7671, with insulation resistance tests carried out using the appropriate d.c. test voltage as specified in Table 61 of BS 7671 (shown here as Table 5.2). The installation will be deemed to conform with the Regulations if, with the main switchboard and each distribution circuit tested separately, with all its final circuits connected but with current using equipment disconnected, it has an insulation resistance not less than that specified in Table 61.

Table 61 Minimum value of insulation resistance		
Circuit nominal voltage (V)	***Test voltage dc (V)***	***Minimum insulation resistance***
SELV and PELV	250	≤0.5 MΩ
Up to and including 500 V with the exception of the above systems	500	≥1.0 MΩ
Above 500 V	1000	≥1.0 MΩ

Table 5.2 Minimum value of insulation resistance (BS 7671)

The tests should be carried out with the main switch off, all fuses in place, switches and circuit breakers closed, lamps removed, and fluorescent and discharge luminaires and other equipment disconnected. Where the removal of lamps and/or the disconnection of current using equipment is impracticable, the local switches controlling such lamps and/or equipment should be open.

Simple installations that contain no distribution circuits should be tested as a whole; however, to perform the test in a complex installation it may need to be subdivided into its component parts. Although an insulation resistance value of not less than 1.0MΩ complies with the Regulations, where an insulation resistance value of less than 2 MΩ is recorded, the possibility of a latent defect exists. If this is the case, each circuit should be separately tested and its measured insulation resistance should be greater than 2 megohms.

Test 1 – Insulation resistance between line conductors

- Single-phase circuits – Test between the line and neutral conductors at the appropriate switchboard.
- Three-phase circuits – Before this test is carried out you must ensure that the incoming neutral has been disconnected so there is no connection with earth.

Now make a series of tests between line conductors in turn at the appropriate switchboard as follows:

- between Brown phase and to the Black phase, Grey phase and Neutral (Blue) grouped together
- between Black phase and to the Grey phase and Neutral (Blue) grouped together
- between Grey phase and Neutral (Blue).

Where it is not possible to group conductors in this way, conductors may be tested singly. Figure 5.10 illustrates the testing of insulation resistance on a single-phase lighting circuit. But remember on a two-way circuit to operate the two-way switch and re-test to make sure that you have tested all of the strappers!

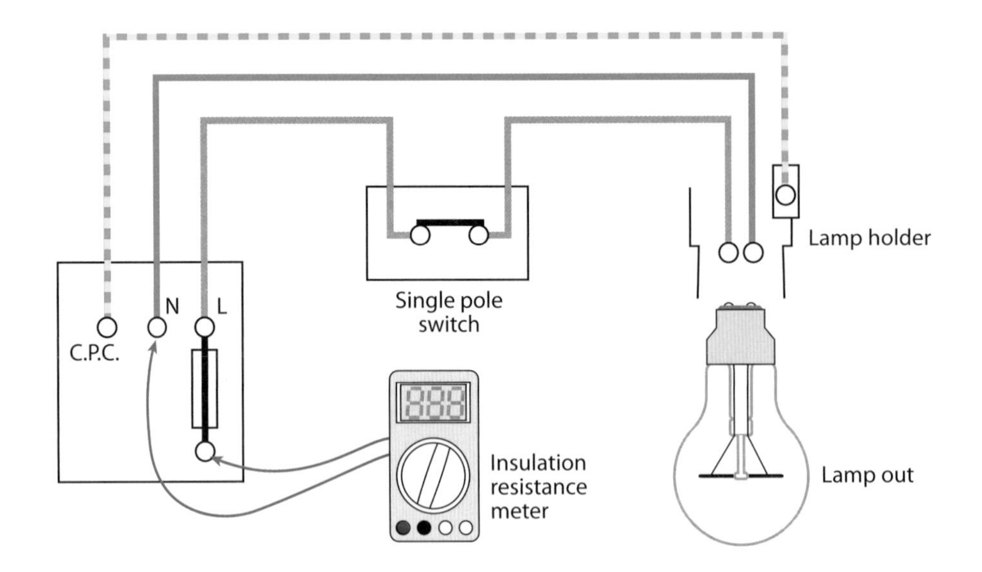

Figure 5.10 Insulation resistance test on single-phase lighting circuit

Test 2 – Insulation resistance from earth to line and neutral connected together

- Single-phase circuits – Test between the line and neutral conductors and earth at the appropriate distribution board. Where any circuits contain any two-way switching, the two-way switches will require to be operated and another insulation resistance test carried out, including the two-way strapping wire which was not previously included in the test.
- Three-phase circuits – Measure between all line conductors and neutral bunched together, and earth. Where a low reading is obtained (less than 2 megohms) it may be necessary to test each conductor to earth separately. Figure 5.11 shows the test of insulation resistance between earth to line and neutral connected together on a socket outlet circuit.

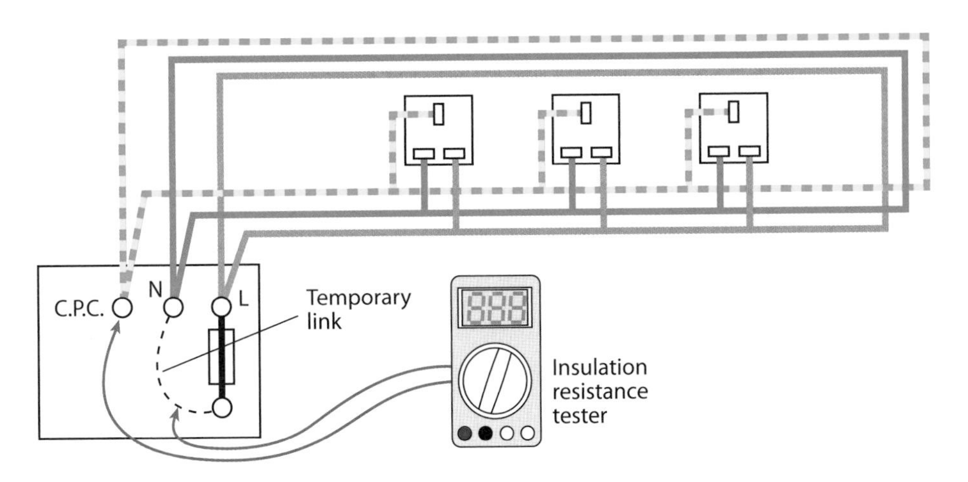

Figure 5.11 Insulation resistance test between earth to line and neutral connected together on socket outlets

Some electricians prefer, or find it quicker, to test between individual conductors rather than to group them together. As an example, Figure 5.12 shows the ten readings that would need to be taken for a three-phase circuit.

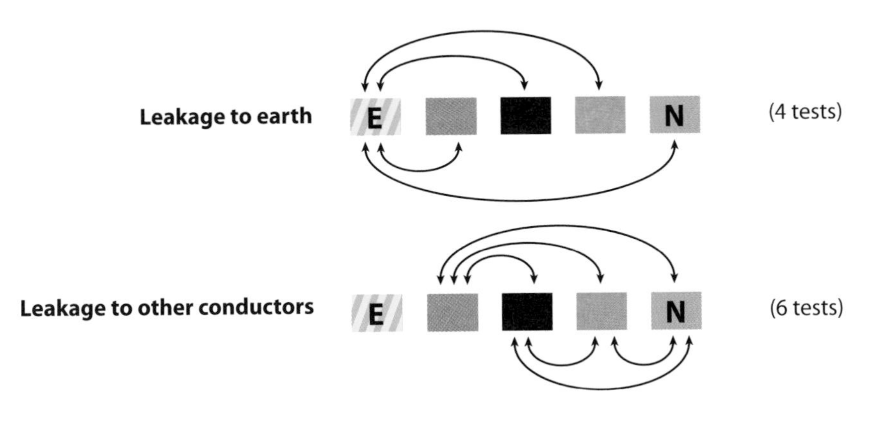

Figure 5.12 Three-phase circuit test

Polarity

A test needs to be performed to check the polarity of all circuits. This must be done before connection to the supply, with either an ohmmeter or the continuity range of an insulation and continuity tester.

The purpose of a polarity test is to verify that:

- each fuse and single-pole control and protective device is connected in the line conductor only
- with the exception of ES14 and ES27 lampholders to BS EN 60238, in circuits that have an earthed neutral, centre contact bayonet and ES lampholders have the outer or screwed contacts connected to the neutral conductor
- wiring has been correctly connected to socket outlets and similar accessories.

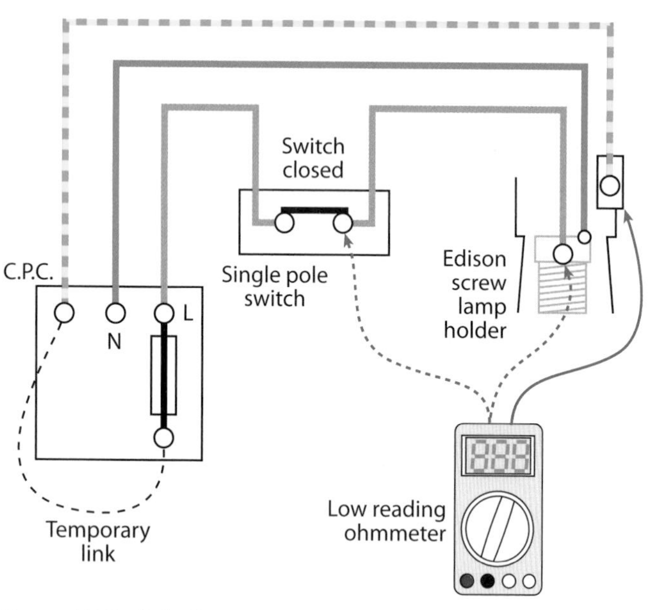

Figure 5.13 Polarity test of lighting circuit using continuity tester

In essence, having established the continuity of the cpc in an earlier test, we now use this as a long test lead, temporarily linking it out with the circuit line conductor at the distribution board and then making our test across the phase and earth terminals at each item in the circuit under test. Remember to close lighting switches before carrying out the test.

Figure 5.13 shows the test for polarity of a lighting circuit using a continuity tester.

For ring circuits, if the tests required by Regulation 713-03 ring circuit continuity have been carried out, the correct connections of line, neutral and cpc conductors will have been verified and no further testing is required. For radial circuits the $R_1 + R_2$ measurements should also be made at each point, using this method.

Earth electrode resistance

Where the earthing system incorporates an earth electrode as part of the system, the electrode resistance to earth needs to be measured. In previous years, the metal pipes of water mains were used, but the change to more plastic pipe work means this practice can no longer be relied upon.

Some of the types of accepted earth electrode are:

- earth rods or pipes
- earth tapes or wires
- earth plates
- underground structural metalwork embedded in foundations
- lead sheaths or other metallic coverings of cables
- metal pipes.

The resistance to earth depends upon the size and type of electrode used, and remember that we want as good a connection to earth as possible. The connection to the electrode must be made above ground level.

Measurement by standard method

When measuring earth electrode resistances to earth where low values are required, as in the earthing of the neutral point of a transformer or generator, test method 1 below may be used, using an earth electrode resistance tester.

Test method 1

Before this test is undertaken, the earthing conductor to the earth electrode must be disconnected either at the electrode or at the main earthing terminal. This will ensure that all the test current passes through the earth electrode alone. However, as this will leave the installation unprotected against earth faults, switch off the supply before disconnecting the earth.

The test should be carried out when the ground conditions are at their least favourable, i.e. during a period of dry weather, as this will produce the highest resistance value.

The test requires the use of two temporary test electrodes (spikes) and is carried out in the following manner:

- Connect the earth electrode to terminals C_1 and P_1 of a four-terminal earth tester. To exclude the resistance of these test leads from the resistance reading, individual leads should be taken from these terminals and connected separately to the electrode. However, if the test lead resistance is insignificant, the two terminals may be short-circuited at the tester and connection made with a single test lead, the same being true if you are using a three-terminal tester.
- Connection to the temporary 'spikes' is now made as shown in Figure 5.14. The distance between the test spikes is important. If they are too close together, their resistance areas will overlap. In general, you can expect a reliable result if the distance between the electrode under test and the current spike is at least ten times the maximum length of the electrode under test, e.g. 30 m away from a 3 m long rod electrode. (Resistance area is important where livestock are concerned, as the front legs of an animal may be outside, and the back legs of the same animal inside the resistance area, thus creating a potential difference. As little as 25 V can

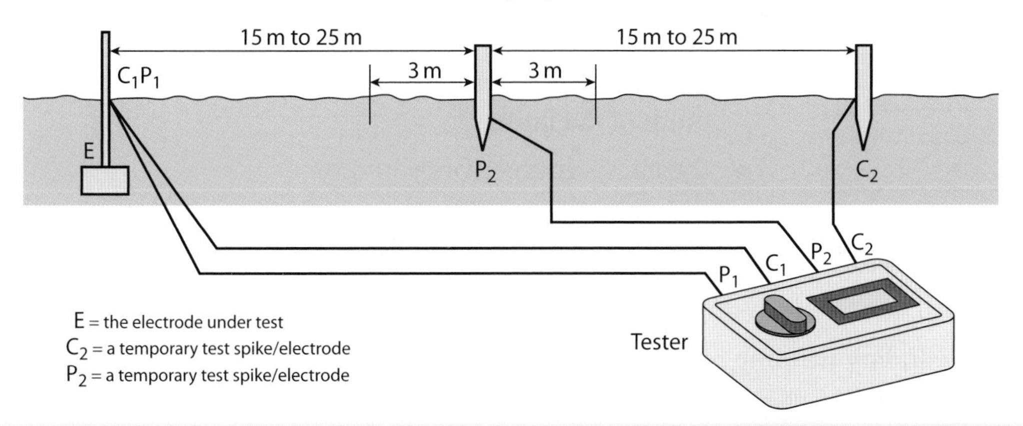

Figure 5.14 Earth electrode test

be lethal to them, so it's important to ensure that all of the electrode is well below ground level and RCD protection is used).

- We then take three readings:
  1. first with the potential spike initially midway between the electrode and current spike
  2. second at a position 10 per cent of the electrode-to-current spike distance back towards the electrode
  3. third at a position 10 per cent of the distance towards the current spike.
- By comparing the three readings, a percentage deviation can be determined. We do this calculation by taking the average of the three readings, finding the maximum deviation of the readings from this average in ohms, and expressing this as a percentage of the average. The accuracy of the measurement using this technique is on average about 1.2 times the percentage deviation of the readings. It is difficult to achieve an accuracy better than 2 per cent, and you should not accept readings that differ by more than 5 per cent. To improve the accuracy of the measurement to acceptable levels, the test must be repeated with larger separation between the electrode and the current spike.

Once the test is completed, make sure that the earthing conductor is re-connected.

Test method 2

Guidance Note 3 to BS 7671 lists a Test method 2. However, this is an alternative method for use on RCD protective TT installations only.

Earth-fault loop impedance

When designing an installation, it is the designer's responsibility to ensure that, should a phase-to-earth fault develop, the protection device will operate safely and within the time specified by BS 7671. Although the designer can calculate this in theory, it is not until the installation is complete that the calculations can be checked.

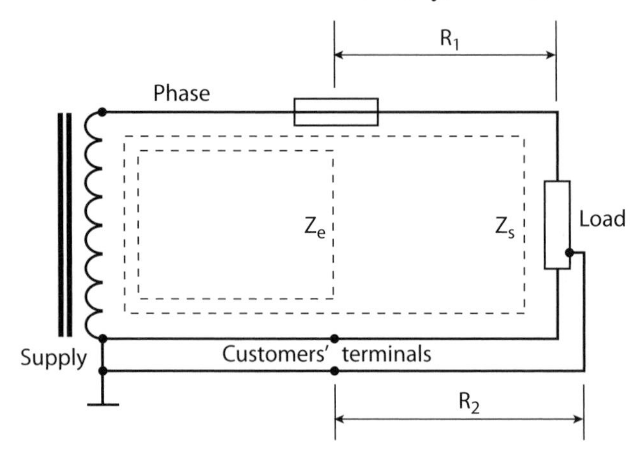

Figure 5.15 Earth-fault loop

It is necessary therefore to determine the earth-fault loop impedance (Z_s) at the furthest point in each circuit and to compare the readings obtained with either the designer's calculated values or the values tabulated in BS 7671 or the *Unite Guide to Good Electrical Practice.*

The earth-fault loop is made up of the following elements:

- the phase conductor from the source of the supply to the point of the fault
- the circuit protective conductor
- the main earthing terminal and earthing conductor
- the earth return path (dependent on the nature of the supply, TN-S, TN-C-S etc.)
- the path through the earthed neutral of the supply transformer
- the secondary winding of the supply transformer.

The value of earth-fault loop impedance may be determined by:

- direct measurement of Z_s
- direct measurement of Z_e at the origin of the circuit and adding to this the value of $R_1 + R_2$ measured during continuity tests: $Z_s = Z_e + (R_1 + R_2)$
- obtaining the value of Z_e from the electricity supplier and adding to this the value of $R_1 + R_2$ as above. However, where the value of Z_e is obtained from the electricity supplier and is not actually measured, a test must be carried out to ensure that the main earthing terminal is in fact connected to earth using an earth loop impedance tester or an approved test lamp.

Direct measurement of Z_s

Direct measurement of earth-fault loop impedance is achieved by use of an earth-fault loop impedance tester, which is an instrument designed specifically for this purpose. The instrument operates from the mains supply and therefore can only be used on a live installation.

The instrument is usually fitted with a standard 13 A plug for connecting to the installation directly through a normal socket outlet, although test leads and probes are also provided for taking measurements at other points on the installation.

In order to eliminate any parallel earth return paths, the main equipotential bonding conductors are disconnected for the duration of the test. This will ensure that the reading is not distorted by the presence of gas or water service pipes acting as part of the earth return path. Precautions must be taken, however, to ensure that the main equipotential bonding conductors are reconnected after the test has taken place.

Earth-fault loop impedance testers are connected directly to the circuit being tested and care must be taken to prevent danger. Should a break have occurred anywhere in the protective conductor under test then the whole of the earthing system could become live. It is essential therefore that protective conductor continuity tests be carried out prior to the testing of earth-fault loop impedance. Communication with other users of the building and the use of warning notices and barriers is essential.

Measurement of Z_e

The value of Z_e can be measured using an earth fault loop impedance tester at the origin of the installation. However, as this requires the removal of covers and the exposure of live parts, extreme care must be taken and the operation must be supervised at all times. The instrument is connected using approved leads and probes between the phase terminal of the supply and the means of earthing with the main switch open or with all sub-circuits isolated. In order to remove the possibility of parallel paths, the means of earthing must be disconnected from the equipotential bonding conductors for the duration of the test. With the instrument correctly connected and the test button pressed, the instrument will give a direct reading of the value of Z_e. It must be remembered to re-connect all earthing connections on completion of the test.

Verification of test results

The measured values of earth fault loop impedance obtained (Z_S) should be less than the values stated in BS 7671 Chapter 4, tables 41.2, 41.3 and 41.4, reproduced below.

Please note that these values are for conductors at their normal operating temperature. If the conductors are at a different temperature at the point of test, then a de-rating factor should be applied in line with the guidance given in Appendix 14 of BS 7671.

Table 41.2 of BS 7671 (see Regulation 411.4.6)						
(a) General purpose (gG) fuses to BS 88-2.2 and BS 88.6						
Rating (amperes)	6	10	16	20	25	32
Z_S (ohms)	8.52	5.11	2.70	1.77	1.44	1.04
(b) Fuses to BS 1361						
Rating (amperes)	5	15	20	30		
Z_S (ohms)	10.45	3.28	1.70	1.15		
(c) Fuses to BS 3036						
Rating (amperes)	5	15	20	30		
Z_S (ohms)	9.58	2.55	1.77	1.09		
(d) Fuses to BS 1362						
Rating (amperes)	3	13				
Z_S (ohms)	16.4	2.42				

Note: The circuit loop impedances given in the table should not be exceeded when the conductors are at their normal operating temperature. If the conductors are at a different temperature when tested, the reading should be adjusted accordingly. See Appendix 14.

Table 5.4 Maximum earth fault impedance (Z_S) for fuses, for 0.4 s disconnection time with U_O of 230 V

Table 41.3 of BS 7671														
(a) Type B circuit-breakers to BS EN 60898 and the overcurrent characteristics of RCBOs to BS EN 61009-1														
Rating (amperes)	3	6	10	16	20	25	32	40	50	63	80	100	125	I_n
Z_S (ohms)	15.33	7.67	4.60	2.87	2.30	1.84	1.44	1.15	0.92	0.73	0.57	0.46	0.37	46/ I_n
(b) Type C circuit-breakers to BS EN 60898 and the overcurrent characteristics of RCBOs to BS EN 61009-1														
Rating (amperes)		6	10	16	20	25	32	40	50	63	80	100	125	I_n
Z_S (ohms)		3.83	2.30	1.44	1.15	0.92	0.72	0.57	0.46	0.36	0.29	0.23	0.18	23/ I_n
(c) Type D circuit-breakers to BS EN 60898 and the overcurrent characteristics of RCBOs to BS EN 61009-14														
Rating (amperes)		6	10	16	20	25	32	40	50	63	80	100	125	I_n
Z_S (ohms)		1.92	1.15	0.72	0.57	0.46	0.36	0.29	0.23	0.18	0.14	0.11	0.09	11.5/ I_n

Table 5.5 Maximum earth fault loop impedance (Z_S) for circuit-breakers with U_o of 230 V, for instantaneous operating giving compliance with the 0.4 s disconnection time of Regulation 411.3.2.2 and 5 s disconnection time of Regulation 411.3.2.3

Table 41.4								
(a) General purpose (gG) fuses to BS 88-2.2 and BS 88-6								
Rating (amperes)	6	10	16	20	25	32	40	50
Z_S (ohms)	13.5	7.42	4.18	2.91	2.30	1.84	1.35	1.04
Rating (amperes)		63	80	100	125	160	200	
Z_S (ohms)		0.82	0.57	0.42	0.33	0.25	0.19	
(b) Fuses to BS 1361								
Rating (amperes)	5	15	20	30	45	60	80	100
Z_S (ohms)	16.4	5.00	2.80	1.84	0.96	0.70	0.50	0.36
(c) Fuses to BS 3036								
Rating (amperes)	5	15	20	30	45	60	100	
Z_S (ohms)	17.7	5.35	3.83	2.64	1.59	1.12	0.53	
(d) Fuses to BS 1362								
Rating (amperes)	3	13						
Z_S (ohms)	23.2	3.83						

Table 5.6 Maximum earth fault loop impedance (Z_S) for 5 s disconnection time with U_O of 230 V (see Regulation 411.4.8)

Operation of residual current devices (RCDs)

Although the majority of residual current devices (RCDs) in use are designed to operate at earth leakage currents not exceeding 30 mA, other values of operating current are available for use in special circumstances. All RCDs are electromechanical devices that must be checked regularly to confirm that they are still in working order. This can be done at regular intervals by simply pressing the test button on the front of the device.

Where an installation incorporates an RCD, Regulation 514.12.2 requires a notice to be fixed in a prominent position at or near the origin of the installation.

The integral test button incorporated in all RCDs only verifies the correct operation of the mechanical parts of the RCD and does not provide a means of checking the continuity of the earthing conductor, the earth electrode or the sensitivity of the device. This can only be done effectively by use of an RCD tester specifically designed for testing RCDs as described below.

Test method

The test must be made on the load side of the RCD between the phase conductor of the protected circuit and the associated circuit protective conductor. The load being supplied by the RCD should be disconnected for the duration of the test. The test instrument is usually fitted with a standard 13 A plug top, and the easiest way of making these connections, wherever possible, is by plugging the instrument into a suitable socket outlet protected by the RCD under test.

The test instrument operates by passing a simulated fault current of known value through the RCD and then measures the time taken for the device to trip. Although different types of RCD have different requirements (time delays etc.), for general purpose RCDs the test criteria are as follows:

- With an earth leakage current equivalent to 50 per cent of the rated tripping current of the RCD, the device should not open (this is to ensure that nuisance tripping due to leakage currents less than the rated tripping current will not occur).
- For general purpose RCDs to BS 4293 an earth leakage current equivalent to 100 per cent of the rated tripping current of the RCD, should cause the device to open in less than 200 milliseconds.
- Individual RCD-protected socket outlets should also meet the above criteria.
- For general purpose RCDs to BS EN 61008 or RCBOs to BS EN 61009 an earth leakage current equivalent to 100 per cent of the rated tripping current of the RCD should cause the device to open in less than 300 milliseconds unless it is of 'Type S' which incorporates an intentional time delay in which case it should trip within the time range of 130 to 500 milliseconds.
- Where an RCD with an operating current not exceeding 30 mA is used to provide supplementary protection against direct contact, a test current of five times the

rated trip current (e.g. 5 × 30 = 150 mA) should cause the RCD to open in less than 40 milliseconds.

A further mechanical function check should be carried out to ensure that the operating switch of the residual current device is functioning correctly. Results of all the above tests should be recorded on the appropriate test results schedule.

Safety

Under certain circumstances these tests can result in potentially dangerous voltages appearing on exposed and extraneous conductive parts within the installation and therefore suitable precautions must be taken to prevent contact by persons or livestock with any such part. Other building users should be made aware of the tests being carried out and warning notices should be posted as necessary.

Electrical test instruments

BS EN 61010 covers basic safety requirements for electrical test instruments, and all instruments should be checked for conformance with this standard before use. Older instruments may have been manufactured in accordance with BS 5458 but, provided these are in good condition and have been recently calibrated, there is no reason why they cannot be used.

Instruments may be analogue (i.e. fitted with a needle that gives a direct reading on a fixed scale) or digital, where the instrument provides a numeric digital visual display of the actual measurement being taken. Insulation and continuity testers can be obtained in either format whilst earth-fault loop impedance testers and RCD testers are digital only.

Calibration and instrument accuracy

To ensure that the reading being taken is reasonably accurate, all instruments should have a basic measurement accuracy of at least 5 per cent. In the case of analogue instruments a basic accuracy of 2 per cent of full-scale deflection should ensure the required accuracy of measured values over most of the scale.

All electrical test instruments should be calibrated on a regular basis. The time between calibrations will depend on the amount of usage that the instrument receives, although this should not exceed 12 months in any circumstances. Instruments have to be calibrated in laboratory conditions against standards that can be traced back to national standards; therefore this usually means returning the instrument to a specialist test laboratory.

On being calibrated the instrument will have a calibration label attached to it stating the date the calibration took place and the date the next calibration is due (see Figure 5.16 overleaf). It will also be issued with a calibration certificate detailing the tests that have been carried out and a reference to the equipment used. The user of the instrument should always check to ensure that the instrument is within calibration before being put to use.

Safety tip

Prior to RCD tests being carried out, it is essential for safety reasons that the earth loop impedance of the installation has been tested to check that the earth-return path is sound and that all the necessary requirements have been met. BS 7671 stipulates the order in which tests should be carried out

Did you know?

The basic standard covering the performance and accuracy standards of electrical test instruments is BS EN 61557, which incidentally also requires compliance with the safety requirements of BS EN 61010

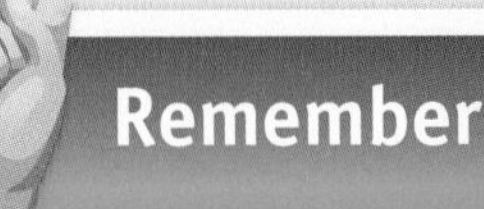

Remember

Although older instruments often generated their own operating voltage by use of an in-built hand-cranked generator, almost all modern instruments have internal batteries for this purpose. It is essential therefore that the condition of the batteries is checked regularly, including the absence of corrosion at the battery terminals

Remember

When using an instrument out on site, the accuracy of the instrument will probably not be as good as the accuracy obtained under laboratory conditions. Operating accuracy is always worse than basic accuracy and can be affected by battery condition, generator cranking speed, ambient temperature, instrument alignment or loss of calibration

Remember

Records of calibrations carried out and copies of all calibration certificates issued should be retained in a safe place in case it is necessary to validate any test results carried out at some later date

A further adhesive label (example shown in Figure 5.17) is often placed over the joint in the instrument casing stating that the calibration is void should the seal be broken. A broken seal will indicate whether anyone has deliberately opened the instrument and possibly tampered with the internal circuitry.

Instrument serial No. : ....

Date tested : ....

Date next due : ....

Figure 5.16 Adhesive calibration label

Calibration void if seal is broken

Figure 5.17 Calibration seal

Instruments that are subject to any electrical or mechanical misuse (e.g. if the instrument is subject to an electrical short circuit or is dropped) should be returned for re-calibration before being used again.

Electrical test instruments are relatively delicate and expensive items of equipment and should be handled in a careful manner. When not in use they should be stored in clean, dry conditions at normal room temperature. Care should also be taken of instrument leads and probes to prevent damage to their insulation and to maintain them in a good, safe working condition.

Types of instrument

Low-resistance ohmmeters

Where low resistance measurements are required when testing earth continuity, ring circuit continuity and polarity, then a low reading ohmmeter is required. This may be a specialised low-reading ohmmeter or the continuity scale of a combined insulation and continuity tester. Whichever type is used it is recommended that the test current should be derived from a source of supply not less than 4 V and no greater than 24 V with a short circuit current not less than 200 mA (instruments manufactured to BS EN 61557 will meet the above requirements).

Errors in the reading obtained can be introduced by contact resistance or by lead resistance. Although the effects of contact resistance cannot be eliminated entirely and may introduce errors of 0.01 ohm or greater, lead resistance can be eliminated either by clipping the leads together and zeroing the instrument before use, where this facility is provided, or alternatively measuring the resistance of the leads and subtracting this from the reading obtained.

Insulation resistance ohmmeters

Insulation resistance should have a high value and therefore insulation resistance meters must have the ability to measure high resistance readings. The test voltage required for measuring insulation resistance is given in BS 7671 Table 61 as shown in Table 5.8.

Remember

BS 7671 requires that the instruments used for measuring insulation resistance must be capable of providing the test voltages stated above whilst maintaining a test current of 1 mA. Instruments that are manufactured to BS EN 61557 will satisfy the above requirements

Circuit Nominal Voltage (volts)	Test Voltage d.c. (volts)	Minimum Insulation Resistance (megohms)
SELV and PELV	250	≥0.5 MΩ
Up to and including 500 V with the exception of the above	500	≥1.0 MΩ
Above 500 V	1000	≥1.0 MΩ

Table 5.8 Minimum values of insulation resistance (BS 7671)

The photograph shows a typical modern insulation and continuity tester that will measure both low values of resistance for use when carrying out continuity and polarity tests and also high values of resistance when used for insulation resistance tests.

Instruments of this type are usually enclosed in a fully insulated case for safety reasons and have a range of switches to set the instrument correctly for the type of test being carried out i.e. continuity or insulation. The instrument also has a means of selecting the voltage range required e.g. 250 V, 500 V, 1000 V.

Other features of this particular type of instrument are the ability to lock the instrument in the 'on' position for hands-free operation and an automatic nulling device for taking account of the resistance of the test leads.

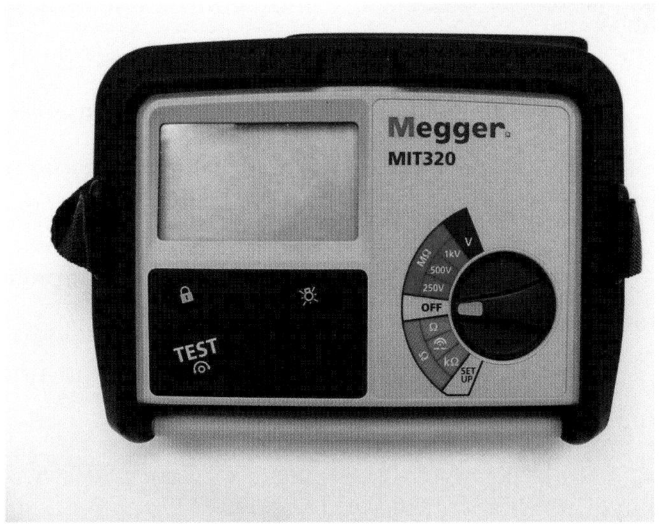

Modern insulation and continuity tester

Earth-loop impedance testers

Earth-loop impedance testers of the type shown in the photograph have the capability to measure both earth-loop impedance and also prospective short-circuit current, depending on which function is selected on the range selection switch. The instrument also has a series of LED warning lights to indicate whether the polarity of the circuit under test is correct or not.

The instrument gives a direct digital read-out of the value of the measurement being taken at an accuracy of plus or minus 2 per cent.

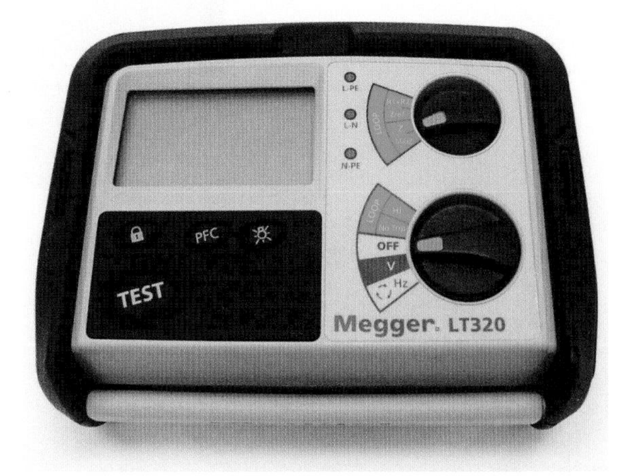

Earth-loop impedance tester

RCD testers

Instruments for testing residual current devices, such as the one shown in the photograph, have two selection switches. One switch that should be set to the rated tripping current of the RCD (e.g. 30 mA, 100 mA etc.) and the other set to the test current required i.e. 50 per cent or 100 per cent of the rated tripping current or 150 mA for testing 30 mA RCDs when being used for supplementary protection.

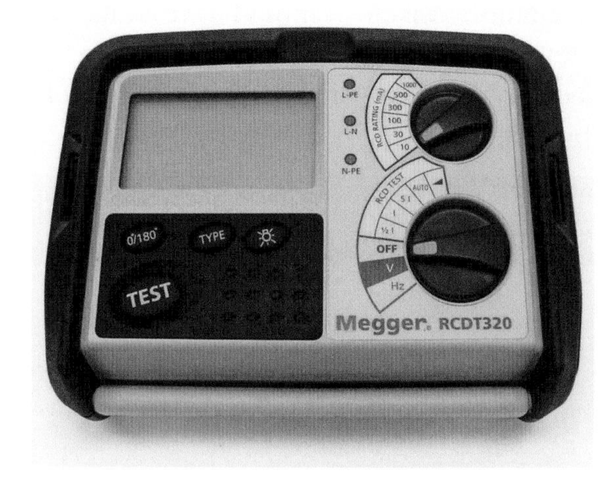

RCD tester

A modern innovation by manufacturers is the production of an 'all in one' instrument that has the ability to carry out the most common tests required by the Regulations. These are:

- continuity tests (including polarity tests)
- insulation resistance tests
- earth-loop impedance tests
- RCD tests
- measurement of prospective short circuit current.

The photograph below shows an example of this type of instrument, which by manipulation of the function and range switches, will perform all of the above tests.

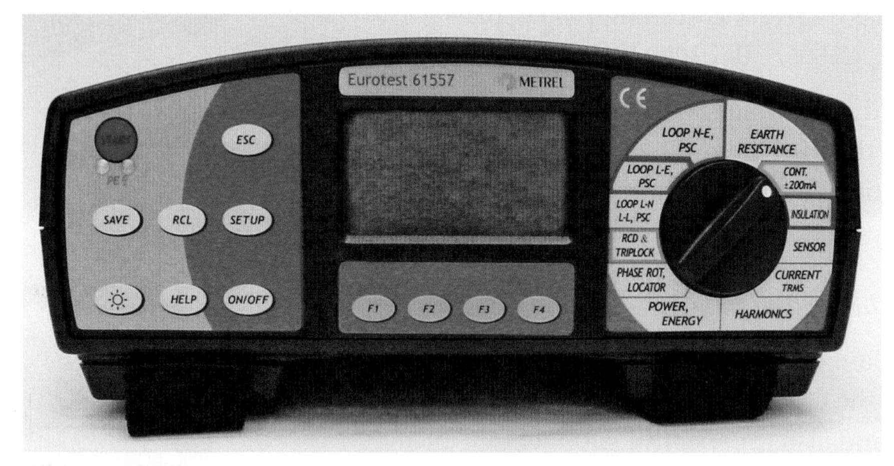

All-in-one RCD tester

Certificates

Electrical Installation Certificate

BS 7671 requires that, following the inspection and testing of all new installations, alterations and additions to existing installations or periodic inspections, an Electrical Installation Certificate together with a schedule of test results shall be given to the person ordering the work.

Examples of such forms are provided in Appendix 6 of BS 7671. Other professional organisations such as the Electrical Contractors Association (ECA) or the National Inspection Council for Electrical Installation Contractors (NICEIC) produce their own versions of these forms for their own members' use. The different types of forms available are as follows.

- Periodic Inspection Report
- Electrical Installation Certificate (Type 1)
- Electrical Installation Certificate (Type 2)
- Inspection Schedule
- Schedule of Test Results
- Electrical Installation – Minor Works Certificate

Periodic Inspection Report

The Periodic Inspection Report is for use when carrying out a routine periodic inspection and test of an existing installation and is not for use when alterations and additions are made. An Inspection Schedule and a Schedule of Test Results should accompany the Periodic Inspection Report. The general requirements for the content and completion of all the above documents are stated in the introduction to Appendix 6 of BS 7671 and are summarised below:

- Electrical Installation Certificates and Electrical Installation – Minor Works Certificates must be completed and signed by a competent person or persons in respect of the design, the construction, and the inspection and test of the installation.
- Periodic Inspection Reports must be completed and signed by a competent person in respect of just the inspection and test of the installation.
- Competent persons must have a sound knowledge and relevant experience of the type of work being undertaken and of the technical requirements of BS 7671. They must also have a sound knowledge of the inspection and testing procedures contained in the Regulations and must use suitable testing equipment.
- Electrical Installation Certificates must identify who is responsible for the design, construction, and inspection and testing, whether this is new work or an alteration or addition to an existing installation. As explained below this may be one person or it may be a number of different organisations.

- Minor Works Certificates must also identify who is responsible for the design, construction, and inspection and testing of the work carried out.
- Periodic Inspection Reports must indicate who is responsible for carrying out the periodic inspection and testing of the existing installation.
- A Schedule of Test Results must be issued with all Electrical Installation Certificates and Periodic Inspection Reports.
- When signing the above forms on behalf of a company or other business organisation, individuals should state for whom they are acting.
- For larger or more complex installations, additional documentation such as inspection checklists may be required for clarification purposes.

The design, construction, and inspection and testing of an installation may be the responsibility of just one person, in which case the shortened version of the Electrical Installation Certificate may be used.

On larger or more complex installations the design, construction, and inspection and test may be the responsibility of a number of different organisations, in which case all the parties concerned should be identified and must sign the certificate in the appropriate place. This may be the case where a design consultant is employed to carry out the design of the installation and an electrical contractor is employed to carry out the construction.

Many larger companies now employ dedicated inspectors whose job it is to inspect and test the work of other electricians within the company. In this case the certificate may be signed by the person carrying out the installation work as the constructor and the person carrying out the inspection and test as the inspector.

Provided that each party is clearly identified in the appropriate place, this is quite acceptable. In other cases the inspection and testing of the installation may be sub-contracted to a second contractor, in which case they should sign the certificate for that section of the work.

The certificate also provides accommodation for the identification and signatures of two different design organisations. This is for use where different sections of the installation are designed by different parties e.g. the electrical installation may be designed by one consultant and the fire-alarm system or other specialist service may be designed by another. If this should be the case then the details and signatures of both parties are required. BS 7671 provides advice notes for the completion of each type of form and also provides guidance notes for the person receiving the certificate. These are particularly helpful for non-technical persons.

Electrical Installation Certificate (Type 1)

For use when inspecting and testing a new installation, a major alteration or addition to an existing installation where the design, construction, and inspection and testing of the installation are the responsibility of different organisations. See Figure 5.18.

For Design
I/We being the person(s) responsible for the design of the Electrical Installation (as indicated by my/our signatures) particulars of which are described above, having exercised reasonable skill and care when carrying out the design hereby CERTIFY that the design work for which I/we have been responsible is to the best of my/our knowledge and belief in accordance with BS7671 amended to .... except for the departures (if any) detailed below.

Details of departures from BS7671(if any)
....
....
....

The extent of liability of the signatory is limited to the work described above.
For the DESIGN of the installation.

Signature : .... Date :.................... Name : .... Designer No.1

Signature : .... Date :.................... Name : .... Designer No.2

For Construction
I/We being the person(s) responsible for the construction of the Electrical Installation (as indicated by my/our signatures) particulars of which are described above, having exercised reasonable skill and care when carrying out the construction hereby CERTIFY that the construction work for which I/we have been responsible is to the best of my/our knowledge and belief in accordance with BS7671 amended to .... Except for the departures (if any) detailed below.

Details of departures from BS7671 (if any)
....
....
....

The extent of liability of the signatory is limited to the work described above.
For the CONSTRUCTION of the Installation.

Signature : .... Date :.................... Name : .... Constructor

For Inspection & testing
I/we being the person(s) responsible for the inspection and testing of the Electrical Installation (as indicated by my/our signatures) particulars of which are described above, having exercised reasonable skill and care when carrying out the inspection and testing hereby CERTIFY that the work for which I/we have been responsible is to the best of my/our knowledge and belief in accordance with BS7671 amended to .... except for the departures detailed below.

Details of departures from BS7671 (if any)
....
....
....

The extent of liability of the signatory is limited to the work described above.
For the INSPECTION & TEST of the installation

Signature : .... Date :.................... Name : .... Inspector

Figure 5.18 Electrical Installation Certificate (Type 1)

As stated previously, the full certificate requires not only the signatures of those persons responsible for the design, construction, and inspection and testing of the installation but also their full details including the company they represent. See Figure 5.19.

Particulars of the Signatories to the Electrical Installation Certificate

Designer No.1
Name : .... Company : ....
Address : ....
....
Post Code : .... Tel No : ....

Designer No.2
(if applicable)
Name : .... Company : ....
Address : ....
....
Post Code : .... Tel No : ....

Constructor
Name : .... Company : ....
Address : ....
....
Post Code : .... Tel No : ....

Inspector
Name : .... Company : ....
Address : ....
....
Post Code : .... Tel No : ....

Figure 5.19 Signatories to full certificate

Electrical Installation Certificate (Type 2)

A shortened version of the Type 1 form used when inspecting and testing a new installation, major alteration or addition to an existing installation where one person is responsible for the design, construction, and inspection and testing of the installation. See Figure 5.20.

This certificate is intended for use with new installations or alterations or additions to an existing circuit where new circuits have been installed. The original certificate is to be given to the person ordering the work and a copy retained by the contractor. This certificate is not valid unless accompanied by a Schedule of Test Results.

The signatures on the certificate should be those of the persons authorised by the companies carrying out the work of design, construction, and inspection and testing.

The recommended time to the next inspection should be stated, taking into account the frequency and quality of maintenance that the installation is likely to receive.

The prospective fault current should be the greater of either that measured between live conductors or that between phase conductor(s) and earth.

Client Details : ....

Address of Installation : ....

.... **Postcode :** ....

Description & Extent of Installation :	**New Installation**
Description :	**Addition**
Extent :	**Alteration**

For Design, Construction and Inspection & Testing

I being the person responsible for the Design, Construction and Inspection & Testing of the Electrical Installation (as indicated by my signature below) and particulars of which are described above, having exercised reasonable skill when carrying out the Design, Construction and Inspection & Test, hereby CERTIFY that the said work for which I have been responsible is to the best of my knowledge and belief in accordance with BS7671, amended to .... (date) except for the departures, if any, detailed below.

Departures from BS7671 (if any)

....

The extent of liability of the signatory is limited to the work described above as the subject of this certificate.

Name : .... Position : ....
Signature : .... Date : ....
For and on behalf of : ....
Address : ....
.... Tel No : ....

Next Inspection

I recommend that this installation is further inspected and tested after an interval of not more than : ....

Supply Characteristics and Earthing Arrangements

Earthing Arrangements	Number & Type of Live conductors	Nature of Supply	Protective device Characteristics
TN-S	1 phase – 2 wire	Nominal Voltage :	Type : ....
TN-C-S	1 phase – 3 wire	Frequency :	Current Rating : ....
TT	2 phase – 3 wire	Prospective Fault Current :	
	3 phase – 4 wire	External loop impedance :	

Guidance Notes (for those receiving the certificate)

- This certificate has been issued to confirm that the electrical installation work to which it relates has been designed, constructed, and inspected and tested in accordance with BS 7671 (*IEE Wiring Regulations*).
- As the person ordering the work you should have received the original certificate and the contractor should have retained a copy. However, if you are not the user of the installation (the occupier) then you should pass the certificate or a full copy (including schedules) to the user immediately.
- The original certificate should be retained in a safe place to be shown to anyone carrying out further work on the installation at a later date (this includes periodic inspection). If you later vacate the property, this certificate will demonstrate that the installation did comply with BS 7671 at the time the certificate was issued. The Construction (Design and Management) (CDM) Regulations require that copies of the Installation Certificate together with any attached schedules are retained in the project Health and Safety File.
- To ensure the safety of the installation it will need to be inspected at regular intervals throughout its life. The recommended period between inspections is stated on the certificate.
- This certificate is only intended to be used for new installations or for new work such as an alteration or addition to an existing installation. For the routine inspection and testing of an existing installation a 'Periodic Inspection Report' should be issued.
- This certificate is only valid if it is accompanied by the appropriate Schedule of Test Results.

Particulars of the Installation

Means of Earthing	Maximum Demand : .... Amperes per Phase ....
Suppliers facility	Details of Earth Electrode (where applicable)
Earth Electrode	Type : .... Location : .... Resistance to Earth : ....

Main Protective Conductors

Earthing Conductor : Material : .... csa : .... Connection verified : Yes/No

Main Equipotential Bonding Conductors : Material : .... csa : .... Connection verified : Yes/No

Bonding to water service : . yes/no Bonding to gas service : yes/no Bonding to oil service : yes/no

Bonding to structural steel : yes/no Bonding to other incoming service : yes/no (state details) ....

Main Switch or Circuit Breaker

BS. Type : .... No of Poles :Current Rating : A Voltage Rating : V

Location : .... Fuse Rating or setting : A

RCD operating current : .... mA RCD operating time : .... mS

Comments on Existing Installation : (in cases of alteration or additions)

....

....

....

....

Schedules

The attached Inspection Checklists and Schedule of Test Results form part of this certificate.

Attached : .... Inspection Checklists and .... Schedule of Test Results

Figure 5.20 Electrical Installation Certificate (Type 2 Short form)

Inspection Schedule

The Inspection Schedule provides confirmation that a visual inspection has been carried out as required by Part 6 of BS 7671 and covers all the inspection requirements listed in Regulation 611.3. The completed Inspection Schedule is attached to and forms part of the Electrical Installation Certificate.

Each item on the inspection schedule (Figure 5.21) should be checked and either ticked as satisfactory or ruled out if not applicable. In almost all cases, protection against direct contact will be by the presence of insulation and the enclosure of live parts, and protection against indirect contact will be by earthed equipotential bonding and automatic disconnection of supply (together with the presence of earthing and bonding conductors). Therefore other means of protection against either direct or indirect contact referred to below can be ruled out as not applicable.

Some example points are given below.

- SELV is an extra low-voltage system which is electrically separate from earth and from other systems. Where this method is used the particular requirements of Regulation 417 of BS 7671 must be checked.
- Protection against direct contact by the use of obstacles, or placing out of reach, are usually only employed in special circumstances but where they are used the particular requirements of Regulation 417 of BS 7671 must be met.
- For the requirements of systems using the protective measure electrical separation, see Regulation 413 of BS 7671.

On completion of the Inspection Schedule it should be signed and dated by the person responsible for carrying out the inspection.

1. Connection of conductors
2. dentification of conductors
3. Routing of cables in safe zones or protected against damage
4. Selection of conductors for current and voltage drop
5. Connection of single pole devices in phase conductor only
6. Correct connection of socket outlets and lampholders
7. Presence of fire barriers and protection against thermal effects
8. Method of protection against shock

8a. Protection against both direct and indirect contact
SELV
Limitation of discharge of energy

8b. Protection against indirect contact
Insulation of live parts
Barriers or enclosures
Obstacles
Placing out of reach
PELV

8c. Protection against indirect contact

8c. (i) Earthed equipotential bonding and automatic disconnection
Presence of earthing conductors

8c. (iii) Non conducting locations

8c. (iv) Earth-free local equipotential bonding
Presence of earth free equipolentia bonding conductors

8c. (v) Electrical separation

9. Prevention of mutual detrimental influence
Proximity of non-electrical services
Separation of Band 1 and Band 2 circuits
Separation of fire alarm and emergency lighting circuits
10. Presence of appropriate devices for isolation & switching
11. Presence of undervoltage protection devices
12. Choice and setting of protective and monitoring devices
(a) Residual current devices
(b) Overcurrent device
13. Labelling of protective devices switches and terminals
14. Selection of equipment suitable for external influences
15. Adequate access to switchgear and equipment
16. Presence of danger notices and other warning signs
17. Presence of diagrams & instructions

Figure 5.21 Inspection Schedule

Schedule of Test Results

The Schedule of Test Results is a written record of the results obtained when carrying out the electrical tests required by Part 6 of BS 7671. The following notes give guidance on the compilation of the Schedule, which should be attached to the Electrical Installation Certificate. See Figure 5.22.

Contractor : .... Address /Location of distribution board .... Type of Supply TN-S /TN-C-S/TT .... Instruments: ....
Date of Test : .... Ze at origin : .... Continuity : ....
Signature : .... PFC: .... Insulation : ....
Method of protection against indirect contact : .... Loop Impedance: .... RCD Tester : ....

Description of Work :

Circuit Description	Overcurrent Device		Wiring			Test Results									
	Short Circuit Capacity KA		Ref Method	Conductors		Continuity			Insulation Resistance		P	Zs	Functional Testing		Remarks
	Type	Rating		Live	cpc	R1+R2	R2	Ring	Live-Live	Live-Earth			RCD (Time)	Other	
1															
2															
3															
4															
5															
6															
7															
8															
9															
10															
11															
12															

Note 1 Any equipment vulnerable to testing : ....

Note 2 Any deviations from BS7671 : ....

Figure 5.22 Schedule of Test Results

1. The type of supply should be determined from either the supply company or by inspection of the installation.
2. Z_e should be measured with the main bonding disconnected. If Z_e is determined from information provided by the supply company rather than measured then the effectiveness of the earth must be confirmed by testing.
3. The value of prospective fault current recorded on the results schedule should be the greater of that measured between live conductors (short-circuit current) or line conductor and earth (earth-fault current).
4. In the column headed 'Overcurrent Device', details of the main protective device including its short circuit should be inserted.
5. The column headed 'Reference Method' refers to the method used for the installation of cables.
6. Continuity tests are to include a check for correct polarity of the following:
   - every fuse and single pole device to be connected in the phase conductor only
   - all centre contact lamp holders to have the outer contact connected to the neutral conductor
   - wiring of socket outlets and other accessories correctly connected.
7. Continuity of protective conductors – every protective conductor, including bonding conductors, must be tested to verify that it is sound and correctly connected.

8. Continuity of ring circuit conductors – a test must be carried out to verify the continuity of each conductor including the protective conductor of every ring circuit. A satisfactory reading should be indicated by a tick in the appropriate columns.

9. The sum of the resistances of the phase conductor and the protective conductor ($R_1 + R_2$) should be recorded and, after appropriate correction for temperature rise, may be used to determine Z_s ($Z_s = Z_e + (R_1 + R_2)$). Where Test method 2 is used, the maximum value of R_2 should be recorded.

10. Insulation resistance – where necessary all electronic devices should be disconnected from the installation so that they are not damaged by the tests themselves. Details of such equipment should be recorded on the Schedule of Test Results so that the person carrying out routine periodic testing in the future is aware of the presence of such equipment.

11. The insulation resistance between conductors and between conductors and earth shall be measured and the minimum values recorded in the appropriate columns on the Schedule of Test Results. The minimum insulation resistance for the main switchboard and each distribution board tested separately with all final circuits connected but with all current-using equipment disconnected shall be not less than that given.

12. Following the energising of the installation, polarity should be re-checked before further testing.

13. Earth-loop impedance (Z_s) – this may be determined either by direct measurement at the furthest point of a live circuit or by measuring Z_e at the origin of the circuit and adding to it the measured value of $R_1 + R_2$. Alternatively the value of Z_e may be obtained by inquiry from the electricity supply company. Values of Z_s should be less than the values (corrected where necessary) given in tables 5.3 to 5.7 of this book.

14. Functional testing – the correct operation of all RCDs and RCBOs must be tested by using an instrument that simulates a fault condition. The time taken to trip the device is automatically recorded by the instrument and should be written in the appropriate space on the Schedule of Test Results. At rated tripping current RCDs should operate within 200 milliseconds.

Electrical Installation – Minor Works Certificate

Minor works are defined as an addition to an electrical installation that does not extend to the installation of a new circuit e.g. either the addition of a new socket outlet or of a lighting point to an existing circuit. This certificate includes space for the recording of essential test results and does not require the addition of a Schedule of Test Results.

PART 1 – Description

1. Description of the Minor Works : ....
2. Location/Address : ....
3. Date of completion : ....
4. Departures from BS7671 (if any) : ....

PART 2 - Installation Details

1. Earthing arrangements (where known) : TN-C-S TN-S TT
2. Method of protection against indirect contact : ....
3. Protective device for the modified circuit : Type : .... Rating : ....
4. Comments on existing installation (including adequacy of earthing and bonding arrangements.) ....

PART 3 Essential Tests

1. Earth continuity satisfactory : Yes/No
2. Insulation resistance : Phase /Neutral : .... Megohms ....
   Phase/Earth : .... Megohms ....
   Neutral/Earth : .... Megohms ....
3. Earth fault loop impedance : .... ohms
4. Polarity correct : Yes/No
5. RCD operation (if applicable) : Operating current : .... mA Operating time : .... ms

PART 4 Declaration

I/We CERTIFY that the said works do not impair the safety of the existing installation, that the said works have been designed, constructed and inspected and tested in accordance with BS7671 amended to .... and that the said works have to the best of my/our knowledge and belief, at the the time of my/our inspection complied with all the requirements of BS7671 except as detailed in Part 2.

2. Name : .... 3. Signature : ....
For and on behalf of : .... Position : ....
Address : .... Date : ....

Figure 5.23 Minor Works Certificate

Scope

The Minor Works Certificate shown in Figure 5.23 is only to be used for additions to an electrical installation that do not include the introduction of a new circuit e.g. the addition of an extra socket outlet or lighting point to an existing circuit (see BS 7671 Regulations 631.3 and 633.1).

Part 1 – Description

- Description of the minor works – this requires a full and accurate description in order that the work carried out can be readily identified.
- Location/address – state the full address of the property and also where the work carried out can be located.
- Date of completion.
- Departures from BS 7671 (if any) – departures from the Regulations would only be expected in the most unusual circumstances.

Part 2 – Installation details

- Earthing arrangements – the type of supply should be indicated by a tick i.e. TN-C-S, TN-S, TT etc.
- Method of protection against indirect contact – the method of protection against indirect contact should be clearly identified e.g. earthed equipotential bonding and automatic disconnection of supply using fuse/circuit breaker or RCD.
- Protective device – state the type and rating of the protective device used.
- Comments on existing installation – if the existing installation lacks either an effective means of earthing or adequate main equipotential bonding conductors, this must be clearly stated. Any departures from BS 7671 may constitute non-compliance with either the *Electricity Supply Regulations* or the *Electricity at Work Regulations* and it is important that the client is informed in writing as soon as possible.

Part 3 – Essential tests

- Earth continuity – it is essential that the earth terminal of any new accessory is satisfactorily connected to the main earth terminal.
- Insulation resistance – the insulation resistance of the circuit that has been added to must comply with the requirements of BS 7671 (Table 61).
- Earth-fault loop impedance – the earth-fault loop impedance must be measured to establish that the maximum permitted disconnection time is not exceeded.
- Polarity – a check must be made to ensure that the polarity of the additional accessory is correct.
- RCD operation – if the additional work is protected by an RCD, then it must be checked to ensure that it is working correctly.

Part 4 – Declaration

- The certificate must be completed and signed by a competent person in respect of the design, construction, and inspection and testing of the work carried out (in the case of minor works this will invariably be the responsibility of one person).
- The competent person must have a sound knowledge and experience relevant to the work undertaken and to the technical standards laid down in BS 7671. This person should also be fully aware of the inspection and testing procedures contained in the Regulations and must use suitable test instruments.
- When signing the form on behalf of a company or other organisation, individuals should state clearly whom they represent.

Guidance notes (for those receiving the certificate)

- This certificate has been issued to confirm that the work to which it relates has been designed, constructed, and inspected and tested in accordance with BS 7671.
- You should have received the original certificate and the contractor should have retained a duplicate. If you are not the owner of the installation then you should pass the certificate or a copy of it to the owner.
- The Minor Works Certificate is only to be used for additions, alterations or replacements to an installation that do not include the provision of a new circuit. This may include the installation of an additional socket outlet or lighting point to an existing circuit or the replacement or relocation of an accessory such as a light switch. A separate certificate should be issued for each existing circuit on which work of this nature has been carried out. The Minor Works Certificate is not valid where more extensive work is carried out, when an Electrical Installation Certificate should be issued.
- The original certificate should be retained in a safe place and be shown to any person inspecting the installation or carrying out further work in the future. If you later vacate the property, this certificate will demonstrate to the new owner that the minor works covered by the certificate met the requirements of the Regulations at the time the certificate was issued.

Initial inspection

The following text provides a detailed description of the procedures required to carry out an initial inspection of an electrical installation. Substantial reference has been made to the *IEE Wiring Regulations* (BS 7671), the *IEE On-Site Guide* and *IEE Guidance Note No.3* and it is recommended that wherever possible these documents are referred to should clarification be required.

The most important considerations prior to carrying out any inspection and test procedure are that all the required information is available, the person carrying out the procedure is competent to do so and that all safety requirements have been met. Forward planning is also a major consideration and it is essential that suitable inspection checklists have been prepared and that appropriate certification document is available for completion. It is also important to realise that a large proportion of any new installation will be hidden from view once the building fabric has been completed and therefore it is preferable to carry out a certain amount of visual inspection throughout the installation process, e.g. conduit, cable tray or trunking is often installed either above the ceiling or below the floor and once the ceiling or floor tiles have been fitted it is difficult and often expensive to gain access for inspection purposes.

The same applies to testing and it may be advisable to carry out tests such as earth continuity during construction rather than after the building has been completed.

It must be remembered, however, that when visual inspection and/or tests are carried out during the construction phase, the results must be recorded on the appropriate checklists or test certificates.

Remember

Although the major part of any inspection will be visual, other human senses may be employed e.g. a piece of equipment with moving parts may generate an unusual noise if it is not working correctly, or an electrical device which overheats will be hot to touch as well as giving off a distinctive smell. The senses of hearing, touch and smell will assist in detecting these

Information

Before carrying out the inspection and test of an installation it is essential that the person carrying out the work be provided with the following information:

1. The maximum demand of the installation expressed in amperes per phase together with details of the number and type of live conductors both for the source of energy and for each circuit to be used within the installation (e.g. single-phase two-wire a.c. or three-phase four-wire a.c. etc.).
2. The general characteristics of the supply such as:
   - the nominal voltage (U_0)
   - the nature of the current (I) and its frequency (Hz)
   - the prospective short circuit current at the origin of the installation (kA)
   - the earth-fault loop impedance (Z_e) of that part of the system external to the installation
   - the type and rating of the overcurrent device acting at the origin of the installation.

If this information is not known it must be established either by calculation, measurement, inquiry or inspection.

3. The type of earthing arrangement used for the installation e.g. TN-S, TN-C-S, TT etc.
4. The type and composition of each circuit (i.e. details of each sub-circuit, what it is feeding, the number and size of conductors and the type of wiring used).
5. The location and description of all devices installed for the purposes of protection, isolation and switching (e.g. fuses/circuit breakers etc.).
6. Details of the method selected to prevent danger from shock in the event of an earth fault. (This will invariably be protection by earthed equipotential bonding and automatic disconnection of the supply.)
7. The presence of any sensitive electronic device which may be susceptible to damage by the application of 500 volts d.c. when carrying out insulation resistance tests.

This information may be gained from a variety of sources such as the project specification, contract drawings, as fitted drawings or distribution board schedules. If such documents are not available, then the person ordering the work should be approached.

Scope

BS 7671 states that as far as reasonably practicable, an inspection shall be carried out to verify that:

- all equipment and materials used in the installation are of the correct type and comply with the appropriate British Standards or acceptable equivalent
- all parts of the installation have been correctly selected and installed
- no part of the installation is visibly damaged or otherwise defective
- the installation is suitable for the surrounding environmental conditions.

Inspection requirements

In order to meet the above requirements the inspection process should include the checking of at least the relevant items from the following list.

1. Connection of conductors

Every connection between conductors or between conductors and equipment must be electrically continuous and mechanically sound. We must also make sure that all connections are adequately enclosed but accessible where required by the Regulations.

2. Identification of conductors

A check should be made that each conductor is identified in accordance with the requirements of BS 7671 Table 51. Although numbered sleeves or discs may be used in special circumstances, the most common form of identification is by means of coloured insulation or sleeving. It should be noted in particular that only protective conductors shall be identified by a combination of the colours green and yellow.

3. Cable routes

Cables should be routed out of harm's way and protected against mechanical damage where necessary. Permitted cable routes are clearly defined or alternatively cables should be installed in earthed metal conduit or trunking.

4. Current carrying capacity

Where practicable the size of cable used should be checked for current carrying capacity and voltage drop based upon information provided by the installation designer.

5. Verification of polarity

A check must be made that all single pole devices are connected in the phase conductor only.

6. Accessories and equipment

Accessories and equipment should be checked to ensure they have been connected correctly including correct polarity.

7. Selection and erection to minimise the spread of fire

A check must be made (preferably during construction) that fire barriers, suitable seals and/or other means of protection against thermal effects have been provided as necessary to meet the requirements of the Regulations.

8. Protection against electric shock

A check must be made that the requirements of BS 7671 Chapter 41 have been met for the protection method being used.

9. Prevention of mutual detrimental influence

Account must be taken of the proximity of other electrical services of a different voltage band and of non-electrical services and influences, e.g. fire alarm and emergency lighting circuits must be separated from other cables and from each other, and Band 1 and Band 2 circuits must not be present in the same enclosure or wiring system unless they are either segregated or wired with cables suitable for the highest voltage present. Mixed categories of circuits may be contained in multicore cables, subject to certain requirements.

Band 1 circuits are circuits that are nominally extra-low voltage i.e. not exceeding 50 volts a.c. or 120 volts d.c., e.g. telecommunications or data and signalling. Band 2 circuits are circuits that are nominally low voltage, i.e. exceeding extra-low voltage but not exceeding 1000 volts a.c. between conductors or 600 volts a.c. between conductors and earth.

10. Isolating and switching devices

BS 7671 requires that effective means suitably positioned and ready to operate shall be provided so that all voltage may be cut off from every installation, every circuit within the installation and from all equipment, as may be necessary to prevent or remove danger. This means that switches and/or isolating devices of the correct rating must be installed as appropriate to meet the above requirements. It may be advisable where practicable to carry out an isolation exercise to check that effective isolation can be achieved. This should include switching off, locking-off and testing to verify that the circuit is dead and no other source of supply is present.

11. Under voltage protection

Sometimes referred to in starters as no-volt protection, suitable precautions must be taken where a loss or lowering of voltage or a subsequent restoration of voltage could cause danger. The most common situation would be where a motor driven machine stops due to a loss of voltage and unexpectedly restarts when the voltage is restored unless precautions such as the installation of a motor starter containing a contactor are employed. Regulations require that where unexpected restarting of a motor may cause danger, the provision of a motor starter designed to prevent automatic restarting must be provided.

12. Labelling

A check should be carried out to ensure that labels and warning notices as required by BS 7671 have been fitted e.g. labelling of circuits, MCBs, RCDs, fuses and isolating devices, periodic inspection notices advising of the recommended date of the next inspection, and warning notices referring to earthing and bonding connections.

13. Selection of equipment appropriate to external influences

All equipment must be selected as suitable for the environment in which it is likely to operate. Items to be considered are ambient temperature, presence of external heat sources, presence of water, likelihood of corrosion, ingress of foreign bodies, impact, vibration, flora, fauna, radiation, building use and structure.

14. Access to switchgear and equipment

BS 7671 requires that every piece of equipment that requires operation or attention must be installed so that adequate and safe means of access and working space are provided.

15. Presence of diagrams, charts and other similar information

All distribution boards should be provided with a distribution board schedule that provides information regarding types of circuits, number and size of conductors, type of wiring etc. These should be attached within or adjacent to each distribution board.

16. Erection methods

Correct methods of installation should be checked, in particular fixings of switchgear, cables, conduit etc., which must be adequate and suitable for the environment.

Inspection Schedule

All the previous items should be inspected and the results noted on an Inspection Schedule, an example of which is given in Figure 5.24. Where items are not applicable to the type of installation being inspected (e.g. different methods of protection against direct and indirect contact), these should either be ruled out or marked N/A (not applicable).

Inspection Schedule

No.	Item	☐
1	Correction of conductors	☐
2	Identification of conductors	☐
3	Routing of cables in safe zones or protected against damage	☐
4	Selection of conductors for current and voltage drop	☐
5	Connection of single-pole devices in phase conductor only	☐
6	Correct connection of socket outlets and lamp holders	☐
7	Presence of fire barriers and protection against thermal effects	☐
8	Method of protection against shock	☐
8a	Protection against both direct and indirect contact	
	SELV	☐
	Limitation of discharge of energy	☐
8b	Protection against direct contact	
	Insulation of live parts	☐
	Batteries or enclosures	☐
	Obstacles	☐
	Placing out of reach	☐
	PELV	☐
8c	Protection against indirect contact	☐
8d	(i) Earthed equipotential bonding and automatic disconnection	☐
	Presence of bonding conductors	☐
	Presence of earthing conductors	☐
	Presence of protective conductors	☐
8d	(ii) Use of class 2 equipment	☐
8d	(iii) Non-conducting locations	☐
8d	(iv) Earth-free local equipotential bonding	☐
	Presence of earth-free equipotential bonding conductors	☐
8d	(v) Electrical Separation	☐
9	Prevention of mutual detrimental influence	
	Priority of non-electrical services	☐
	Separation of Band 1 and Band 2 circuits	☐
	Separation of fire alarm and emergency lighting circuits	☐
10	Presence of appropriate devices for isolation and switching	☐
11	Presence of under voltage protection devices	☐
12	Choice and setting of protective and monitoring devices (a) residual current devices (b) overcurrent device	☐
13	Labelling of protective devices switches and terminals	☐
14	Selection of equipment suitable for external devices	☐
15	Adequate access to switchgear and equipment	☐
16	Presence of danger notices and other warning signs	☐
17	Presence of diagrams and instructions	☐
18	Erection methods	☐
19	Requirements for special locations	☐

Signed: ____________________

Date: ____________________

Figure 5.24 Inspection Schedule

Inspection checklists

To ensure that all the requirements of the Regulations have been met, inspection checklists should be drawn up and used appropriate to the type of installation being inspected. Examples of suitable checklists are given in Table 5.9 which follows.

Joint Boxes (tick if satisfactory)	
All joint boxes comply with the appropriate British Standard.	
Joints accessible for inspection where required.	
All conductors correctly connected.	
Joints protected against mechanical damage.	

Wiring Accessories (General Requirements) (tick if satisfactory)	
All accessories comply with the appropriate British Standard.	
Boxes and other enclosures securely fastened.	
Metal boxes and enclosures correctly earthed.	
Flush boxes not projecting above surface of wall.	
No sharp edges which could cause damage to cable insulation.	
Non-sheathed cables not exposed outside box or enclosure.	
Conductors correctly identified.	
Bare protective conductors sleeved green and yellow.	
All terminals tight and contain all strands of stranded conductor.	
Cord grips correctly used to prevent strain on terminals.	
All accessories of adequate current rating.	
Accessories suitable for all conditions likely to be encountered.	

Socket Outlet (tick if satisfactory)	
Comply with appropriate British Standard.	
Mounting height above floor or working surface is correct.	
All sockets have correct polarity.	
Sockets not installed in bathroom or shower room (unless shaver type socket).	
Sockets not within 2.5 m of a shower in a room other than a bathroom.	
Sockets controlled by a switch if the supply is direct current.	
Sockets correctly protected where floor mounted.	
Circuit protective conductor connected directly to the earthing terminal of the socket outlet on a sheathed wiring installation.	

Table 5.9 Inspection checklists (continued overleaf)

Socket Outlet continued (tick if satisfactory)	
Earthing tail provided from the earthed metal box to the earthing terminal of the socket outlet.	
Socket outlets not used to supply a water heater with un-insulated elements.	

Lighting Controls (tick if satisfactory)	
Light switches comply with appropriate British Standard.	
Switches suitably located.	
Single-pole switches connected in phase conductor only.	
Correct colour coding of conductors.	
Correct earthing of metal switch plates.	
Switches out of reach of a person using bath or shower.	
Switches for inductive circuits (discharge lamps) de-rated as necessary.	
Switches labelled to indicate purpose where this is not obvious.	
All switches of adequate current rating.	
All controls suitable for their associated luminaire.	

Lighting Points (tick if satisfactory)	
All lighting points correctly terminated in suitable accessory or fitting.	
Ceiling roses comply with appropriate British Standard.	
No more than one flexible cord unless designed for multiple pendants.	
Devices provided for supporting flex used correctly.	
All switch wires identified.	
Holes in ceiling above ceiling rose made good to prevent spread of fire.	
Ceiling roses not connected to supply exceeding 250 V.	
Flexible cords suitable for the mass suspended.	
Lamp holders comply with appropriate British Standard.	
Luminaire couplers comply with appropriate British Standard.	

Conduits (general) (tick if satisfactory)	
All inspection fittings accessible.	
Maximum number of cables not exceeded.	
Solid elbows used only as permitted.	
Conduit ends reamed and bushed.	
Adequate number of boxes.	

Table 5.9 Inspection checklists (continued overleaf)

Conduits (general) continued (tick if satisfactory)	
All unused entries blanked off.	
Lowest point provided with drainage holes where required.	
Correct radius of bends to prevent damage to cables.	
Joints and scratches in metal conduit protected by painting.	

Circuit Protective Conductors (enter circuit details from specifications)	
1.	
2.	
3.	
4.	
5.	
6.	
7.	
8.	

Table 5.9 Inspection checklists (continued)

Where an item is not relevant to a specific installation N/A should be placed in the tick box

Periodic inspection and testing

General requirements

The regular inspection and testing of electrical installations is necessary because over a period of time the condition of all installations will deteriorate to some extent. This may be due to normal wear and tear, accidental damage, corrosion or other effects due to environmental influences, normal ageing or deterioration due to excessive electrical loading.

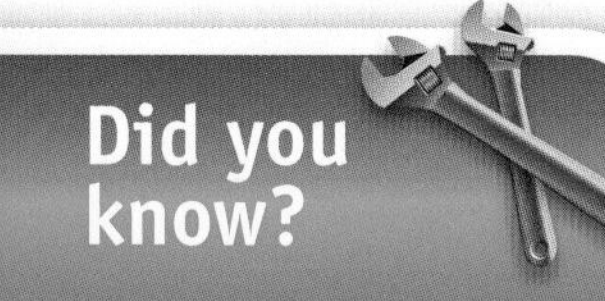

Did you know?

The requirements for periodic inspection and testing are given in BS 7671 Chapter 62

The Electricity at Work Regulations (1989) state that: 'As may be necessary to prevent danger, all systems shall be maintained so as to prevent, as far as is reasonably practicable, such danger'.

This means that all electrical installations must be maintained in a safe condition, and regular inspection and testing (periodic inspection) is an essential part of any such preventative maintenance programme. In addition to statutory requirements other bodies such as licensing authorities, insurance companies, mortgage lenders etc. may also require periodic inspection and testing to be carried out on a regular basis.

Other reasons for carrying out periodic inspection and testing are:

- to confirm compliance with BS 7671
- on a change of ownership of the premises
- on a change of use of the premises
- on a change of tenancy of the premises
- on completion of alterations or additions to the original installation
- following any significant increase in the electrical loading of the installation
- where there is reason to believe that damage may have been caused to the installation.

In the case of an installation that is under constant supervision while in normal use, such as a factory or other industrial premises, periodic inspection and testing may be replaced by a system of continuous monitoring and maintenance of the installation provided that adequate records of such maintenance are kept.

When carrying out the design of an electrical installation, and particularly when specifying the type of equipment to be installed, the designer should take into account the likely quality of the maintenance programme, and the periods between periodic inspection and testing to be specified on the Electrical Installation Certificate.

Both Section 6 of the Health and Safety at Work Act and the *Construction (Design and Management) Regulations* require information on the requirements for routine checks and periodic inspections to be provided. The advice of the Health and Safety

Executive in their *Memorandum of Guidance on the Electricity at Work Regulations* indicates that practical experience of an installation's use may indicate a need for an adjustment to the frequency of checks and inspections i.e. more often or less frequent depending on the likely deterioration of the installation during its normal use. This would be a matter of judgement for the duty holder.

Routine checks

Electrical installations should still be routinely checked in the intervening time between periodic inspection and testing. In domestic premises it is likely that the occupier will soon notice any damage or breakages to electrical equipment and will take steps to have repairs carried out. In commercial or industrial installations a suitable reporting system should be available for users of the installation to report any potential danger from deteriorating or damaged equipment.

In addition to this, a system of routine checks should be set up to take place between formal periodic inspections. The frequency of these checks will depend entirely on the nature of the premises and the usage of the installation. Routine checks are likely to include activities such as those listed in Table 5.10.

Activity	Check
Defect reports	Check that all reported defects have been rectified and that the installation is safe
Inspection	Look for: Breakages Wear or deterioration Signs of overheating Missing parts (covers/screws) Switchgear still accessible Enclosure doors secure Labels still adequate (readable) Loose fittings
Operation	Check operation of: Switchgear (where reasonable) Equipment (switch off and on) RCD (using test button)

Table 5.10 Routine checks

The recommended period between both routine checks and formal inspections are given in Section 7 of the *Unite Guide to Good Electrical Practice*. The requirements for such inspections are stated in BS 7671 Chapter 62 and specify that all inspections should provide careful scrutiny of the installation without dismantling or with only partial dismantling where absolutely necessary. It is considered that the unnecessary dismantling of equipment or disconnection of cables could produce a risk of introducing faults that were not there in the first place.

The frequency of periodic inspection and testing should aim to provide, as far as reasonably possible, the following:

- the safety of persons and livestock against the effects of electric shock or burns
- protection against damage to property by fire or heat arising from an installation defect
- confirmation that the installation has not been damaged and has not deteriorated to the extent that it may impair safety
- the identification of any defects in the installation or non-conformity with the current edition of the Regulations that may cause danger.

In summary, the inspection should ensure that:

- the installation is safe
- the installation has not been damaged
- the installation has not deteriorated so as to impair safety
- any items that no longer comply with the Regulations or may cause danger are identified.

In practical terms the inspector is carrying out a general inspection to ensure that the installation is safe. However, the inspector is required to record and make recommendations with respect to any items that no longer comply with the current edition of the Regulations.

As with all inspections the person carrying out the work must be competent and have sufficient knowledge and experience of the type of installation to be inspected and tested. Enquiries should be made to the person responsible for the installation with regard to the provision of charts and/or diagrams to indicate the type of circuits, means of isolation and switching, and types and ratings of protective devices including a written record of all previous inspection and test results.

Statutory and non-statutory documentation

1. *The Electricity Supply Regulations (1988).*
2. *The Electricity at Work Regulations.*
3. BS 5266 Pt.1 Code of Practice for emergency lighting systems (other than cinemas). Other Regulations and intervals cover testing of batteries and generators.
4. BS 5839 Pt.1 Code of Practice for the design, installation and servicing of fire alarm systems.
5. Local authority conditions of licence.
6. SI 1995 No 1129 (clause 27) *The Cinematography (Safety) Regulations.*

Guidance Note 3 requirements for inspection and testing

Guidance Note 3 3.8.1 General Procedure

In old installations where information such as drawings, distribution board schedules, charts etc. are not available, some exploratory work may be necessary to enable inspection and testing to be carried out safely and without damage to existing equipment. A survey should be carried out to identify all items of switchgear and control gear and their associated circuits.

During the survey a note should be made of any known changes in either the structure of the building, the environmental conditions or of any alterations or additions to the installation which may affect the suitability of the wiring or the method of installation.

A careful check should be made of the type of equipment on site so that, where required, electronic or other equipment that may be damaged by high-test voltages can be either disconnected or short-circuited to prevent damage. If computer equipment is to be disconnected for the purpose of testing then two important issues need to be considered:

- The user of the equipment must be informed in order that data within the computer can be backed up and stored on disc if necessary.
- Many commercial computer installations have an emergency back-up electrical supply that will automatically energise if the mains supply is disconnected. This means that circuits that you assume are isolated may well be kept live from a different source of supply.

If the inspection and testing cannot be carried out safely without the use of drawings, diagrams or schedules etc. then the person ordering the work should be informed.

Remember

All information in this section is based on Guidance Note 3 as written for the 16th edition of the IEEE Regulations. Much of the information will remain applicable and therefore only references to table numbers have been corrected to align with the 17th editon.

Guidance Note 3 3.8.2

The requirements of BS 7671 for Periodic Inspection and Testing are for a thorough **Inspection** of the installation supplemented by **Testing** where necessary. For safety reasons a visual inspection must be carried out before opening any enclosures, removing covers or carrying out any tests.

To comply with the Electricity at Work Regulations (1989) the inspection should be carried out wherever possible with the supply de-energised. A thorough visual inspection should be carried out of all electrical equipment that isn't concealed and, where damage or deterioration has occurred, this must be recorded on the inspection schedule. The inspection should include a check on the condition of all equipment and materials used in the installation with regard to the following:

- safety
- damage
- external influences
- wear and tear
- overload
- suitability
- corrosion
- age
- correct operation.

Detailed inspection requirements

Anticipation of danger

As stated previously, when carrying out the inspection the opportunity should be taken to identify any equipment that may be damaged if subjected to high-test voltages. As well as computer equipment this may include safety systems such as fire or intruder alarms that could well have electronic components susceptible to test voltages.

Joints and connections

It may be impossible to inspect every joint and termination in an electrical installation. Therefore, where necessary, a sample inspection should be made. Provided the switchgear and distribution boards are accessible as required by the Regulations, then a full inspection of all conductor terminations should be carried out and any signs of overheating or loose connections should be investigated and included in the report. For lighting points and socket outlets a suitable sample should be inspected in the same way.

Conductors

The means of identification of every conductor including protective conductors should be checked and any damage or deterioration to the conductors, their insulation, protective sheathing or armour should be recorded. This inspection should include each conductor at every distribution board within the installation and a suitable sample of lighting points, switching points and socket outlets.

Flexible cables and cords

Where flexible cables or cords form part of the fixed installation the inspection should include:

- examination of the cable or cord for damage or deterioration
- examination of the terminations and anchor points for any defects
- checking the correctness of the installation with regard to additional mechanical protection or the application of heat resistant sleeving where necessary.

Switches

The *IEE Guidance Notes 3 (Inspection & Testing)*, recommends that a random sample of at least 10 per cent of all switching devices be given a thorough internal visual inspection to assess their electrical and mechanical condition. Should the inspection reveal excessive wear and tear or signs of damage due to arcing or overheating then, unless it is obvious that the problem is associated with that particular switch, the inspection should be extended to include all remaining switches associated with the installation.

Protection against thermal effects

Although sometimes difficult due to the structure of the building, the presence of fire barriers and seals should be checked wherever reasonably practicable.

Basic and fault protection

Separated Extra Low Voltage (SELV) is commonly used as a means of protection against both basic and fault protection. When inspecting this type of system, the points to be checked include the use of a safety isolating transformer, the need to keep the primary and secondary circuits separate and the segregation of exposed conductive parts of the SELV system from any connection with the earthing of the primary circuit or from any other connection with earth.

Basic protection

Inspection of the installation should confirm that all the requirements of the Regulations have been met with regard to protection against direct contact with live conductors. This means checking to ensure there has been no damage or deterioration of any of the insulation within the installation, no removal of barriers or obstacles and no alterations to enclosures that may allow access to live conductors.

Remember

An RCD must not be used as the sole means of protection against direct contact with live parts

Fault protection

The protective measure must be established and recorded on the inspection schedule. Where ADS is used, the requirements of BS 7671 must be met in terms of protective earthing (Regulation 411.3.1.1) and protective equipotential bonding (Regulation 411.3.1.2).

Protective devices

A check must be made that each circuit is adequately protected with the correct type, size and rating of fuse or circuit breaker. A check should also be made that each protective device is suitable for the type of circuit it is protecting and the earthing system employed, e.g. will the protective device operate within the disconnection time allowed by the Regulations and is the rating of the protective device suitable for the maximum prospective short circuit current likely to flow under fault conditions?

Enclosures and mechanical protection

The enclosures of all electrical equipment and accessories should be inspected to ensure that they provide protection not less than IP2X or IPXXB, and where horizontal top surfaces are readily accessible they should have a degree of protection of at least IP4X. IP2X represents the average finger of 12.5 mm diameter and 80 mm in length and can be tested by a metal finger of these dimensions. IP4X provides protection against entry by strips greater than 1.0 mm or solid objects exceeding 1.0 mm in diameter.

Marking and labelling

Labels should be applied adjacent to every fuse or circuit breaker indicating the size and type of fuse or the nominal current rating of the circuit breaker and details of the circuit they protect. Other notices and labels required by the Regulations are as follows:

1. At the origin of every installation as shown in Figure 5.25.

IMPORTANT
This installation should be periodically inspected and tested and a report on its condition obtained as prescribed in BS7671 (formally the IEE Wiring Regulations for Electrical Installations) published by the Institute of Electrical Engineers.

Date of last inspection : ....

Recommended date of next inspection : ....

Figure 5.25 Label for origin of installation

2. Where different voltages are present:
   - in equipment or enclosures within which a voltage exceeding 250 volts exists but where such a voltage would not be expected
   - terminals between which a voltage exceeding 250 volts exists, which although contained in separate enclosures are within arm's reach of the same person
   - the means of access to all live parts of switchgear or other live parts where different nominal voltages exist.
3. Earthing and bonding connections: a label with the words shown in Figure 5.26 shall be permanently fixed in a visible position at or near the following points:
   - the point of connection of every earthing conductor to an earth electrode
   - the point of connection of every bonding conductor to an extraneous conductive part
   - the main earth terminal of the installation where separate from the main switchgear.

Safety Electrical Connection – Do Not Remove

Figure 5.26 Label for earthing and bonding connections

4. Residual Current Devices: where RCDs are fitted within an installation, a suitable permanent durable notice marked in legible type no smaller than the example shown in BS 7671 shall be permanently fixed in a prominent position at or near the main distribution board. The notice shall contain the following words:

   'This installation, or part of it, is protected by a device that automatically switches off the supply if an earth-fault develops. Test quarterly by pressing the button marked 'T' or 'Test'. The device should switch off the supply and should then be switched on to restore the supply. If the device does not switch off the supply when the button is pressed, seek expert advice.'

Caravan installations: all touring (mobile) caravans shall have a notice fixed near the main switch giving instructions on the connection and disconnection of the caravan installation to the electricity supply. Details of the wording required on the notice are given in full in BS 7671.

Periodic testing

Periodic testing is supplementary to periodic inspection and the same level of testing as for a new installation is not necessarily required.

When the building being tested is unoccupied, isolation of the supply for testing purposes will not cause a problem, but where the building is occupied, then testing should be carried out to cause as little inconvenience as possible to the user. Where isolation of the supply is necessary then the disconnection should take place for as short a time as possible and by arrangement with the occupier. It may be necessary to complete the tests out of normal working hours in order to keep disruption to a minimum.

Sequence of tests

Because it may not be possible to disconnect the installation under test from the supply, the recommended sequence of carrying out periodic tests is different from the sequence of tests for a new installation and is given below:

- continuity of all protective conductors (including equipotential bonding conductors and continuity of ring circuit conductors where required)
- polarity
- earth-fault loop impedance
- insulation resistance
- operation of devices for isolation and switching
- operation of residual current devices
- operation of circuit breakers.

Detailed test procedures

Continuity of protective conductors and equipotential bonding conductors

Where the installation can be safely isolated from the supply, then the circuit protective conductors and equipotential bonding conductors can be disconnected from the main earthing terminal in order to verify their continuity.

Where the installation cannot be isolated from the supply, the circuit protective conductors and the equipotential bonding conductors must not be disconnected from the main earthing terminal as under fault conditions extraneous metalwork could become live. Under these circumstances a combination of inspection, continuity testing and earth loop impedance testing should establish the integrity of the circuit protective conductors.

When testing the effectiveness of the main bonding conductors or supplementary bonds, the resistance value between any service pipe or extraneous metalwork and the main earthing terminal should not exceed 0.05 ohms.

Polarity

Polarity tests should be carried out to check that:

- polarity is correct at the intake position and the consumer unit or distribution board. Single pole switches or control devices are connected in the line conductor only
- socket outlets and other accessories are connected correctly
- with the exception of ES14 and ES27 lampholders to BS EN 60238, in circuits that have an earthed neutral, centre contract bayonet and ES lampholders have the outer or screwed contacts connected to the neutral conductor
- all multi-pole devices are correctly installed.

Where it is known that no alterations or additions have been made to the installation since its last inspection and test, then the number of items to be tested can be reduced by sampling. It is recommended that at least 10 per cent of all switches and control devices should be tested together with any centre contact lamp holders and 100 per cent of socket outlets. However if any cases of incorrect polarity are found then a full test should be made of all accessories in that part of the installation and the sample of the remaining should be increased to 25 per cent.

Earth-fault loop impedance

Earth-fault loop impedance tests should be carried out at:

- the origin of each installation and at each distribution board
- all socket outlets
- at the furthest point of each radial circuit.

Results obtained should be compared with the values documented during previous tests, and where an increase in values has occurred these must be investigated.

Insulation resistance

Insulation resistance tests can only be carried out where it is possible to safely isolate the supply. All electronic equipment susceptible to damage should be disconnected or alternatively the insulation resistance test should be made between line and neutral conductors connected together and earth. Where practicable the tests should be carried out on the whole of the installation with all switches closed and all fuse links in place. Where this is not possible, the installation may be subdivided by testing each separate distribution board one at a time.

BS 7671 Table 61 states a minimum acceptable resistance value of 1 MΩ. However, if the measured value is less than 2 megohms then further investigation is required to determine the cause of the low reading.

Where individual items of equipment are disconnected for these tests and the equipment has exposed conductive parts, then the insulation resistance of each item of equipment should be checked. In the absence of any other requirements the minimum value of insulation resistance between live components and the exposed metal frame of the equipment should be not less than 1.0 MΩ.

Operation of main switches and isolators

All main switches and isolators should be inspected for correct operation and clear labelling and to check that access to them has not been obstructed. Where the operation of the contacts of such devices is not visible it may be necessary to connect a test lamp between each line and the neutral on the load side of the device to ensure that all supply conductors have been broken.

Operation of Residual Current Devices

RCDs should be tested for correct operation by the use of an RCD tester to ensure they trip out in the time required by BS 7671 as well as by use of the integral test button. A check should also be made that the tripping current for the protection of a socket outlet to be used for equipment outdoors should not exceed 30 mA.

Operation of circuit breakers

All circuit breakers should be inspected for visible signs of damage or damage caused by overheating. Where isolation of the supply to individual sub-circuits will not cause inconvenience, each circuit breaker should be manually operated to ensure that the device opens and closes correctly.

Periodic Inspection Report

BS 7671 requires that the results of any periodic inspection and test should be recorded on a Periodic Inspection and Test Report of the type illustrated in Figure 5.27. The report should include the following:

- a description of the extent of the inspection and tests and what parts of the installation were covered
- any limitations (e.g. portable appliances not covered)
- details of any damage, deterioration or dangerous conditions which were found
- any non-compliance with BS 7671
- schedule of test results.

Details of the Client

Name : ....
Address : ....
....
....

Purpose of Report : ....

Details of the Installation
Occupier : ....
Installation address (if different from above) : ....
....
Description of Premises : Domestic ☐ Commercial ☐ Industrial ☐ Other ☐
Estimated age of the installation : ....Years. Evidence of alterations or additions : ....Yes/No
Date of last inspection : .... Records available : ....Yes/No

Extent & Limitations of the Inspection
Extent of the installation covered by this report : ....
....
....
Limitations : ....
....
....
This inspection has been carried out in accordance with BS7671 (I.E.E Wiring Regulations). Cables concealed within trunking and conduit or cables concealed under floors, buried underground, installed in roof spaces or generally hidden within the fabric of the building have not been inspected.

Next Inspection
I/We recommend that this installation should be further inspected and tested after an interval of not more than.............. Months/years, providing that any observations requiring urgent attention are attended to without delay.

Declaration
Inspected & tested by

Name : .... Signature : ....
For and on behalf of : .... Position : ....
Address : ....
....
.... Date : ....

Figure 5.27 Periodic Inspection Report

If any items are found which may cause immediate danger these should be rectified immediately. If this is not possible then they must be reported to a responsible person without delay.

When inspecting older installations, which may have been installed in accordance with a previous edition of the *IEE Wiring Regulations*, provided that all items which do not conform to the present edition of BS 7671 are reported, the installation may still be acceptable provided that no risk of shock, fire or burns exists.

Supply characteristics

Earthing Arrangements	**Number and Type of Live Conductors**	**Nature of Supply**	**Protective Device Details**
TN-C ☐	a.c 1-phase-2 wire ☐	Nominal voltage :	Type : . . . .
TN-S ☐	1-phase-3 wire ☐	V	. . . .
TN-C-S ☐	2-phase-3 wire ☐	Nominal frequency :	. . . .
TT ☐	3-phase-3 wire ☐	. . . . .Hz	Nominal current rating
Alternative (details to	3-phase-4 wire ☐	Prospective fault	. . . .
be given on attached	d.c 2-pole ☐	Current : . . . . .kA	. . . . .A
schedules).	3-pole ☐	External loop	
		Impedance : . . . .ohms	

Particulars of the Installation

Means of Earthing	Details of the installation earth electrode		
	Type (rod – tape etc)	Location	Resistance to
Suppliers facility ☐			
earth ☐			
Earth electrode . . . . .ohms			

Main Protective Conductors

Main earthing conductor : material . . . . .csa

Main equipotential bonding conductor : material . . . . .csa

To incoming water service . . . .☐	To incoming gas service . . . .☐
To incoming oil service . . . .☐	To structural steel . . . .☐
To lightning protection . . . .☐	

To other incoming services (state details) . . . .

Main Switch or Circuit Breaker

BS Type ;	No of poles : . . . .
Current rating : . . . . .A	Voltage rating : V
Location :	Fuse rating or setting : . . . . .A
Residual current device – operating current : . . .mA	Operating time : . . . . .ms

Figure 5.28 Supply characteristics and earthing arrangements

Summary of periodic testing

Type of test	Recommendations
Continuity	Tests to be carried out between: • all main bonding connections • all supplementary bonding connections Note! When an electrical installation cannot be isolated, protective conductors including bonding conductors must not be disconnected
Polarity	Tests to be carried out: • origin of installation • all socket outlets • 10 per cent of control devices (including switches) • 10 per cent of centre contact lamp holders Note! If incorrect polarity is found then a full test should be made on that part of the installation and testing on the remainder increased to 25 per cent. If further faults are found the complete installation must be tested
Earth-fault loop impedance	Tests to be carried out at: • origin of installation • each distribution board • each socket outlet • extremity of every radial circuit
Insulation resistance	If this test is to be carried out then test: • the whole installation with all protective devices in place and all switches closed • where electronic devices are present, the test should be carried out between phase and neutral conductors connected together and earth
Functional	Activities to be carried out: • all isolation and switching devices to be operated • all labels to be checked • all interlocking mechanisms to be verified • all RCDs to be checked both by test instrument and by test button • all manually operated circuit breakers to be operated to verify they open and close correctly

Table 5.11 Summary of periodic testing

On the job: Periodic inspection

You are asked to carry out a periodic inspection at a small workshop. All goes well until you come to check the office that is attached to the factory, where you notice that about 20 staff are working on computers. No installation drawings exist for the office, as it turns out the factory was built as an extension to the offices many years ago.

1 Can you identify the problems you will encounter in this job?

2 How would you deal with the situation?

(A suggested answer can be found on page 333)

Unsatisfactory test results

Continuity

When testing the continuity of circuit protective conductors or bonding conductors we should always expect a very low reading, which is why we must always use a low-reading ohmmeter. Main and supplementary bonding conductors should have a reading of not more than 0.05 ohms whilst the maximum resistance of circuit protective conductors can be estimated from the value of $(R_1 + R_2)$ given in the *Unite Guide to Good Electrical Practice*. These values will depend upon the cross-sectional area of the conductor, the conductor material and its length.

A very high (end of scale) reading would indicate a break in the conductor itself or a disconnected termination that must be investigated. A mid-range reading may be caused by the poor termination of an earthing clamp to the service pipe e.g. a service pipe which is not cleaned correctly before fitting the clamp or corrosion of the metal service pipe due to its age and damp conditions.

Polarity

Correct polarity is achieved by the correct termination of conductors to the terminals of all equipment. This may be main intake equipment such as isolators, main switches and distribution boards or accessories such as socket outlets, switches or lighting fittings. Polarity is either correct or incorrect; there is nothing in between. Incorrect polarity is caused by the termination of live conductors to the wrong terminals and is corrected by re-connecting all conductors correctly.

Did you know?

A reading only slightly higher than the required reading may be possible to correct by replacing the conductor with one of a larger cross-sectional area

Insulation resistance

The value of insulation resistance of an installation will depend upon the size and complexity of the installation and the number of circuits connected to it. When testing a small domestic installation you may expect an insulation resistance reading in excess of 200 megohms whilst a large industrial or commercial installation with many sub-circuits, each providing a parallel path, will give a much smaller reading if tested as a whole. It is recommended that, where the insulation resistance reading is less than 2 megohms, individual distribution boards or even individual sub-circuits be tested separately in order to identify any possible cause of poor insulation values.

An extremely low value of insulation resistance would indicate a possible short circuit between line conductors or a bare conductor in contact with earth at some point in the installation, either of which must be investigated. A reading below 1.0 MΩ would suggest a weakness in the insulation, possibly due to the ingress of dampness or dirt in such items as distribution boards, joint boxes or lighting fittings etc. Although PVC insulated cables are not generally subject to a deterioration of insulation resistance due to dampness (unless the insulation or sheath is damaged), mineral insulated cables can be affected if dampness has entered the end of a cable before the seal has been applied properly. Other causes of low insulation resistance can be the infestation of equipment by rats, mice or insects.

Earth-fault loop impedance

As explained previously the earth-fault loop path is made up of those parts of the supply system external to the premises being tested (Z_e) and the phase conductor and circuit protective conductor within the installation ($R_1 + R_2$), the total earth-fault loop impedance being $Z_s = Z_e + (R_1 + R_2)$.

Should the value of impedance measured be higher than that required by the design of the installation, then as we have no influence on the external value of impedance (Z_e) we can only reduce the value of Z_s by installing circuit protective conductors of a larger cross-sectional area or, if aluminium conductors have been used, by changing these to copper. If the value were still too high to guarantee operation of the circuit protective device in the time required by BS 7671, then consideration would have to be given to changing the type of protective device (i.e. fuses to circuit breakers).

Remember

Further information and advice on the interpretation of test results and fault finding is given in Chapter 6 Fault diagnosis and rectification

Residual Current Devices (RCDs)

Where a Residual Current Device (RCD) fails to trip out when pressing the integral test button this would indicate a fault within the device itself, which should therefore be replaced.

Where a Residual Current Device fails to trip out when being tested by an RCD tester then it would suggest a break in the earth return path, which must be investigated. If the RCD does trip out but not within the time specified then a check should be made that the test instrument is set correctly for the nominal tripping current of the device under test.

FAQ

Q When should I prove that my test instrument is working correctly?

A You can never tell when something will stop working properly. Therefore, you should always prove that your test instrument is working correctly both before you start to test and after you have completed the test. Only then can you be sure that your readings are valid.

Q Why do I need to do a ring circuit continuity test?

A The purpose of this test is to establish that a ring exists and that it has been correctly connected. As the IEE Regulations require that each leg of the ring be tested, it is probably sensible to mark each conductor at the distribution board as being either an Incoming leg from the last socket, or an Outgoing leg going to the first.

FAQ

Q When carrying out a ring circuit continuity test, why is it that a higher reading at a socket indicates a spur exists?

A We know when we do this test, that for a correctly connected circuit, the reading at each socket should be roughly the same. Try to imagine a ring circuit for what it actually is, i.e. each conductor of the ring circuit leaving the distribution board terminal, travelling out around the various socket outlets, thus forming a big circle (ring) and then returning to that same distribution board terminal.

If you think about it, it doesn't matter where you are on a circle, the distance all the way around it back to where you are is the same, no matter where you start measuring it from. This is why the reading at each socket outlet should be roughly the same. If you are moving from socket to socket, obtaining similar readings and then reach a socket where the reading obtained is higher, you have probably found a spur. This is because the spur is connected to that socket by using additional cable outside the ring and you are now measuring the additional resistance of that cable, hence the higher reading.

Knowledge check

1. How often is it recommended that a hospital has an electrical inspection?
2. There are four reasons why we carry out an initial inspection. What are they?
3. Name five things that should be covered by an initial inspection.
4. Name all the initial tests that should be carried out on an electrical installation, placing them in the correct sequence and indicating which are the 'live' tests.
5. Describe what is meant by the term 'fault protection'.
6. When carrying out an insulation resistance test, why is it important to remove all GLS lamps before applying the test?
7. Why is it unacceptable to only test an RCD by pressing the test button located on the front of the RCD?
8. A 30 mA RCD is providing supplementary protection against direct contact. A test current of what should cause the RCD to trip within how many milliseconds?
9. What is a Minor Works Certificate and when should it be used?
10. Describe a TN-C-S earthing system.

Chapter 6

Fault diagnosis and rectification

OVERVIEW

Electricians should have the ability to recognise when something is not up to standard or is not functioning correctly, such as when a piece of equipment is not suitable (e.g. a metal-clad switch that has been installed in a damp environment) or a circuit is not operating correctly (e.g. a two-way lighting circuit that is only operating from one switch). Not all faults are easily visible; some are concealed and some may develop over a period of time. Regular inspection, tests and maintenance checks will limit such faults. Testing is not only needed at the completion of works but should also take place during the installation process of the wiring system. This chapter will cover the following areas:

- **Safe working practices for fault diagnosis**
- **General categories of fault**
- **Specific types of fault**
- **Knowledge and understanding of electrical systems and equipment**
- **Stages of diagnosis and rectification**
- **Factors affecting the fault repair process**
- **Special situations**

Safe working practices for fault diagnosis

Isolation

Remember, that before beginning work on any piece of electrical equipment or circuit you should make sure that it is completely isolated from the supply by following recognised procedures.

One such procedure has been drawn up by the Joint Industry Board for the Electrical Contracting Industry and sets the standard for safe systems of working in the industry. The JIB procedure has two parts to it as shown in Figures 6.1 and 6.2:

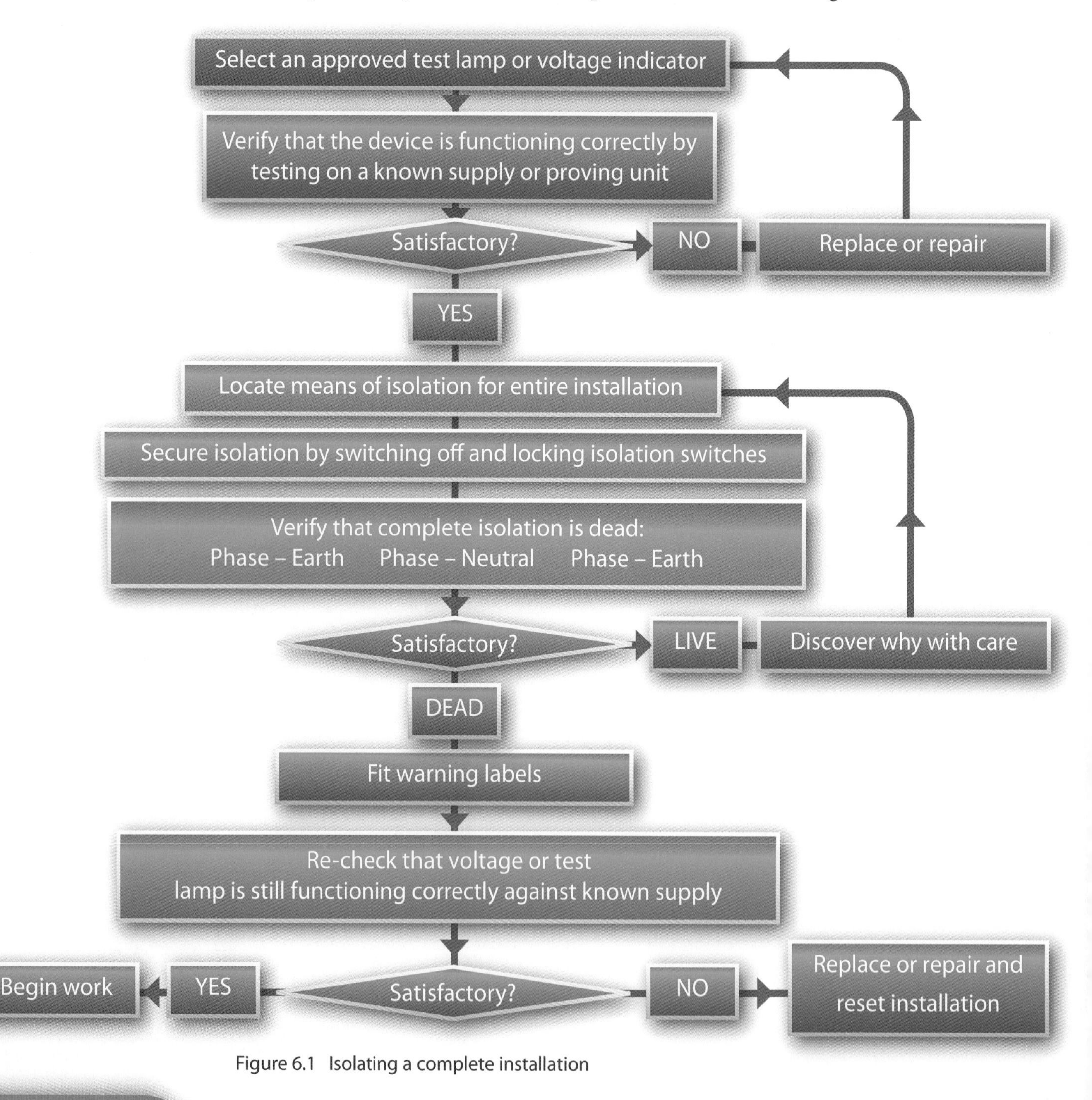

Figure 6.1 Isolating a complete installation

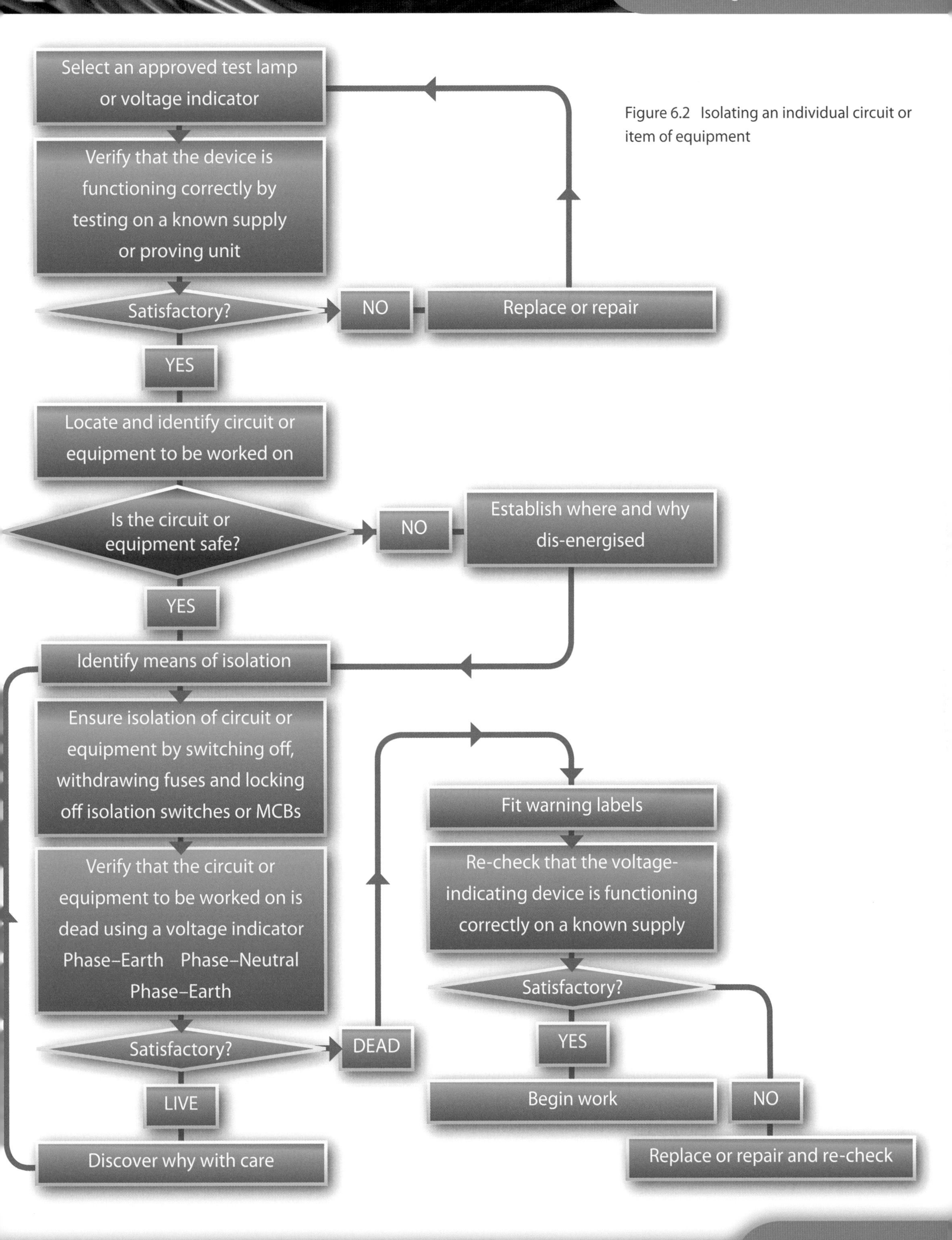

Figure 6.2 Isolating an individual circuit or item of equipment

Remember

You must ensure that your test equipment has a current calibration certificate, which indicates that the instrument is working properly and providing accurate readings. If you do not do this, test results could be void

Test instruments

All test equipment must be regularly checked to make sure it is in good and safe working order.

If you are checking the equipment yourself, the following points should be noted and, if you have any doubt about an instrument or its accuracy, ask for assistance, as test instruments are very expensive and any unnecessary damage caused by ignorance should be avoided.

Guidance Note GS 38

Published by the Health and Safety Executive, GS 38 is for electrical test equipment used by electricians and gives guidance to electrically competent people involved in electrical testing, diagnosis and repair. Electrically competent people may include electricians, electrical contractors, test supervisors, technicians, managers or appliance repairers.

Voltage indicating devices

Instruments used solely for detecting a voltage fall into two categories:

- Detectors that rely on an illuminated lamp (test lamp) or a meter scale (test meter): test lamps are fitted with a 15 watt lamp and should not give rise to danger if the lamp is broken. A guard should also protect it.
- Detectors that use two or more independent indicating systems (one of which may be audible) and limit energy input to the detector by the circuitry used. An example is a two-pole voltage detector, i.e. a detector unit with an integral test probe, an interconnecting lead and a second test probe.

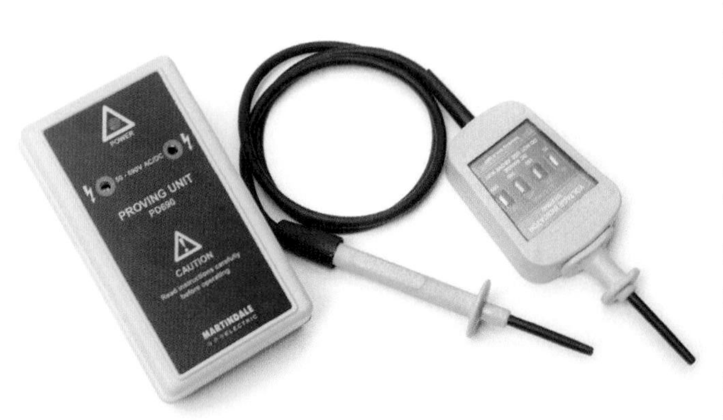

Voltage indicating device

Both these detectors are designed and constructed to limit the current and energy that can flow into the detector. This limitation is usually provided by a combination of the circuit design using the concept of protective impedance, and current-limiting resistors built into the test probes. The detectors are also provided with in-built test features to check the functioning of the detector before and after use. The interconnecting lead and second test probes are not detachable components. These types of detector do not require additional current-limiting resistors or fuses to be fitted provided that they are made to an acceptable standard and the contact electrodes are shrouded.

Test lamps and voltage indicators are recommended to be clearly marked with the maximum voltage which may be tested by the device and any short time rating for the device if applicable. This rating is the recommended maximum current that should pass through the device for a few seconds, as these devices are generally not designed to be connected for more than a few seconds.

Operating devices

It is not an adequate precaution to simply isolate a circuit electrically before commencing work on it. **It is vital to ensure that once a circuit or item of electrical equipment has been isolated, it cannot be inadvertently switched back on**.

Figure 6.3 Device locked off with padlock

A good method of providing full electrical and mechanical isolation is to lock off the device (or distribution board containing the device) with a padlock (see Figure 6.3), with the person working on the isolated equipment keeping the key in their pocket. A skilled person should be the only one allowed to carry out this or similar responsible tasks concerning the electrical installation wiring.

Many items of electrical equipment when installed are provided with a local means of switching or isolation. This is not only convenient for the user (as control is readily accessible) but it is a safety feature and indeed a requirement of BS 7671 *IEE Wiring Regulations* (Chapter 53) that 'equipment should be readily controlled, protected and isolated'. A good example of local isolation is when a motor is provided with a control unit as this can include the starter, switch and lock-off facility.

The types of equipment used in circuits to provide switching and isolation of the circuits, and indeed complete installations, can be categorised as having one or more of the following functions:

- control
- isolation
- protection.

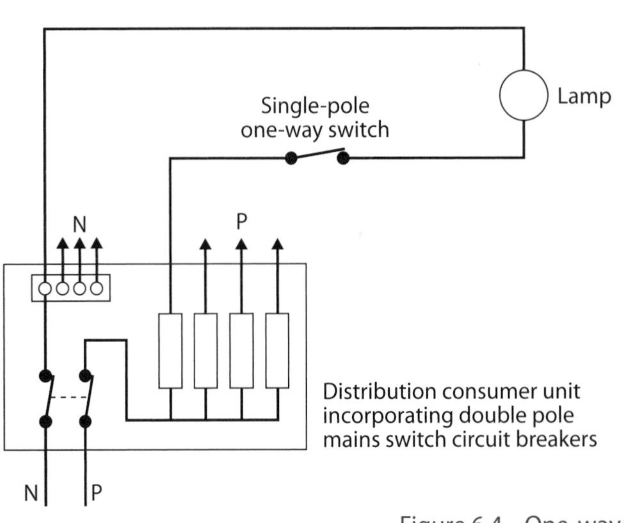

Figure 6.4 One-way lighting circuit with mains switch and breakers

A simple example of the control, isolation and protection functions can be represented in any basic simple circuit. As an example, Figure 6.4 shows a one-way lighting circuit supplied from a distribution board having a mains switch and circuit breakers.

From Figure 6.4 we can see that:

- The distribution consumer unit combines all three functions of control, protection and isolation.
- The main double-pole switch can provide the means for switching off the supply and, when locked off, can provide complete isolation of the installation.
- The circuit breakers in the unit provide protection against faults and overcurrents in the final circuits and, when switched off and locked off, the circuit breakers can provide isolation of each individual circuit.
- The one-way switch has only one function, which is to control the circuit enabling the luminaire to be switched 'on' or 'off'.

On and off-load devices

Not all devices are designed to switch circuits 'on' or 'off'. It is important to know that when a current is flowing in a circuit, the operation of a switch (or disconnector) to break the circuit will result in a discharge of energy across the switch terminals.

You may well have had this happen when entering a dark room and switching on the light, where for an instant you may see a blue flash from behind the switch plate. This is actually the arcing of the current as it dissipates and makes contact across the switch terminals. A similar arcing takes place when circuits are switched off or when protective devices operate thus breaking fault current levels.

Fundamentally, an isolator is designed as an off-load device and is usually only operated after the supply has been made dead and there is no load current to break. An on-load device can be operated when current is normally flowing and is designed to make or break load current.

Remember

A plug is not designed to make or break load current and that the appliance should be switched off before removing the plug

An example of an on-load device could be a circuit breaker, which is not only designed to make and break load current but has been designed to withstand make and break high levels of fault current. Remembering the three previous functions, it is important to install a device that meets the needs at a particular part of a circuit or installation. Some devices can meet the needs of all three functions. However, some devices may only be designed to meet a single function.

All portable appliances should be fitted with the simplest form of isolator, a fused plug. This, when unplugged from the socket outlet, provides complete isolation of the appliance from the supply.

Restoration of the supply

After the fault has been rectified, which may have resulted in either parts being replaced or simple reconnection of conductors, it is important that the circuit is tested for functionality. These tests may be simple manual rotation of a machine or the sequence of tests as prescribed in the *IEE Wiring Regulations*. For example, a simple continuity test will check resistance values, open and closed switches and their operation.

General categories of fault

In this section you will recognise different types of faults and understand that consequences of the fault depend on its location within the installation or in a specific circuit. You should also understand the need for care when installing electrical systems and how poor, or lack of, maintenance could lead to such faults. The essential points to be covered are:

- position of faults (complete loss of supply at the origin of the installation and localised loss of supply)
- operation of overload and fault current devices
- transient voltages
- insulation failure
- plant, equipment or component failure
- faults caused by abuse, misuse and negligence
- prevention of faults by regular maintenance.

Position of faults

The knowledge of fault finding and the diagnosis of faults can never be complete because no two situations are exactly the same. Also, as technology advances and the systems we install are being constantly improved, then the faults that develop become more complicated to solve. Therefore an understanding of the electrical installation and the equipment we install becomes paramount.

To diagnose faults an electrician will adopt basic techniques and these can usually solve the most common faults that occur. However, there are occasions when it may be impractical to rectify a fault. This could be due to costs involved in down time, or the cost of repair being more than the cost of replacement, in which case it could mean the renewal of wiring and/or equipment or components. Such situations and outcomes must be monitored, and the client should be kept informed and be made party to the decision to repair or replace.

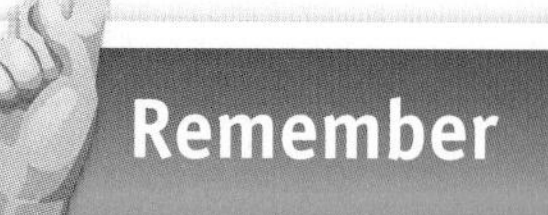

Remember

Always keep the client informed

Almost all faults should never exist. A fault can be compared to an accident, where careful planning and thought can prevent accidents. It is therefore part of a designer's job to build into the design of the electrical installation fault protection and damage limitation, such as:

- Installing more than one circuit. When a fault occurs on one of the two circuits the fault can be limited to that circuit.
- Installing fuses and protective devices to disable the fault and limit its effect.
- Ensuring the ability to access, maintain and repair. Good maintenance will limit faults, and access to the installation allows for maintenance and repair.

It is easier to find faults on installations where there are plenty of circuits. Indeed it is a requirement of the *IEE Wiring Regulations* BS 7671 Part 3 under **Installation circuit arrangements**, where Regulation 314.1 states 'Every installation shall be divided into circuits as necessary to:

i) avoid danger and minimise inconvenience in the event of a fault

ii) facilitate safe operation, inspection, testing and maintenance.'

We can see that compliance with this Regulation will help us to locate faults more easily, as usually by a process of elimination (operating each fuse individually) or simply looking at the device to see which one has operated we can identify a defective circuit or device.

Meltdown of insulation due to overcurrent

Meltdown of plastic connector due to overcurrent

Protective devices and simple fuses are designed to operate when they detect large currents due to excess temperature. In the case of a short circuit fault, high levels of fault current can develop, causing high temperatures and breakdown of insulation. Such faults, if allowed to develop, can cause fires.

When considering the consequences of a fault, it is important to know that the location of the fault can limit its severity with regard to disruption and inconvenience

Figure 6.5 could be a typical layout of equipment for a small industrial or commercial installation where a three-phase supply is required and the load is divided on the three final consumer units, and will assist you in understanding simple installation layouts and the consequence of a located fault.

Key to Figure 6.5

A	Supply company's service with incoming supply cable and fuses.
A1	Supply company's cables feeding KWh meter.
B	Supply company's metering equipment KWh meter.
B1	Consumer's meter tails.
C	Consumer's main switch – typical TP&N 100 amp.
C1	Sub final circuit conductors brown phase.
C2	Sub final circuit conductors black phase.
C3	Sub final circuit conductors grey phase.
D	Consumer's unit with DP main switch and protective devices.
E	Consumer's unit with DP main switch and protective devices.
F	Consumer's unit with DP main switch and protective devices.
D1	Consumer's final circuits i.e. sockets and lighting.
E1	Consumer's final circuits i.e. sockets and lighting.
F1	Consumer's final circuits i.e. sockets and lighting.

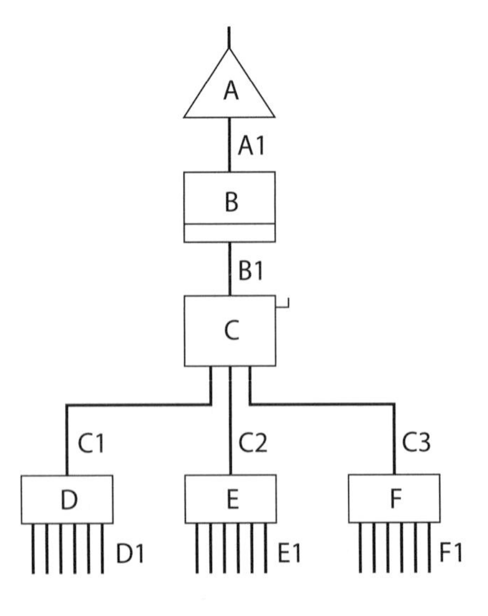

Figure 6.5 Typical three-phase supply for small installation

Complete loss of supply at the origin of the installation

With reference to Figure 6.5 it can be seen that a fault at points A to B1 could result in the complete loss of supply to the rest of the installation. The equipment at this point is mainly the property of the supply company, and faults that appear on this equipment are usually the result of problems with the supply, such as an underground cable being severed by workmen.

However, faults on the installation wiring would result in the supply company's protective devices operating but this is unusual and is only a consequence of poor design, misuse of the equipment or overloading of the supply company's equipment.

As an example of this, if the supply company's protective device or fuses were rated at 100 amperes per phase and the designer loaded each phase up at 200 amperes, this would result in a 100 per cent overload and the consequence of this would be the operation of the mains protective devices and the whole installation being without power.

Localised loss of supply

Looking again at Figure 6.5 we can recognise what happens when parts of the installation lose their supply. If a fault appeared on Cable C1, then the consumer unit (D) and all the final circuits from it would be dead. If a fault occurred on any one final circuit then only that circuit's protective device should have operated and only that circuit would be dead.

If a fault occurred on any one final circuit it should not affect any other final circuit or any other cable leading up to that consumer unit.

Because the system in Figure 6.5 is three phase, this should mean that a fault on any one phase should only result in that phase being affected, again limiting the consequence of a fault. This further highlights the reason for Regulation 314-01-01 that we mentioned earlier, which stated 'Every installation shall be divided into circuits as necessary to:

i) avoid danger and minimise inconvenience in the event of a fault

ii) facilitate safe operation, inspection, testing and maintenance.'

Operation of overload and fault current devices

When a fault is noticed, it is usually because a circuit or piece of equipment has stopped working and this is usually because the protective device has done its job and operated. The rating of a protective device should be greater than, or at least equal to, the rating of the circuit or equipment it is protecting, e.g. 10 x 100 watt lamps equate to a total current use of 4.35 amperes. Therefore a device rated at 5 or 6 amperes could protect this circuit.

A portable domestic appliance which has a label rating of 2.7 kW equates to a total current of 11.74 amperes. Therefore a fuse rated at 13 amperes should be fitted in the plug.

Definitions

Overload current – an overcurrent occurring in a circuit which is electrically sound

Overcurrent – a current exceeding the rated value

Earth fault current – a fault current which flows to earth

Short circuit current – an overcurrent resulting from a fault of negligible impedance between live conductors

Protective devices are designed to operate when an excess of current (greater than the design current of the circuit) passes through it. The fault current's excess heat can cause a fuse element to rupture or the device mechanism to trip, dependent on which type of device is installed. These currents may not necessarily be circuit faults, but short-lived overloads specific to a piece of equipment or outlet. The Regulations categorise these as **overload current** and **overcurrent**.

For conductors, the rated value is the current-carrying capacity. Most excess currents are, however, due to faults, either **earth faults** or **short circuit** type which cause excessive currents.

Whichever type of fault occurs the designer should take account of its effect on the installation wiring and choose a suitable device to disconnect the fault quickly and safely. The fundamental effect of any fault is a rise in current and therefore a rise in temperature.

High temperature destroys the properties of insulation, which in itself could lead to a short circuit. High currents damage equipment, and earth fault currents are dangerous to the body from electric shock. Some examples of each are given below.

Overload faults

- adaptors used in socket outlets exceeding the rated load of circuit
- extra load being added to an existing circuit or installation
- not accounting for starting current on a motor circuit.

Wrong type of starter in fluorescent tube has melted due to excess power demand

Short circuit faults

- insulation breakdown
- severing of live circuit conductors
- wrong termination of conductors energised before being tested.

Earth faults

- insulation breakdown
- incorrect polarity
- poor termination of conductors.

Fuse holder has melted due to overloading

Poor termination of (cpc) circuit protective conductors

Transient voltages

A transient voltage can be defined as a variance or disturbance to the normal voltage state, the normal voltage state being the voltage band that the equipment is designed to operate.

It is the installation designer's responsibility to stipulate adequate sizes for the circuit conductors. But this sizing procedure will only prevent voltage drop during normal circuit conditions. Designers cannot prevent transient voltages as they are outside their control, but they can compensate by including equipment that will protect against such voltage variations and disturbances.

It is now becoming common practice to install filter systems to IT circuits, which provide stabilised voltage levels. These devices can also reduce the effect of voltage spikes by suppressing them. If transient voltages are not recognised they can cause damage to equipment and lead to a loss of data that could prove costly, especially if the user was an organisation such as a bank.

Some common causes of transient voltages are:

- supply company faults
- electronic equipment
- heavy current switching (causing voltage drops)
- earth-fault voltages
- lightning strikes.

Transient voltages are becoming increasingly problematic with the expanding use of IT equipment, not only in commerce and industry, but also in the home

If transient voltages are thought to exist on installations, monitoring equipment can be installed to record events over periods of time. The results of the monitoring can usually trace the cause

Insulation failure

Insulation is designed primarily to separate conductors and to ensure their integrity throughout their life. Insulation is also used to protect the consumer against electric shock by protecting against direct contact and is also often used as a secondary protection against light mechanical damage, such as in the case of PVC/PVC twin and cpc cables. However, as previously mentioned, the breakdown of electrical insulation can lead to short circuit or earth-faults.

When insulation of conductors and cables fail, it is usually due to one or more of the following:

- poor installation methods
- poor maintenance
- excessive ambient temperatures
- high fault current levels
- damage by third party.

Poorly connected 13 A plug top resulting in dangerous working conditions

Plant, equipment and component failure

It is said that nothing lasts forever and this is certainly true of electrical equipment. There will be some faults that you will attend that will be the result of a breakdown simply caused by wear and tear, although it must be said that planned maintenance systems and regular testing and inspections can extend the life of equipment. Some common failures on installations and plant are:

- switches not operating – due to age
- motors not running – new brushes required
- lighting not working – lamp's life expired
- fluorescent luminaire not working – new lamp or starter needed
- outside PIR not switching – ingress of water causing failure
- corridor socket outlet not working due to poor contacts created by excessive use/age.

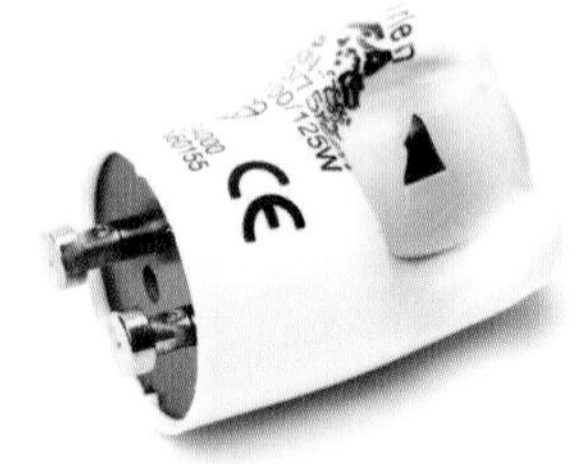

Burnt out starter

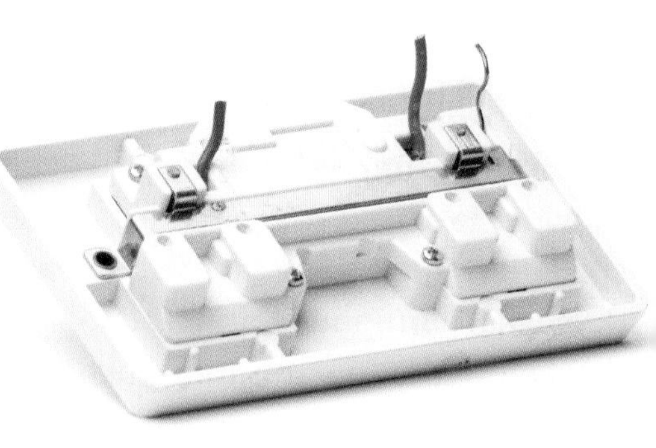

Poorly wired socket outlet

Faults caused by misuse, abuse and negligence

A common reason for faults on any electrical system or equipment is misuse, where the system or the equipment is simply not being used as it was designed. Every item of electrical equipment comes with user instructions that usually cover procedures and precautions, and such instructions should be read carefully as misuse can lead to invalidation of the guarantee.

When an electrical installation is completed, a manual is normally handed over to the client that includes all product data and installation schedules and test results. This data will help the client when additions to the installation are made, inspections and tests are carried out or to assist in maintaining the installation.

Some faults are caused by carelessness during the installation process, and simple faults caused by poor termination and stripping of conductors can then lead to serious short circuits or overheating. Therefore good mechanical and electrical processes should always be carried out and every installation should be tested and inspected prior to being energised. Some examples of faults arising from misuse are given below.

Remember

If you are called out to repair/rectify a fault and you find that misuse of the system has led to the fault, then the guarantee for the installation may be invalidated by the fault. This may lead to the client having to pay for the repair costs

User misuse

- using an MCB as a switch where constant use could lead to breakdown
- unplugging on-load portable appliances thus damaging socket terminals
- damp accessories, for instance, due to hosing walls in dry areas
- MCBs' nuisance-tripping caused by connecting extra load to circuits.

MCB

Installer misuse

- poor termination of conductors – overheating due to poor electrical contact
- loose bushes and couplings – no earth continuity – electric shock risk!
- wrong size conductors used – excessive voltage drop – excess current which could lead to an inefficient circuit and overheating of conductors
- not protecting cables when drawing them in to enclosures, causing damage to insulation
- overloading conduit and trunking capacities, causing overheating and insulation breakdown.

Poor termination of conductors causing electric shock risk

Prevention of faults by regular maintenance

A well designed electrical installation should serve the consumer efficiently and safely for many years, whereas an installation that isn't regularly inspected and tested could lead to potential faults and areas of weakness not being discovered. Even a visual inspection can uncover simple faults which, if left, could lead to major problems in the future.

Large organisations in the commercial sector and industry have realised the need for regular inspections and planned maintenance schedules. Some major supermarkets choose to have regular maintenance take place when the stores are closed to the public, as they realise a store closure or part closure to carry out corrective electrical work could cost sales. A typical example of maintenance in shops is lamp cleaning, changing and repairs and servicing to refrigeration and air conditioning. Large manufacturing companies usually carry out maintenance during planned holiday periods when the workforce is on annual leave and machinery, equipment and the electrical installation wiring can be accessed without loss of production.

Carrying out repairs and servicing in this way gives electricians and engineers the opportunity to test installations, carry out remedial work and to do general servicing on equipment and machinery.

Specific types of fault

Remember

It is part of our job to realise and understand why items of equipment and wiring types are being installed and to know they are suitable

The type of equipment we install as electricians is normally chosen by a designer in line with the needs of the client's specification. We therefore often take it for granted that the equipment and wiring we fit is right for the job, and the designer has not only met the needs of the client but also the environment.

The manufacturer and the types of cables and accessories that you install must comply with BS and BSEN Standards and must be installed in accordance with BS 7671 *IEE Wiring Regulations*. Part 5 of the Regulations covers Selection and Erection of Equipment, in particular:

- selection of type of wiring system
- external influences
- current carrying capacity of conductors
- cross-sectional area of conductors
- voltage drop
- electrical connections
- minimising spread of fire
- proximity to other services
- selection and erection in relation to maintainability.

The items listed in this area of the Regulations should have been accounted for in the initial design but it is the responsibility of the installer to enforce them. This is achieved by adopting good practice when installing, and understanding the consequences (faults) that would occur if such good practice and Part 5 were ignored.

Cable interconnections

Cable interconnections are used on many occasions in electrical installations. Their use is generally seen to be poor planning, and good design would limit their use. However, they are used in one or more of the following ways:

- lighting circuit joint boxes for line, neutral, switch wire and strappers
- power circuits, ring final socket outlet wiring and spurs etc.
- street lighting and underground cables where long runs are needed
- general alterations and extensions to circuits when remedial work is being done
- rectification of faults or damage to wiring.

Where they must be used, they should be mechanically and electrically suitable. They must also be accessible for inspection as laid down by Regulation 526.3, which states that 'every connection and joint shall be accessible for inspection, testing and maintenance'. There is an exception to this Regulation when any of the following is used:

- a compound or encapsulated joint (commonly underground)
- a connection between a cold tail and heating element (under floor)
- a joint made by welding, soldering, brazing or compression tool.

The joints in non-flexible cables shall be made by soldering, brazing, welding, mechanical clamps or of a compression type. The devices used should relate to the size of the cable and be insulated to the voltage of the system used. Connectors used must be of the appropriate British Standard and the temperature of the environment must be considered when choosing the connector.

Where cables having insulation of dissimilar characteristics are to be jointed, for example XLPE thermosetting with general purpose PVC, the insulation of the joint must be to the highest temperature of the two insulators. The most common types of terminating devices used are:

- plastic connectors
- porcelain connectors
- soldered joints
- Screwits
- uninsulated connectors
- compression joints
- junction boxes.

Screwit

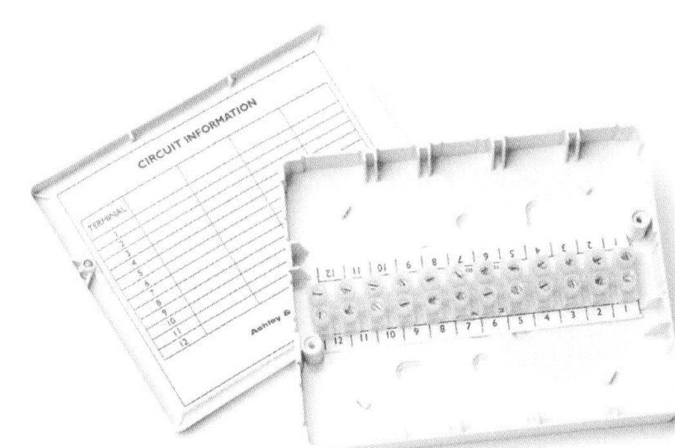

RB4 Junction box

Cable interconnections are usually seen as the first point to investigate in the event of a fault. This is because they are seen as the weak link in the wiring system and because they are usually readily accessible.

The mechanical and electrical connection when joining two conductors together relies on good practice by the installer to ensure that the connection is made soundly. The terminating device used should:

- be the correct size for the cross-section of the cable conductor
- be at least the same current rating as the circuit conductor
- have the same temperature rating as the circuit conductor
- be suitable for the environment.

The most common fault at this point would probably be due to a poor/loose connection, which would produce a resistive joint and excessive heat that could lead to insulation breakdown or eventually fire. When a fault occurs at a cable interconnection it may not necessarily be due to the production of heat but to the lack of support or strain being placed on the conductors, which may lead to the same outcome as above.

Remember

Follow the safe isolation procedures before removing any cover or lid from any termination point

Cable terminations, seals and glands

Terminations

The same rules as for cable interconnections apply to the termination of cables, i.e. good mechanical and electrical connections. The same consequences will occur if they are not adhered to, namely:

- Care must be taken not to damage the wires.
- BS 7671 requires that a cable termination of any kind should securely anchor all the wires of the conductor that may impose any appreciable mechanical stress on the terminal or socket.
- A termination under mechanical stress is liable to disconnect, overheat or spark.
- When current is flowing in a conductor a certain amount of heat is developed and the consequent expansion and contraction may be sufficient to allow a conductor under stress, particularly tension, to be pulled out of the terminal or socket.
- If the PF improvement capacitor became disconnected in a luminaire, the circuit current would increase.
- A fault caused at a poorly connected terminal would be known as a high resistance fault.
- One or more strands or wires left out of the terminal or socket will reduce the effective cross-sectional area of the conductor at that point. This may result in increased resistance and probably overheating.
- Poorly terminated conductors in circuits that continue to operate correctly are commonly known as latent defects.

Types of terminals

There are a wide variety of conductor terminations. Typical methods of securing conductors in accessories include pillar terminals, screw heads, nuts and washers, and connectors as shown in Figures 6.6, 6.7 and 6.8.

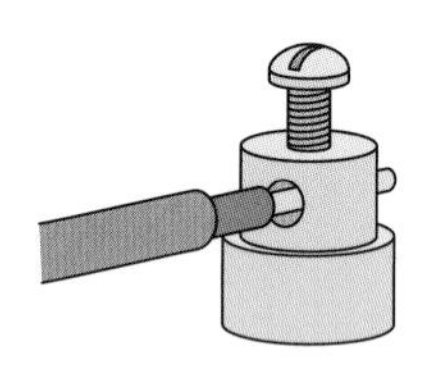

Figure 6.6 Pillar terminal

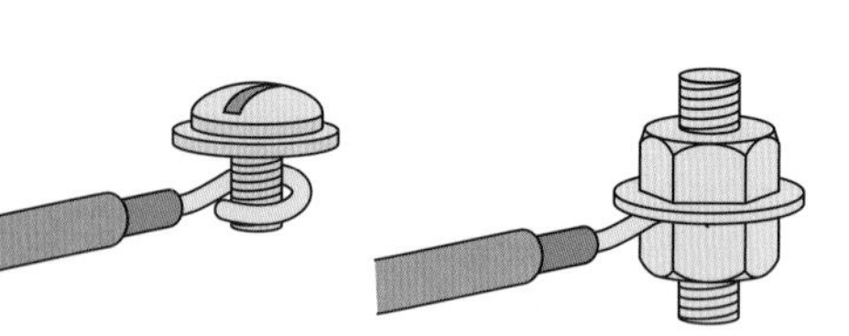

Figure 6.7 Screwhead, nut and washer terminals

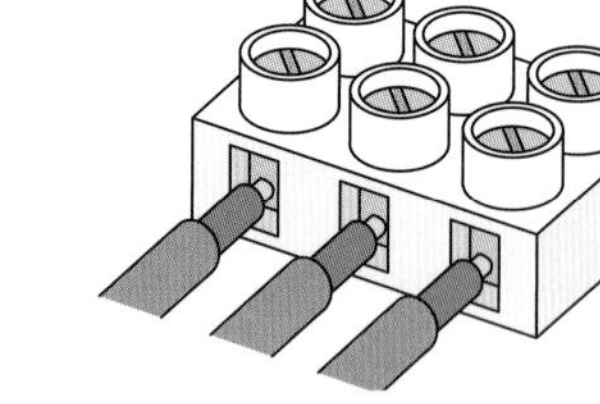

Figure 6.8 Strip connectors

Seals and entries

Where a cable or conductor enters an accessory or piece of equipment, the integrity of the conductor's insulation and sheath, earth protection and the enclosure or accessory should be maintained. Some cables and wiring systems have integrated mechanical protection i.e.:

- PVC/PVC twin and cpc
- PVC/steel wire armoured
- PVC covered mineral insulated copper cable (MICV)
- FP200.

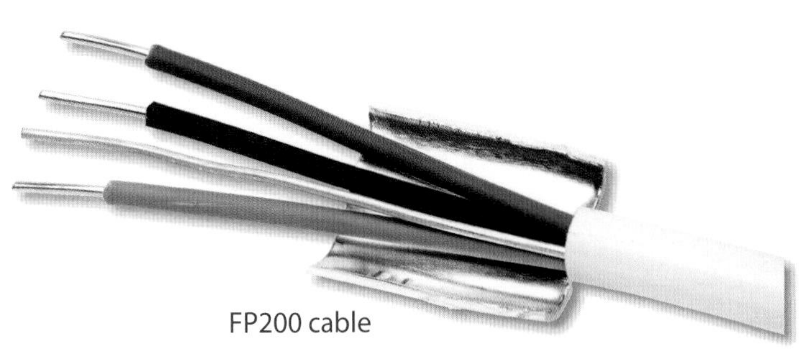

FP200 cable

Twin and cpc PVC insulated and sheathed cable

MICV cable with pot and seal

When these and similar cables are installed, their design capabilities should not be degraded and special glands and seals, which are produced by the makers, should be used. Where PVC/PVC cables enter accessories, the accessory itself should have no loss of integrity and there should be no damage to any part of the cable.

When carrying out a visual inspection of an installation, either at the completion of works or at a periodic inspection, the checking of terminations of cables and conductors is an integral part of the inspection. Some items listed for checking in Regulation 611.3 include:

- connection of conductors
- identification of conductors
- routing of cables
- selection of conductors in accordance with the design
- correct connection of accessories and equipment.

These checks are made to comply with Regulation 611.2, which states 'The inspection shall be made to verify that the installed electrical equipment is not visibly damaged or defective so as to impair safety'.

We should also routinely check:

- for correct entry of cable into the accessory
- that the correct type of gland and seal has been used
- that accessory or enclosure seals and building structure/cable routes sealed.

Accessories including switches, control equipment, contactors, electronic and solid state devices

Remember

Operating mechanisms do not normally break down and where contact points can be serviced and the build up of carbon removed, this will extend the life of the accessory. However, many items are not serviceable and when a breakdown occurs through constant use, a repair is neither a choice nor cost-effective

Faults can appear on most items of equipment that we install. The most common fault is due to wear of the terminal contacts, which constantly make and break during normal operation. All items of equipment should be to BS or BSEN and must be type tested by the manufacturer.

Some accessories break down due to excessive use, and through experience the electrician can usually recognise them. During regular inspections and maintenance of an installation the inspector will check 10 per cent of accessories visually, which will usually be items that are constantly being used such as:

- entrance hall switches
- socket outlets in kitchens for kettles
- cleaner's socket
- any item which is regularly being switched on or off.

Control equipment

This is also termed switchgear and is found at the origin of the supply after the metering equipment. It can take many forms; a few examples are:

- domestic installation – double pole switch in consumer unit
- industrial installation – single-phase double pole switch fuse or three-phase triple pole and neutral switch fuse
- large industrial installation moulded case circuit breakers – polyphase
- switch rooms industrial/commercial – as above, also oil and air blast types.

The switchgear is categorised not only in normal current ratings for load in amperes, but also for fault current ratings in 1000 amperes or kA. If we refer back to Figure 6.5 on page 194, item C is a typical example of switchgear control equipment.

If a fault occurred on the outgoing conductors from these devices then the level of fault current would be dependent on the location. The basic rule for fault current level is that the nearer to the supply that the fault occurs, then the greater the level of fault current. This is due to there being less impedance (resistance) to impede the current flow at that point, whereas the further away from the origin of the supply the greater the impedance (resistance) to impede the current flow.

The operation of this type of control gear, either manually or automatically, due to a fault would leave all circuits fed from such equipment dead. Manual operation of such devices is usually for maintenance reasons and it should be noted that permission of the client by prior arrangement would need to be obtained before commencing such work.

Protective devices are usually an integral part of switchgear, which will be one of the following:

- high breaking capacity fuses HBC or HRC
- moulded case circuit breakers MCCB.

Specialist information and manufacturer's instructions are needed for replacing fuses that have operated and when re-setting tripped circuit breakers.

In older type installations you may come across switchgear that is prone to fault due to lack of maintenance. For example oil type circuit breakers have their contact breaker points submerged in a special mineral-based oil. The oil, which quenches arcing, assists in breaking high levels of load and fault currents. If the oil viscosity and level is not regularly checked, this could lead to high levels of fault current circulating, leading to equipment damage and insulation breakdown or even fire.

Contactors

Contactors are commonly used in electrical installations. They can be found in:

- motor control circuits
- control panels
- electronic controllers
- remote switching.

The basic principle of operation of a contactor is the use of the solenoid effect, in which a magnetic coil is used in the energised or de-energised mode and spring loaded contacts for auxiliary circuits are either made or broken when the coil has been energised or de-energised.

Operating coils used in motor control circuits are energised when the start button is operated, with the most common fault associated with their use being coils burning out. This can be due to age, but can be due to the wrong voltage type being used, i.e. a 230 volt coil being used across a 400 volt supply.

Remote switching can utilise the contactor with great effect and convenience because the contacts of large, load switching contactors can be used to carry large loads, i.e. distribution boards and large lighting loads such as used in car parks, floodlights etc. The benefit of the system is that a local 5 A switch in an office or reception area can be operated when entering or leaving, thus energising the coil of the contactor and allowing heavy loads to be carried through the contactor's contacts.

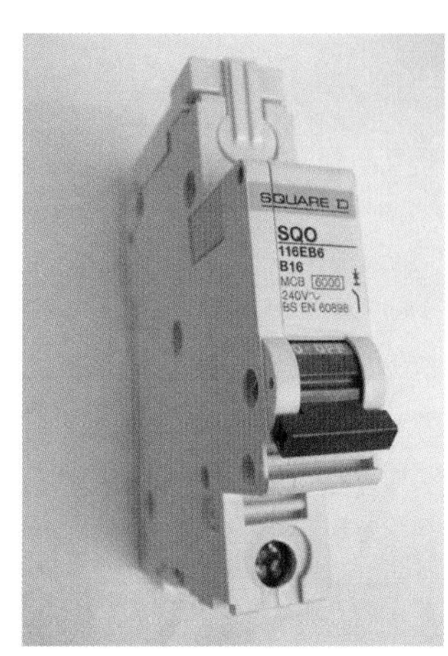

Moulded case circuit breaker

Did you know?

As contacts are regularly being made and broken, a build-up of carbon appears on the contact points. On larger contactors these can simply be cleaned periodically using fine emery paper to remove the residue and make the electrical contact point more efficient

Remember

On smaller contactors used in electronic circuits it is not possible to clean contact points, and when breakdown occurs it is normal practice to replace rather than repair

Electronic and solid state devices

Solid state and electronic equipment work within sensitive voltage and current ranges in millivolts and milliamperes, and are consequently sensitive to mains voltage and heat.

In the case of resistors on circuit boards, excessive heat produces more resistance and eventually an open circuit when the resistor breaks down. We can use a low-resistance ohmmeter to check resistor values.

In the case of capacitors, voltages in excess of their working voltage will cause them to break down, which will result in a short circuit. Such equipment obviously requires specialist knowledge, but some basic actions by the electrician when carrying out tests and inspections can prevent damage to this type of equipment. When testing, such equipment should be disconnected prior to applying tests to the circuits. The reason for this is that some test voltages can damage these sensitive components and their inclusion in the circuit would also give an inaccurate reading.

We have discussed transient voltages and these are a major cause of faults on components and equipment of this type. These voltages can arise from supply company variations or lightning strikes, and therefore most large companies protect their equipment from such voltages and employ specialist companies to install lightning protection and filtering equipment.

Instrumentation

Switch room panels have integrated metering equipment. In this part of the installation please note that **large currents are being monitored**. This monitoring allows the consumer to view current and voltage values at particular times of the day. Some panels will have equipment which monitors and records these values, allowing the consumer to plot peak energy times and budget for energy costs.

As these instruments measure large currents, transformers are used to reduce current and voltage values, and this reduces the size of instrument needed. These transformers are known as CTs and VTs. However, great care must be taken when servicing and repairing such equipment as large voltages can exist across the transformer terminals. The common remedial work on such equipment is the replacement of faulty instruments and burnt-out transformers. It is essential that safe isolation procedures are followed.

Protective devices

These devices operate in the event of a fault occurring on the circuit or equipment that the device is protecting. Typical faults include:

- short circuits between live conductors – phase to neutral single-phase, phase to phase three-phase
- earth faults – between any live conductor and earth
- overcurrents.

The most common reason for a device not operating is that the wrong type or rating of device has been used. Therefore, when replacing or re-setting a device after a fault has occurred, it is important to replace the device or repair it with the same type and rating. This should only be done after the fault has been investigated and corrected. The consequence of switching on a device with an un-repaired fault is for the device to operate again. This type of activity will cause damage to the circuit wiring, the equipment and possibly the device.

When a protective device has operated correctly and this device was the nearest device to the fault, **discrimination** is said to have occurred

Luminaires

The most common fault with luminaires is expiry of lamp life, which obviously only requires replacement. Discharge lighting systems employ control gear which on failure will need replacement as they cannot be repaired. Discharge type lighting may have problems with the control circuit.

Common items of equipment in the luminaire control circuit are:

- The capacitor used for power factor correction. If this had broken down it would not stop the lamp operating but would prevent the luminaire operating efficiently.
- The choke or ballast used to create high p.d. to assist in the lamp starting. A common item which could break down and needs replacement.
- The starter, which is used to assist the discharge across the lamp when switched on at the start. This is the usual part to replace when the lamp fails to light.

Many fluorescent luminaires have starter-less electronic control gear, which is not only quick-start with increased efficiency, but requires less maintenance. They have a longer lamp life but when the quick-start unit fails it will need replacement

Flexible cables and cords

This type of conductor is used to connect many items within an electrical installation, e.g.:

- pendant ceiling rose lamp holders
- flex outlet spur units
- fused plugs to portable appliances
- immersion heaters
- flexible connection to fans and motors.

BS 7671 requires that all flexible cables and cords shall comply with the appropriate British and Harmonised Standard. The most common faults that occur with this type of conductor are likely to arise from poor choice and suitability for the equipment and the environment. Common faults relating to flexible cables and cords are:

- poor terminations into accessory – conductors showing etc.
- wrong type installed i.e. PVC instead of increased temperature type
- incorrect size of conductor – usually too small for load
- incorrectly installed when load bearing in luminaires.

Such problems should be identified during the visual inspection stage.

Portable appliances and equipment

Flexible cables and cords are used to connect many items of fixed and portable equipment to the installation wiring, and most faults on cords and flexible cables usually relate to poor choice or installation. Portable appliances come pre-wired with a fused plug to BS 1363.

When supplying equipment with flexible cables and cords it is important that the requirements of the *IEE Wiring Regulations* are followed but specifically:

- correct size conductor used for the load
- correct type used for the environment – temperature – moisture – corrosion
- correct termination method used to ensure good connection and no stress
- correct type and rating of protective device used to protect the cable and appliance.

Knowledge and understanding of electrical systems and equipment

Experience and understanding of electrical installations takes years to learn and a vast amount of knowledge is gathered during your apprenticeship. However, putting that knowledge into practice only really occurs when you take up responsibilities as an electrician.

When analysing faults on electrical installations, even the trained electrician will need to ask questions to help in rectifying faults.

Hopefully by now you should have come across many different types of wiring systems, equipment, enclosures and protective devices. Equally, you should have now used the procedures for safe isolation and locking off procedures many times. However, when a fault occurs, the art of finding information and asking questions can be a daunting task.

Understanding the electrical system, installation and equipment

These can generally be categorised as follows:

- voltage – 230 volt single-phase or 400 volt three-phase
- installation type – domestic, industrial or commercial etc.
- system type – lighting, power, fire alarm system, emergency lighting, heating etc.

If a fault occurs on a system which is in the process of being installed then the information and data for the system should be at hand. However, the big problem starts when you are asked to rectify a fault on a system that you have had no experience or knowledge of. In this situation you will need to access drawings and data to familiarise yourself with layout, distribution boards etc. The type of information that you will require when called out to a fault can be listed as:

- type of supply – single phase or three phase
- nominal voltage – 230 volt or 400 volt
- type of earthing supply system – TT, TN-S, TN-C-S
- type of protective measure – ADOS
- types of protective device – HRC, MCB, RCD etc.
- ratings of devices
- location of incoming supply services
- location of electrical services
- distribution board schedules
- location drawings
- design and manufacturer's data.

Remember

An electrician's knowledge and training can be a great advantage in recognising and tracing faults, but the task will be made easier and the fault will be remedied quicker if the information listed is available. As a first choice, where this information is not readily available you should either seek advice or try to obtain it

Optimum use of personal and other person's experience of systems and equipment

Your own knowledge of an installation can be an asset not only in the fault finding process, but also to your company for future business. If your company can problem-solve quickly and be relied upon by a client when a problem occurs, it will be your company that will be the first to be called.

The fault may not exist on the installation wiring but on auxiliary equipment such as refrigeration and air conditioning. Unless you have specialist-trained knowledge of such equipment, it is not safe to attempt to rectify the faults and your company may not be insured to work on such systems. However, if the fault is on the circuit supplying these systems, the fault may be investigated and rectified. You may also have to work alongside other specialists, assisting each other in commissioning and testing such systems. On no account should you attempt to investigate or rectify faults on their systems without knowledge or experience and training.

Equally, you will often be asked to repair faults on wiring and equipment that have been caused by inexperienced personnel attempting to install or repair circuits or equipment.

When you first arrive at any installation to investigate and rectify a fault, you should always use the personnel present to help you to obtain any background to the fault and relevant information, and if possible seek the specialist knowledge of the person responsible for the electrical installation or general maintenance.

The different people that you will be dealing with to obtain special and essential knowledge and information could be:

- the electrician who may have previously worked on the system
- design engineer
- works engineer
- shift engineer
- maintenance electrician
- machine operator
- home owner
- site foreman
- shop manager
- school caretaker.

These persons may have access to:

- operating manuals
- wiring and connection diagrams
- manufacturer's product data/ information
- maintenance records
- inspection and test results
- installation specifications
- drawings
- design data
- site diary.

Their experience of the installation and day-to-day knowledge is essential. It will help you to solve, replace or repair the fault more efficiently.

Stages of diagnosis and rectification

To be able to competently investigate, diagnose and find faults on electrical installations and equipment is one of the most difficult jobs for an electrician and therefore, if the electrician is to be successful in fault finding and diagnosis, a thorough working knowledge of the installation or equipment involved is essential.

Consequently, the person carrying out the procedure should take a reasoned approach and apply logical steps or stages to the investigation and subsequent remedy. The electrician should also realise their knowledge limitations and seek expert advice and support where necessary.

In an ideal world an electrician should not embark on any testing or fault finding without some forward planning. However, an emergency or dangerous fault, because of its very nature, may allow very little time for planning the remedy or repair. That said, some faults that are visible and straightforward can be easily repaired, and some careful planning and liaison with the client will limit disruption.

Inspection and testing prior to energising a new installation can be important and can save embarrassment by rectifying problems before circuits are energised. As an example, pre-energising tests can be extremely important when checking the function of switching circuits, particularly two-way and intermediate switching.

The electrician should have knowledge of inspection and testing in order to fault find competently – in particular the correct use of test instruments, the choice of instrument for testing, and knowledge of each instrument's range and limitations.

Remember

The *IEE Wiring Regulations* on testing can be a useful guide, but the *Electricity at Work Regulations* and safe use of test equipment should be followed during the process

Remember

Information is a necessity and no fault finding should commence without the relevant information and background to the fault being made available to the person carrying out the work. Some faults can be recognisable by simply analysing test results and on further investigation can be remedied simply

Logical stages of diagnosis and rectification

Some faults can be rectified very easily, especially when the electrician has a working knowledge of the installation. But there are many occasions when you may be called out to a repair and there is no information for you. It is on these occasions that your years of training and knowledge of wiring systems have to be used. However, this knowledge is not always enough and a sequential and logical approach to rectifying a fault and gathering information is needed. Such an approach could be:

- **Identify the symptom**: This can be done by establishing the events that led up to the problem or fault on the installation or equipment.
- **Gather information**: This is achieved by talking to people and obtaining and looking at any available information. Such information could include manufacturer's data, circuit diagrams, drawings, design data, distribution board schedules and previous test results and certification.
- **Analyse the evidence**: Carry out a visual inspection of the location of the fault and cross-reference with the available information. Interpret the collected information and decide what action or tests need to be carried out. Then determine the remedy.

- **Check supply**: Confirm supply status at origin and locally. Confirm circuit or equipment when fault is isolated from supply.
- **Check protective devices**: Check status of protective devices. If they have operated this would determine location of fault on circuit or equipment.
- **Isolation and test**: Confirm isolation prior to carrying out sequence of tests.
- **Interpret information and test results**: By interpreting the test results, the status of protective devices and other information, the fault may be identified and remedied/rectified.
- **Rectify the fault**: This may be done quite simply or parts or replacement may be needed.
- **Carry out functional tests**: Before restoring the supply, the circuit or equipment will need to be tested not only electrically but also for functionality.
- **Restore the supply**: Care must be taken that the device has been reset or repaired to correct current rating. Make sure the circuit or equipment is switched off locally before restoring supply.

In summary, the logical sequence of events would be:

1. Identify the symptoms
2. Gather information
3. Analyse the evidence
4. Check supply
5. Check protective devices
6. Isolation and test
7. Interpret information and test results
8. Rectify the fault
9. Carry out functional tests
10. Restore the supply.

Information required for fault location

We have already said that some information regarding the fault will be gathered by asking people questions relating to the events leading up to the fault. If the fault occurred on the installation when the electrician was working on it then the information will be to hand. Most often the situation regarding faults is when we are called out to a breakdown and asked to carry out repairs. It is when we are called out to a fault or repair that we will need information such as:

- operating manuals
- wiring and connection diagrams

- manufacturer's product data/information
- maintenance records
- inspection and test results
- installation specifications
- drawings
- design data
- site diary.

Interpretation of test results

Some faults can be recognised at the installation stage when the testing and inspection process is carried out. We should be able to recognise typical test results for each non-live test and interpret the type of fault that may exist.

Non-live tests

The *IEE Wiring Regulations* BS 7671 Part 6 lists non-live and live tests. The non-live tests are carried out on the installation wiring circuits before they are ever energised. These tests will confirm the integrity of the circuit and they can be listed as:

- continuity of protective conductors
- continuity of ring final circuit conductors
- insulation resistance
- polarity.

Continuity of protective conductors

Protective conductors can be found in final circuit wiring and from the main earthing terminals to all metalwork that is to be found in the installation. It is important to verify the continuity of such conductors because we rely upon them for safety in the event of an earth fault and the operation of protective devices and fuses. A low-resistance ohmmeter would be used on conductors to verify a low resistance value. The reading obtained will depend on the length and area of the conductor, as resistance is proportional to length and inversely proportional to cross-sectional area.

If you are testing for the continuity of equipotential bonding the test current should be a minimum of 200 mA (Regulation 612.2.1). A typical reading for earth continuity would be between 0.01 to 0.05 ohms. If a reading greater than 0.5 ohms was obtained, although on the upper side of acceptable, this may need investigating. Although it may be acceptable in relation to the conductor type and size, terminations may require checking for effectiveness.

Continuity of ring final circuit conductors

This test, carried out on the circuit wiring, is designed to highlight open circuits and interconnections within the wiring, and a low-resistance ohmmeter is used to carry out this test. Typical readings will be 0.01 to 0.1 ohms dependent on the conductor size and length of circuit wiring. If the circuit is wired correctly, then the readings on the instrument will be the same at each point of test. If variable readings are found at each point of the test, this may indicate an open circuit or interconnections. If these types of fault exist, the consequence can be an overload on part of the circuit wiring.

Insulation resistance

This test is designed to confirm the integrity of the insulation resistance of all live conductors, between each other and earth, and the type of instrument is known as an insulation resistance tester. Typical test results would be around 50 to 100 megohms, i.e. 50 to 100 million ohms.

Table 61 of the *IEE Wiring Regulations* indicates minimum values between one quarter of a million and one million ohms. These are minimum values and, in practice if these values existed, the circuit may need further investigation.

Depending on where the test takes place, variable values of resistance may be recorded. In the case of a test on the supply side of a distribution board (Figure 6.9), a group value reading will indicate circuits connected in parallel (Figure 6.10).

From Figures 6.9 and 6.10, it can be seen that when testing grouped circuit conductors, a parallel reading would indicate a poor reading, but if each circuit were tested individually then the actual reading would be acceptable.

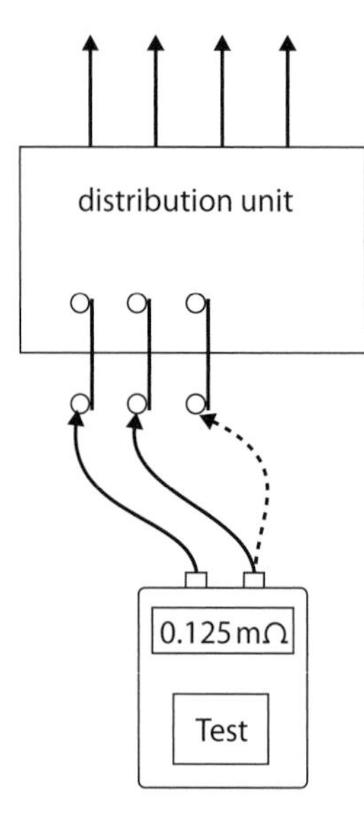

Figure 6.9 Test on supply side of distribution board

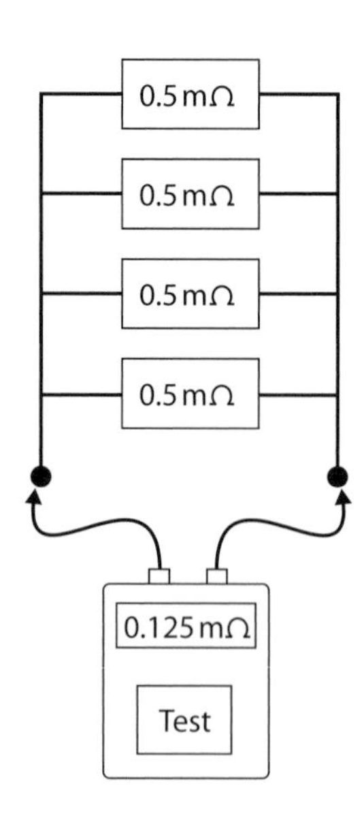

Figure 6.10 Test on circuits connected in parallel

Polarity

The polarity test can be carried out using a low resistance ohmmeter as used for typical continuity testing. This polarity can be confirmed during the continuity test on the circuit and protective conductors. Typical test results will depend on the resistance of the conductors under test but will generally be low, around 0.01 to 0.1 ohms. Values in excess of this may indicate an open circuit or incorrect polarity.

Functional testing

Regulation 612.13.1 states that where fault protection and/or additional protection is to be provided by an RCD, the effectiveness of any test facility incorporated in the device shall be verified.

Such a test would use an appropriate RCD test instrument and the test should confirm the operation of the device independently of the device's integral test button. Residual current circuit breakers or devices are rated in milliamperes and the test should show that the device operates within the milliampere ratings of the device and within the time constraints. Typically if the rated current of the device was 30 mA then the instrument should prove operation at this value within 200 milliseconds.

Regulation 612.13.2 states that assemblies such as switchgear and control gear, drives, controls and interlocks shall be subjected to a functional test, to show that they are properly mounted, adjusted and installed in accordance with relevant requirements of the *IEE Wiring Regulations*. Typical functional tests should be applied to:

- lighting controllers, switches etc.
- motors, fixings, drives, pulleys etc.
- motor controllers
- controls and interlocks
- main switches
- isolators.

Earth-fault loop impedance testing

Overcurrent protective devices must, under earth-fault conditions, disconnect fast enough to reduce the risk of electric shock. This can be achieved if the actual value of the earth loop impedance does not exceed the tabulated values given in BS 7671. The purpose of the test, therefore, is to determine the actual value of the loop impedance (Z_s) for comparison with those values.

The test procedure requires that all main equipotential bonding is in place. The test instrument is then connected by the use of 'flying' leads to the phase, neutral and earth terminals at the remote end of the circuit under test. Press to test and record the results.

Once Z_s and the voltage have been established at the remote point in the circuit, you can divide the voltage by Z_s to give you the fault current to earth. Apply this calculation to the time current characteristic graphs shown in BS 7671 and you will be able to determine the actual disconnection time of the circuit.

Safety tip

Care must be taken when using the external earth probe on this tester because the probe tip is at mains potential

The limitation and range of instruments

The correct choice of instrument for each particular test should now be known, but for those who are still unsure refer to Table 6.1.

Test	Instrument	Range
Continuity of protective conductors	Low resistance ohmmeter	0.005 to 2 ohms or less
Continuity of ring final circuit conductors	Low resistance ohmmeter	0.05 to 0.8 ohms
Insulation resistance	High reading ohmmeter	1 megohm to greater than 200 megohms
Polarity	Ohmmeter Bell/Buzzer	Low resistance None

Table 6.1 Limitation and range of test instruments

Some instruments can provide more than one facility and many manufacturers provide such instruments. Inspectors and testers should know the range of operation of their instruments, be conversant with their operation and understand the instructions on the instrument.

When carrying out testing procedures, and before any instrument is used, some simple checks should be carried out to ensure that the instrument is operating within its range and calibration. A suitable checklist for use prior to testing is given below.

- ✔ Check the instrument and leads for damage or defects.

- ✔ Zero the instrument.

- ✔ Check the battery level.
- ✔ Select the correct scale for testing. If in doubt, ask or select the highest range available.
- ✔ Check the calibration date and record the serial no.

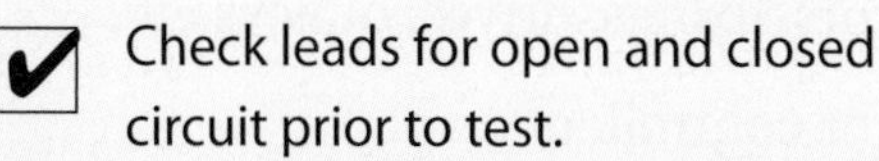

- ✔ Check leads for open and closed circuit prior to test.

- ✔ Record test results.
- ✔ After test, leave selector switches in off position. Some analogue instruments turn off automatically to save battery life.
- ✔ Always store instruments in their cases in secure, dry locations when not in use.

Factors affecting the fault repair process

Having achieved the aim of locating the fault it may be a simple decision by the electrician to eliminate the cause and re-energise the circuit for the equipment or circuit to function again. However, in many cases there are issues that will not allow the power to be switched on to the circuit or equipment under scrutiny. This judgement in many cases cannot be made by the electrician and may need to be discussed by third parties i.e. the client, the manufacturer and the person ordering the work. In this section we will look at the factors which will influence this decision-making process as to whether the equipment or circuit is replaced or repaired and whether the fault can be remedied instantly or at a pre-determined future time.

Cost of replacement

When successfully finding a fault it would not make for good customer relations to go ahead and put the fault right without first considering the consequences to your company or your client. There are points that need consideration such as: 'What will the cost of replacement be?'

This question can only be answered by liaison and consultation with your engineer and the customer, as an agreement must be reached on how to proceed. It may well be that some electrical contractors have an agreement or service contract in place where remedial works are carried out on a day work and contracted basis. This is where a prior contract agreement is signed and the contractor carries out maintenance and repair work at pre-determined rates. However, where this is not the case the question may arise: 'Will the replacement cost be exorbitant?'

It could be that the price quoted for replacement is high, which may result in the customer using your company's quotation to compare with other companies' prices. It may lead to negotiations where only part of a faulty piece of equipment is replaced or repaired, which may reduce the outlay costs to the customer. It is often the case that the customer is given a choice of options to help them decide what to do; they will compare these options and make the decision.

The options may be:

- full replacement
- part replacement
- full repair.

In offering such options it can be seen generally that the time taken to carry out a repair or part replacement may in many cases outweigh the time and cost of a full replacement.

Availability of replacement

As we have said before, in many cases where a complete replacement of equipment or machinery is needed, it is because the fault has arisen due to the life span of the equipment being exceeded. In these cases, the equipment or machinery may have been in operation for so long that a repair may not be possible and the item may even be obsolete. When this happens, a suitable replacement must be found.

The most obvious decision to be made is that of finding a replacement that closely matches the original specification of the faulty item. But this is not always possible and the factor that, in reality, will affect the selection process becomes the availability of the equipment. This is important and will influence the decision to order. It is often the manufacturer who has the replacement item in stock and has a fast delivery that will be given the order to supply, as this will enable the electrician to remedy the fault quickly.

Remember

It is often the case that the original manufacturer is no longer in business and therefore the problem becomes finding a suitable alternative manufacturer. Most manufacturers will have their own design team who will help in the process of finding a suitable alternative piece of equipment

Downtime under fault conditions

It is often the case that the consequence of a fault on a circuit or item of equipment will be costing the client money. This can be due to:

- lost production where manpower and machinery are not working
- lost business where products or services are not being provided to the customer
- data loss because of power failure or disruption during repair.

Sometimes the fault may not be on the electrical installation but on the supply to the consumer's premises. In such cases the rectification of the fault is out of your company's responsibility and control. However, in such circumstances and where the loss of supply is prolonged, it may make good business sense, where possible, to provide a temporary supply to the customer.

Remember

Whatever the cause of the fault, the consequence is often loss of revenue either directly because of output or because of poor service to the customer, which in turn can affect the client's reputation and ultimately result in a future loss of business

Availability of resources and staff

We have already mentioned the availability of equipment due to choice, manufacturer and delivery time, but the availability of staff is often the cause of any delay in rectifying faults and carrying out repairs.

Some faults and repairs may be minor and need little resource in terms of materials and labour. The response time for small faults could be as low as minutes and hours, but in the case of major industrial and commercial repairs the down time and repair time could be weeks or even months. The consequences of some faults may be major contract works such as rewiring or large equipment replacement. In such cases careful planning and decisions must be taken by the contractor's management and design team when deciding to commit the company to carry out such works.

The following are basic considerations that any company should make before commencing such contracts:

- **Availability of staff** – Has the company enough people employed to carry out and complete the work?
- **Competency** – Does the company have staff with relevant experience?
- **Cost implications** – Has the company the monetary capability for the contract?
- **Special plant and machinery** – Has the company this type of resource available?

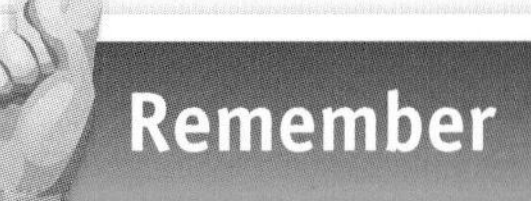

Remember

Repairs must not only be viable for the client but also for the contractor, who would not like to lose money or reputation on installation and repair works

Legal responsibility

The legal responsibilities surrounding the replacement or repair of a fault must also be considered when making the decision to repair or replace. It is usual for a contract to be entered into by both parties as this would help to avoid, or at least assist, in settling future disputes. Such a contract would itemise:

- costs
- time
- guarantee period/warranty
- omissions from contract.

The costs should be agreed and adhered to. Most customers will not pay more than the initial agreed costs of repair unless an allowance was made in the contract for unforeseen work. Unforeseen work could result from something found when dismantling or taking out existing items, such as finding faulty or dangerous wiring underneath an item of heavy plant that is to be repaired.

If the initial costing accounted for repair and replacement of parts, then it is essential that you do what you said you were going to do. For instance, if an inspection is subsequently carried out and it is evident that no work has taken place then you may be liable for legal proceedings to be taken against you for a breach of contract terms.

An accurate cost analysis should be considered when giving an estimate or fixed price for repair work, as it is usual practice to carry out repair work on a daily rate where the customer pays only for the time taken and materials used in rectifying the fault.

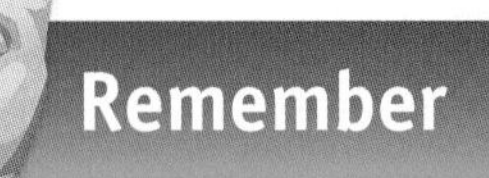

Remember

Most companies will carry insurance to cover against any liabilities in respect of work done

The customer should also be given a guarantee for the work done, both in respect of quality and also for the materials used.

When deciding whether to repair or replace, it may be deemed necessary to avoid writing a guarantee in the contract when the client insists on re-using existing items in order to reduce costs. It is often the case that old industrial or antiquated items would be prone to failure and to avoid legal action such items should not be included in the contract and written out of your guarantee.

Other factors affecting the fault repair process

Careful planning can mean that most work is carried out efficiently with little or no disruption to the client. However, faults on installation equipment and wiring systems are certainly not planned and trained personnel can only minimise the disruption caused by the effects of faults.

In a factory environment, maintenance electricians are on hand to limit the effect of a fault and can usually solve problems and reduce the down time of a machine or circuit to a minimum. This is because of their day-to-day working knowledge of the electrical services and the equipment connected to it.

When an electrician is called out to a fault they may not have the day-to-day knowledge or experience of the maintenance staff. However, through training and knowledge the electrician will use contingency plans to institute the procedures to rectify the fault with the minimum of disruption.

Such contingency plans provide a common format when dealing with fault finding at any customer's premises. Your company will have its own procedures and these may include:

- signing in
- wearing identification badges
- locating supervisory personnel
- locating data drawings etc.
- liaising with the client and office before commencing any work
- following safe isolation procedures.

Special requirements

We have previously looked at the safety procedures and the actions needed for determining causes of faults. If we assume that safety requirements and the location of faults have been met and that the prime objective of the electrical contractor (the electrician) is to locate the fault, rectify and re-commission, then this will be done in liaison with the customer and with minimal disruption. However the following topics are worthy of note.

Access to the system during normal working hours

There is nothing more costly or embarrassing than arriving at a customer's premises to carry out work of any kind, only to find the premises locked or unmanned because no prior arrangements were made for access. Access to the premises and the electrical installation or the specific item of equipment may prove a problem, especially if you are unfamiliar with the customer's premises.

When attending a fault or breakdown we must remember the procedures for ascertaining the information and familiarising ourselves with the installation. Remember our logical sequence was:

1. Identify the symptoms
2. Gather information
3. Analyse the evidence
4. Check supply
5. Check protective devices
6. Isolation and test
7. Interpret information and test results
8. Rectify the fault
9. Carry out functional tests
10. Restore the supply.

If we need to familiarise ourselves with the installation, then access must be readily available not only at the fault repair location but also at the supply intake and isolation point, and this has to be agreed prior to commencing any contracted work, whether it is a repair or a new installation.

Such prior agreement must take into account those personnel who may have their normal day-to-day routine and activities affected. It is not always possible to carry out repair work during normal working hours and so to avoid disruption it is often the case that such work is done outside of normal working hours. It is therefore important that arrangements for out-of-hours work are pre-arranged to account for security and especially access to the premises.

Is there a need for building fabric restoration?

As electricians, we are more often than not being asked to locate and repair faults on existing installations. Consequently, depending on the fault found and the method of rectification needed, we may need to disturb the fabric of the building. It is therefore vitally important to fully discuss with the client all of the implications of the fault diagnosis/rectification process. As part of investigating a fault, we may have to do nothing more than look inside an item of equipment. However, if the fault has been caused by someone putting a nail through a cable, then the repair will involve replacing the cable and that in turn may involve having to damage the building fabric. You should therefore explain these aspects to the client and agree what repair work (e.g. brickwork, plastering, decorating) will be required and which, if any, of these can be carried out satisfactorily by yourself.

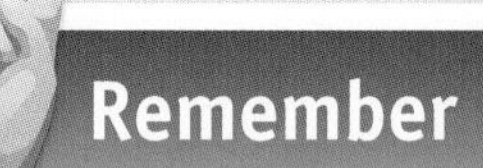

Remember

The expertise of building contractors may be needed and there will be costs associated with this that need to be agreed in advance of any repairs being carried out

Obviously, as with any aspect of fault finding and rectification, always leave the site in a clean and safe condition.

Whether the system can be isolated section by section

After access to the installation has been agreed, the repairs should be carried out using a logical approach. However, if the premises are still occupied and parts of the system are still being used, then it is important that the faulty circuit or plant to be worked on is isolated.

When only a small area or a single circuit or piece of equipment is affected it is not good practice to switch off the mains supply and isolate all the installation. By analysing the installation drawings and data the area or items should be isolated individually to limit disruption.

Remember

All electrical installation systems should have been designed and installed in accordance with the *IEE Wiring Regulations*, thus ensuring that the facility for isolating circuits and sections is inherent in the system

Provision of emergency or stand-by supply

Often the disruption to the customer is not having an electrical supply to an area of the installation. If the work cannot be done outside of normal working hours, a temporary supply could be arranged. Some installations may have a stand-by service in the form of emergency lighting or a stand-by generator, but this is only usual in hospitals or where computers are being used and data is stored.

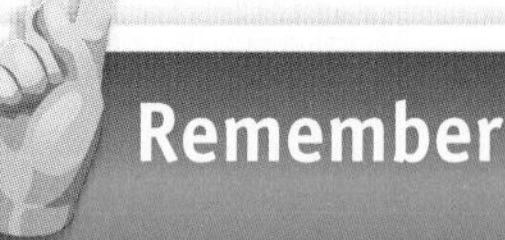

Remember

Any temporary supply which is provided should be wired in accordance with the *IEE Wiring Regulations*

Client demand for a continuous supply

Where the client does not want any interruption in supply and it is not possible to work on the affected area during normal working hours, arrangements should be made for a planned maintenance or shut down in order for the repair to be carried out safely.

Did you know?

Some formal risk assessments are recorded and noted in method statements. These can be seen in your company's Health and Safety Statements

Safety tip

The levels of light carried by fibre optic cabling can be dangerously high so you should never look into the end of a fibre optic cable

Special situations

Special precautions and a risk analysis should always be carried out before any work is commenced. It is a requirement of the Health and Safety at Work Act 1974 that precautions should be accounted for and that any risk is removed or persons carrying out any work activity should be protected from harm. Risk assessments are now commonplace and are part of all our daily work routine. In this section we shall be looking at the risks involved in carrying out fault diagnosis and repair.

Fibre optic cabling

This type of wiring system is becoming increasingly popular in the data and telecommunications industry, although it is unlikely that you will terminate or fault find on such systems. However, your company may be asked to install such systems and you may be asked to assist in their installation.

As its name suggests the cable is made up of high-quality minute glass fibres, which can carry vast amounts of transmitted signals and data using emitted signals of light. The levels of light transmitted can reach dangerously high levels. Therefore YOU SHOULD NEVER LOOK INTO THE END OF FIBRE OPTIC CABLE!

Antistatic precautions

Antistatic precautions are usually prevalent in areas that are hazardous due to the high risk of ignition or explosion. This is usually where flammable, ignitable liquids, gases or powders are stored or dispensed, for instance:

- petrol filling stations
- chemical works
- offshore installations
- paint stores
- flour mills.

Petrol filling station

In such situations we must be aware that NO NAKED FLAMES or equipment which can cause a spark or discharge of energy can be used. There would obviously be a no smoking policy. It is important to adhere to the safety signs and regulations in such areas. When asked to work in such environments you should be given special induction and training to make you aware of the dangers and the risks.

Electro-static discharge

Static is an electric charge, caused by an excess or deficiency of electrons collected on a conducting or non-conducting surface, thus creating a potential difference measured in volts. Static charges are caused in several ways, as described opposite.

Friction

When two surfaces are rubbed together they behave as though electrons were rubbed off one surface and on to the other, leaving one surface with a deficiency of electrons, therefore being positively charged. The other surface acquires and retains electrons, therefore being negatively charged.

Separation

When self-adhesive plastic or cellulose tape is rapidly pulled off a reel it generates a static charge by tearing electrons from one surface to another.

Induction

Static is generated by induction when a charged item, for example a folder of paper, is placed on a desk and on top of that is placed a conductor, say a printed circuit board. If the paper was positively charged, the charge from the paper is immediately induced into the printed circuit board, making it negatively charged.

All materials can be involved in the production of static. In the case of non-conductors (insulators such as paper, plastic and textiles) the charge cannot redistribute itself and therefore remains on the surface.

Static electricity

Did you know?

The reality is that every step, even every shuffle on a seat, generates static charges. Picking up, putting down, wrapping and unwrapping, even just handling an object creates static. So you can see that with all the everyday movements we make, while wearing a lot of clothes made from man-made fibres, we will generate large amounts of static

Protection from electro-static discharge

The main effort must be to protect electro-static sensitive devices by providing special handling areas where they can be handled, stored, assembled, tested, packed and dispatched. In the special handling area there should be total removal of all untreated wrappings, paper, polystyrene, plastics, non-conductive bins, racks and trays, and soldering irons including tips should be earthed. Specifications, drawings and work instructions must be in protective bags, folders or bins. Wrist straps should be worn at all times in the static free area. These straps 'earth' the handler and provide a safe path for the removal of static electricity.

Damage to electronic devices due to over voltage

When testing or inspecting faults we must take account of electronic equipment and their rated voltage. It is usual to isolate or disconnect such equipment to avoid damage due to test voltages exceeding the equipment's rated voltage. Such voltage levels will cause components, control equipment, or data and telecommunication equipment to be faulty. This could be costly not only in repairs but also to the reputation of your company, and could result in the loss of future contracts. The voltage of an insulation resistance tester can reach levels between 250 to 1000 volts.

Avoidance of shut down of IT equipment

On no account should any circuit be isolated where computer equipment is connected. This type of equipment is to be found in many commercial, industrial and even domestic installations. The loss of data is the most common occurrence when systems fail due to faults or power supply problems. Larger organisations will have data protection in the form of uninterruptible power supplies (UPS systems). Whether UPS systems are in place or not, it is still important to plan the isolation of circuits where IT systems are connected. This will give the client time to arrange data storage and download, which will allow systems to be switched off and repairs to be carried out.

Safety tip

On some systems where high frequencies exist, there is a danger of exceptionally high voltages occurring across the terminals. On no account should you as a trainee ever attempt to tamper with capacitive circuits and, if in doubt, seek the assistance of your supervisor

Risk of high frequency on high capacitive circuits

Capacitive circuits could be circuits with capacitors connected to them or some long runs of circuit wiring which may have a capacitive effect. This is usual on long runs of mineral-insulated cables. When working on such circuits no work should commence until the capacitive effect has been discharged. In some cases it would be practical to discharge capacitors manually by shorting out the capacitor.

Danger from storage batteries

Batteries can be found on many installations and provide an energy source in the event of an emergency when there is a power failure. This can be seen where battery panels provide a back-up for the emergency lighting in a building when the power fails.

Batteries can also be found in alarm panels, IT UPS systems and emergency stand-by generators for starting.

Wherever you have cause to work with or near batteries special care should be observed, particularly with regard to:

- Lead acid cells, which contain dilute sulphuric acid. This acid is harmful to the skin and eyes, can rot clothing and is highly corrosive. If there is any contact with the skin it should be immediately diluted and washed with water.
- Lead acid cells also emit hydrogen gas when charging and discharging, which when mixed with air is highly explosive.
- High voltages applied to cell terminals will damage the battery.
- When connecting cells, shorts across the terminals will produce arcing, which could cause an explosion.
- Cells should never be disconnected, as this can also produce arcing, which could cause an explosion.

FAQ

Q Is there any other advice you can give me about fault finding ?

A Apart from using the information given in this chapter, the best advice that I ever received from an electrician was to stay calm and treat fault finding as being a process of elimination and simply 'looking for the differences between what you should see and what you are actually looking at'. Listen carefully to any fault being explained to you, it may help establish the cause more quickly. Wherever possible, look at the fault being demonstrated and if no circuit drawings exist, then make your own. You can then use the drawing to prove wiring and then tick them off on the drawing as you find them and prove their connections.

Q Why all the fuss about safe isolation if they said it was switched off ?

A Never take someone else's word that a circuit or piece of equipment is safe to work on. You should always make sure that, before you start work on any piece of equipment, it is completely isolated from the supply in a safe and approved manner. You really don't want the supply re-energised to something you are working on...do you?

Therefore, when you are visiting a client's premises, investigate the means of isolation thoroughly. It's no good going into a factory to repair a machine where every machine is controlled by a separate supply that is fully interlocked and padlocked if, when given the padlock key for that machine, you then find that every member of the maintenance team has a key to open every padlock.

Equally, never trust circuit markings inside distribution boards. Just because you switched off the MCB that said it controlled the lighting, doesn't mean that it actually does.

Finally, always prove that a supply is not present with a suitable and functioning voltage indicator. Remember that the supply to a piece of equipment might pass through several other bits of equipment on its way to you and if any of them were switched off when you tested for a supply at the equipment, then you would not know whether you had isolated the correct circuit.

On the job: Fault diagnosis

You are asked by your employer to go to a factory unit to investigate a fault with a piece of equipment. When you arrive the customer takes you to the factory area and demonstrates the fault to you. The piece of equipment has a motor, which is fed from a distribution board via a DOL starter/isolator with a 230V coil, which is in turn linked to a remote stop/start control station located some 6 m away in a noise reduction area. The fault is that the motor will not start at all when controlled from the starter/isolator, but at the remote station when the start button is pressed, the motor starts but stops again as soon as you take your finger off the start button. The customer has a wiring diagram for the starter and remote control, so how would you try to solve this problem?

(A suggested answer can be found on page 333)

Knowledge check

1. What is the document published by the HSE that gives guidance for electrical test equipment used by electricians?
2. You are asked to repair a fluorescent lighting fitting that is fed from an MCB distribution board. Describe how you would safely isolate the circuit before starting work on it.
3. Give three means by which an installation designer can influence fault protection and damage limitation.
4. How would you define 'overload current', 'overcurrent', 'earth-fault current' and 'short circuit current'?
5. What is meant by 'transient voltages'?
6. How can regular inspection and maintenance assist in the prevention of faults?
7. A poor or loose connection can lead to what happening?
8. As an installation electrician you may be expected to find faults on a range of systems and equipment. However, what should you generally not attempt to diagnose faults on?
9. Knowledge of an installation is an essential requirement in fault diagnosis. Name four different people that you could approach to gain knowledge of an installation and a fault.
10. Describe the logical stages of fault diagnosis and rectification.

Chapter 7

Motors, generators and motor control

OVERVIEW

In essence there are two categories of motor: those that run on direct current (d.c.) and those that run on alternating current (a.c.).

To understand motors and generators we need to remember the basic principles of magnetic fields, current flow and induced motion and how when used together they can make a motor rotate.

These individual areas were covered in the Science section of Book 1, but we will look at them again in the context of their relationship to motors. This chapter will therefore cover the following:

- Concept revision
- The d.c. motor
- The d.c. and a.c. generator
- The a.c. motor
- Motor construction
- Motor speed and slip calculation
- Synchronous a.c. induction motors
- Motor windings
- Motor starters
- Motor speed control

Concept revision

Magnetic fields

Between the poles of a simple permanent magnet, there is a magnetic field. The direction of the magnetic field moves from the north pole of the magnet to the south pole in closed loops. Like poles repel and unlike poles attract. The magnetic field is represented by lines of force – the magnetic flux (remember the iron filings experiment referred to in chapter 3 of Book 1).

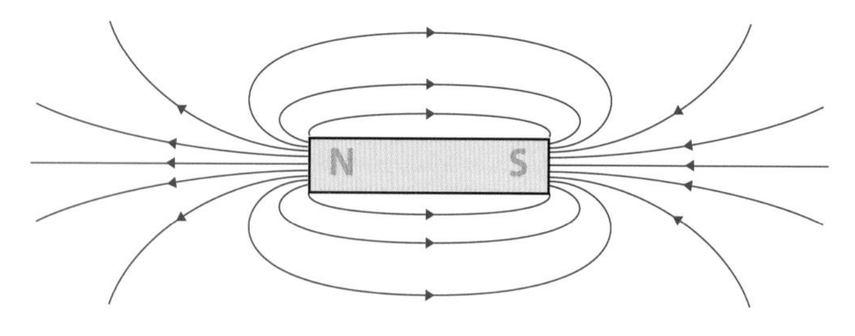

Figure 7.1 Magnetic field around a bar magnet

Current flow and magnetic fields

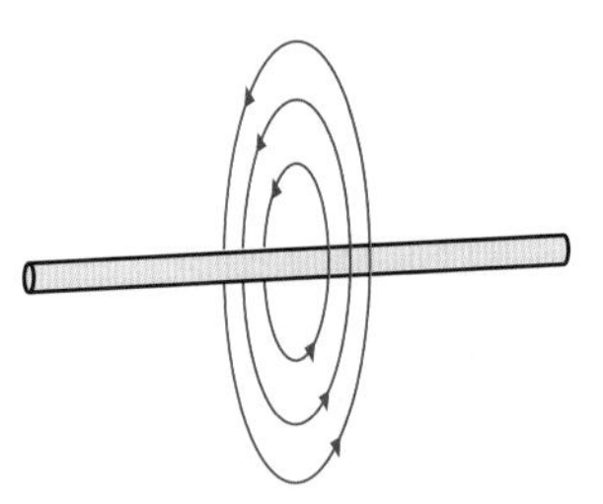

Figure 7.2 Magnetic field around a conductor

Now, let's consider a conductor with an electric current flowing through it. When this happens a magnetic field will be produced in the form of concentric circles along the length of the conductor.

Force on a current-carrying conductor within a magnetic field

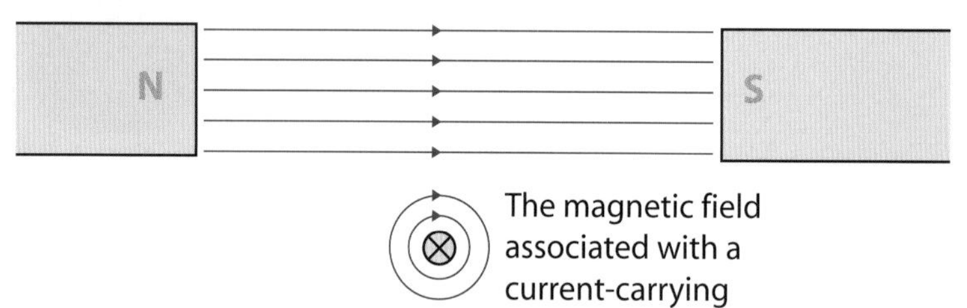

Figure 7.3 The magnetic field associated with two fixed poles

We now know that when a current is present in a conductor a magnetic field is set up around that conductor, always in a clockwise direction, in relation to the direction of current flow. Nearly all motors work on the basic principle that when a current-carrying conductor is placed in a magnetic field it experiences a force. This electromagnetic force is shown in the following diagrams.

If we place the current-carrying conductor into the magnetic field as shown in Figure 7.3, you can see that the current is going away from you; therefore the field is clockwise. This can be best remembered by imagining a corkscrew being twisted into a cork; you need to turn the corkscrew to the right (this symbolises the magnetic field) and the corkscrew is moving away from you, which symbolises the current flowing away from you.

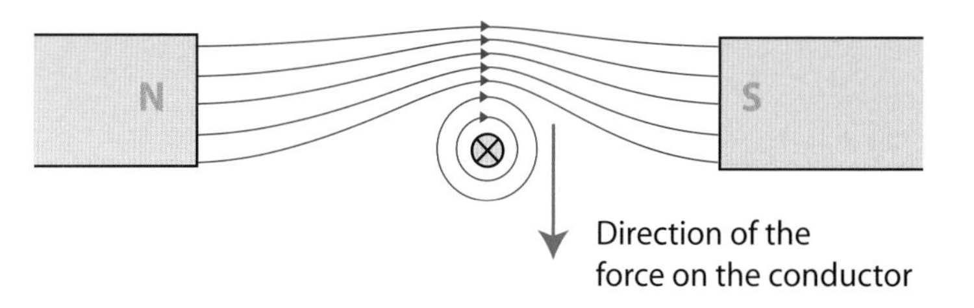

Figure 7.4 The magnetic field when the conductor is placed between the poles

In Figure 7.4 we should now note the following:

- The main field now becomes distorted.
- The field is weaker below the conductor because the two fields are in opposition.
- The field is stronger above the conductor because the two fields are in the same direction and aid each other. Consequently the force moves the conductor downwards.

Remember

⊗ indicates the current flowing away from you

⊙ indicates the current is flowing towards you

In Figure 7.5 we can see that if either the current through the conductor or the direction of the magnetic field between the poles is reversed, the force acting on the conductor now tends to move it in the reverse direction.

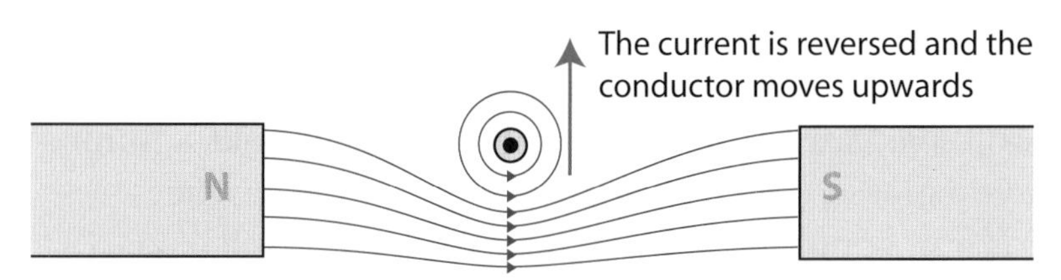

Figure 7.5 Effect of reversing the direction of current or magnetic field

Fleming's left-hand rule

The direction in which a current-carrying conductor tends to move when it is placed in a magnetic field can be determined by **Fleming's left-hand (motor) rule**.

This rule states: if the first finger, the second finger and the thumb of the left hand are held at right angles to each other, as shown in Figure 7.6, then with the first finger pointing in the direction of the Field (N to S) and the second finger pointing in the direction of the current in the conductor, the thumb will now be indicating the direction in which the conductor tends to move.

First finger pointing in the direction of the field (N to S)

Second finger pointing in the direction of the current in the conductor

Thumb points in the direction in which the conductor tends to move

Figure 7.6 Fleming's left-hand (motor) rule

It is this basic principle of a force being exerted on a current-carrying conductor in a magnetic field that is used in the construction of motors and moving coil instruments.

So let's now look at a simple motor that operates using direct current (d.c.), in other words current flowing continuously in the same direction along a conductor, such as that produced by a battery.

Remember

The M in thuMb stands for Motion, the F in First finger stands for Field and the C in seCond finger stands for Current

The d.c. motor

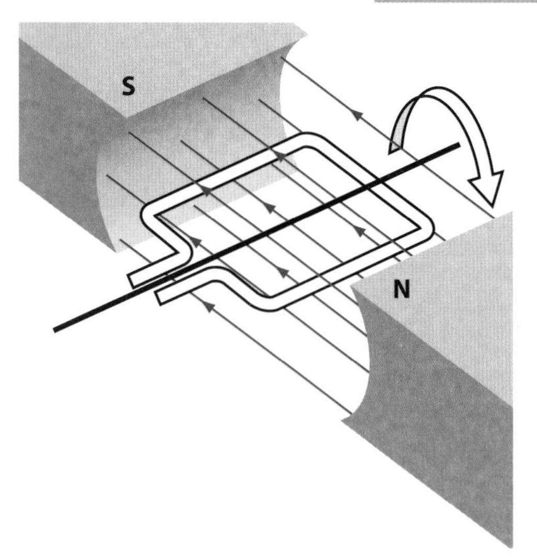

Figure 7.7 Single current-carrying loop in a magnetic field

The operating principle of a d.c. motor is fairly straightforward. Looking at Figure 7.7, if we take a conductor, form it into a loop that is pivoted at its centre, place it within a magnetic field and then apply a d.c. supply to it, a current will pass around the loop. This will cause a magnetic field to be produced around the conductor, and this magnetic field will interact with the magnetic field between the two poles (**pole pair**) of the magnet.

The magnetic field between the two poles (pole pair) becomes distorted and, as mentioned earlier, this interaction between the two magnetic fields causes a bending or stretching of the lines of force. The lines of force behave a bit like an elastic band and consequently are always trying to find the shortest distance between the pole pair. These stretched lines, in their attempt to return to their shortest length, therefore exert a force on the conductor that tries to push it out of the magnetic field, and thus they cause the loop to rotate.

In reality we don't have just one loop, we have many, with them being wound on to a central rotating part of the d.c. motor known as the **armature**. When the armature is positioned so that the loop sides are at right angles to the magnetic field, a turning force is exerted. But we have a problem when the coil has rotated 180°, as the magnetic field in the loop is now opposite to that of the field, and this will tend to push the armature back the way it came, thus stopping the rotating motion.

The solution is to reverse the current in the armature every half rotation so that the magnetic fields will work together to maintain a continuous rotating motion. The device that we use to achieve this switching of polarity is known as the **commutator**.

The commutator

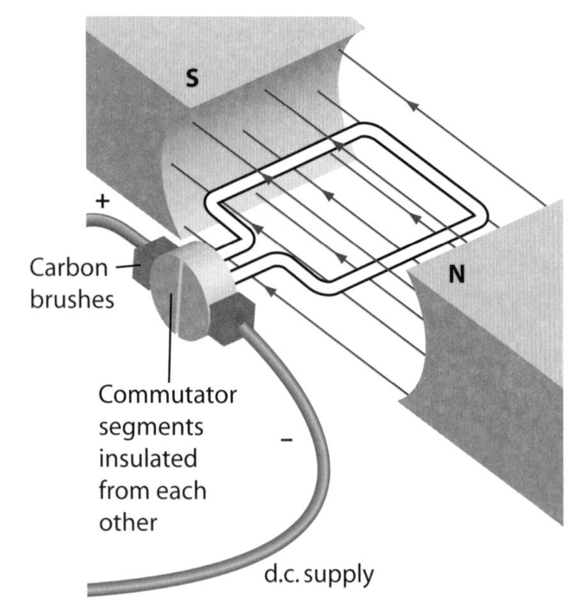

Figure 7.8 Single loop with commutator

Figure 7.8 now gives a different perspective of the same arrangement, showing the device that enables the supply to be connected to the loop, enabling the loop to rotate continuously. This is the commutator. This simplified version has only two segments, connecting to either side of the loop, but in actual machines there may be any number. Large machines would have in excess of 50, and the numbers of loops can be in the hundreds. The commutator is made of copper, with the segments separated from each other by insulation.

Figure 7.9 now shows the direction of current around the loop when connected to a d.c. supply. The brushes remain in a fixed position, against the copper segments of the commutator. Note the direction of current in those sections indicated by X and Y in the diagram.

Figure 7.10 now shows the wire loop having rotated 180°. X and Y have changed positions but, as can be seen from the arrows, current flow remains as it was for the previous diagram. As the poles of the armature electromagnet pass the poles of the permanent magnets, the commutator reverses the polarity of the armature electromagnet. In that instant of switching polarity, inertia keeps the motor going in the proper direction and thus the motor continues to rotate in one direction.

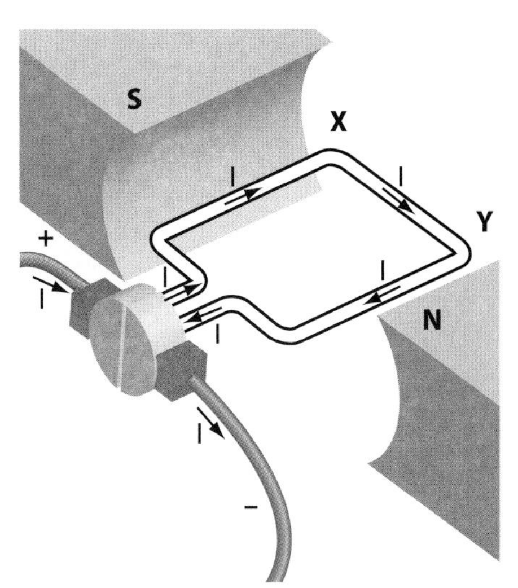

Figure 7.9 Single loop with d.c. flowing

In the previous drawings, we have shown the armature rotating between a pair of magnetic poles. Practical d.c. motors do not use permanent magnets, but use electromagnets instead. The electromagnet has two advantages over the permanent magnet:

1. By adjusting the amount of current flowing through the wire the strength of the electromagnet can be controlled.
2. By changing the direction of current flow the poles of the electromagnet can be reversed.

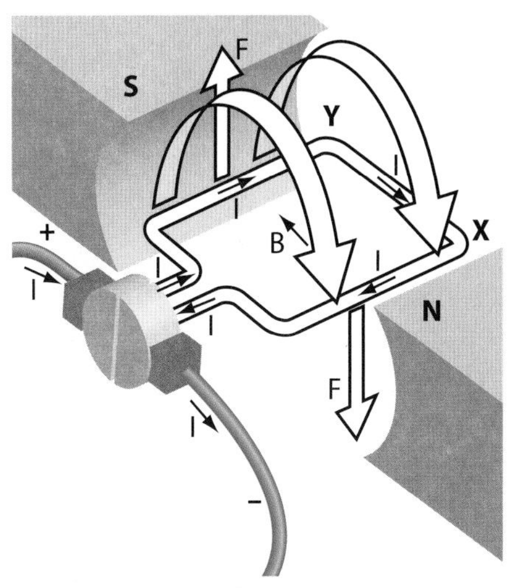

Figure 7.10 Single loop rotated 180°

Reversing a d.c. motor

The direction of rotation of a d.c. motor may be reversed by either:

- reversing the direction of the current through the field hence changing the field polarity.
- reversing the direction of the current through the armature.

Common practice is to reverse the current through the armature, and this is normally achieved by reversing the armature connections only.

Types of d.c. motor

There are three basic forms of d.c. motor:

- series
- shunt
- compound.

They are very similar to look at, the difference being the way in which the field coil and armature coil circuits are wired.

The series motor has the field coil wired in series with the armature. It is also called a universal motor because it can be used in both d.c. and a.c. situations, has a high starting torque (rotational force) and a variable speed characteristic. The motor can therefore start heavy loads, but the speed will increase as the load is decreased.

The shunt motor has the armature and field circuits wired in parallel, and this gives constant field strength and motor speed.

The compound motor combines the characteristics of both the series and the shunt motors and thus has high starting torque and fairly good speed torque characteristics. However, because it is complex to control, this arrangement is usually only used on large bi-directional motors.

The d.c. and a.c. generator

We now know that when a current is present in a conductor a magnetic field is set up around that conductor, always in a clockwise direction, in relation to the direction of current flow.

We also know that when a conductor passes at right angles through a magnetic field, current is induced into the conductor; the direction of the induced current will depend on the direction of movement of the conductor, and the strength of the current will be determined by the speed at which the conductor moves.

It therefore follows that if we were to take this arrangement and connect it to some device that would spin the wire loop within the permanent magnetic field, we would then induce into the wire loop an e.m.f., and were we to connect a load to the armature via the commutator and brushes, a current would flow around the circuit and the load would work. In other words, we would have created a generator.

A variety of sources can be used mechanically to turn the generator's armature, such as steam, wind, water fall or petrol/diesel driven motors. Figure 7.11 below shows just such an arrangement and the voltage output for one complete revolution.

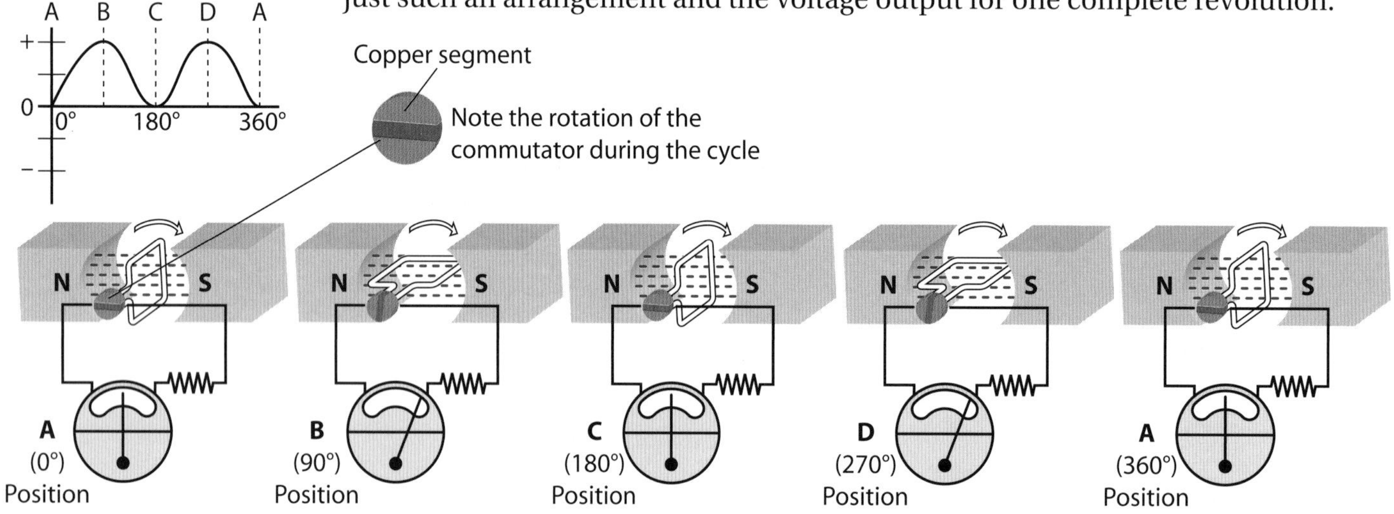

Figure 7.11 Voltage output for one complete revolution

As you can see from Figure 7.11, the output from such a generator has no negative parts in its cycle. This type of generator produces a voltage/current that alternates in magnitude but flows in one direction only: in other words we have a direct current (d.c.).

When a d.c. generator contains only a single coil it provides a pulsating d.c. output, as shown by the wave form in Figure 7.11. Consequently, in general use a number of coils are used to produce a more stable output.

The operating principle of the a.c. generator is much the same as that of the d.c. version. However, instead of the loop ends terminating at the commutator, they are terminated at slip rings. The a.c. generator normally has a stationary armature known as the **stator** and a rotating magnetic field, known as the **rotor**.

The a.c. motor

A simple a.c. motor

As we mentioned earlier, there are three general types of d.c. motor. However, there are many types of a.c. motor, each one having a specific set of operating characteristics such as **torque**, speed, single-phase or three-phase, and this determines their selection for use. We can essentially group them in to two categories: single-phase and three-phase.

In a d.c. motor, electrical power is conducted directly to the armature through brushes and a commutator. Due to the nature of an alternating current, an a.c. motor doesn't need a commutator to reverse the polarity of the current. Whereas a d.c. motor works by changing the polarity of the current running through the armature (the rotating part of the motor), the a.c. motor works by changing the polarity of the current running through the stator (the stationary part of the motor).

The series-wound (universal) motor

We will begin the discussion of a.c. motors by looking at the series-wound (universal) motor because it is different in its construction and operation from the other a.c. motors considered here, and also because it is constructed as we have previously discussed in the Types of d.c. motor section, having field windings, brushes, commutator and an armature.

As can be seen in Figure 7.12, because of its series connection, current passing through the field windings also passes through the armature. The turning motion (torque) is produced as a result of the interaction between the magnetic field in the field windings and the magnetic field produced in the armature.

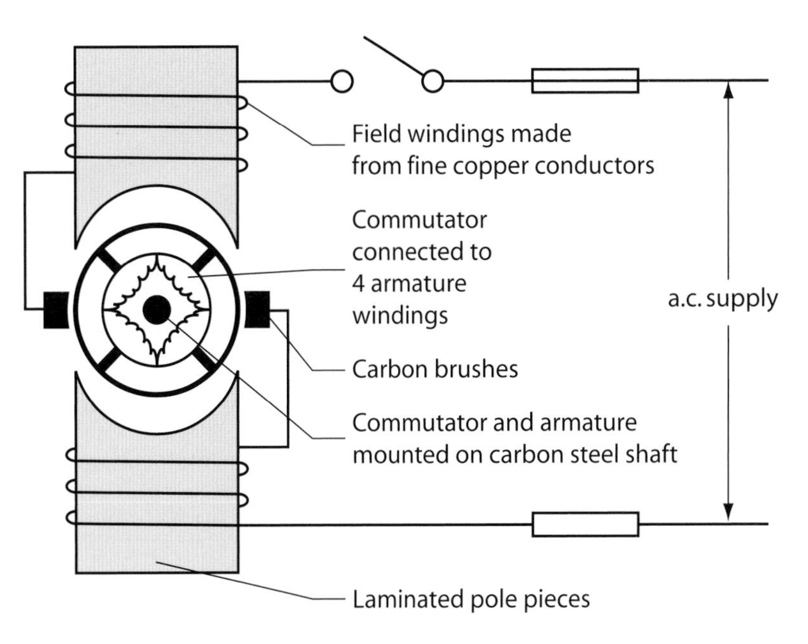

Figure 7.12 Series universal motor

For this motor to be able to run on an a.c. supply, modifications are made both to the field windings and armature **formers**. These are heavily laminated to reduce eddy currents and I^2R losses, which reduces the heat generated by the normal working of the motor, thus making the motor more efficient.

This type of motor is generally small (less than a kilowatt) and is used to drive small hand tools such as drills, vacuum cleaners and washing machines.

A disadvantage of this motor is that it relies on contact with the armature via a system of carbon brushes and a commutator. It is this point that is the machine's weakness, as much heat is generated through the arcing that appears across the gap between the brushes and the commutator. The brushes are spring-loaded to keep this gap to a minimum, but even so the heat and friction eventually cause the brushes to wear down and the gap to increase. These then need to be replaced, otherwise the heat generated as the gap gets larger will eventually cause the motor to fail.

The advantages of this machine are:

- more power for a given size than any other normal a.c. motor
- high starting torque
- relative cheapness to produce.

Three-phase induction motors

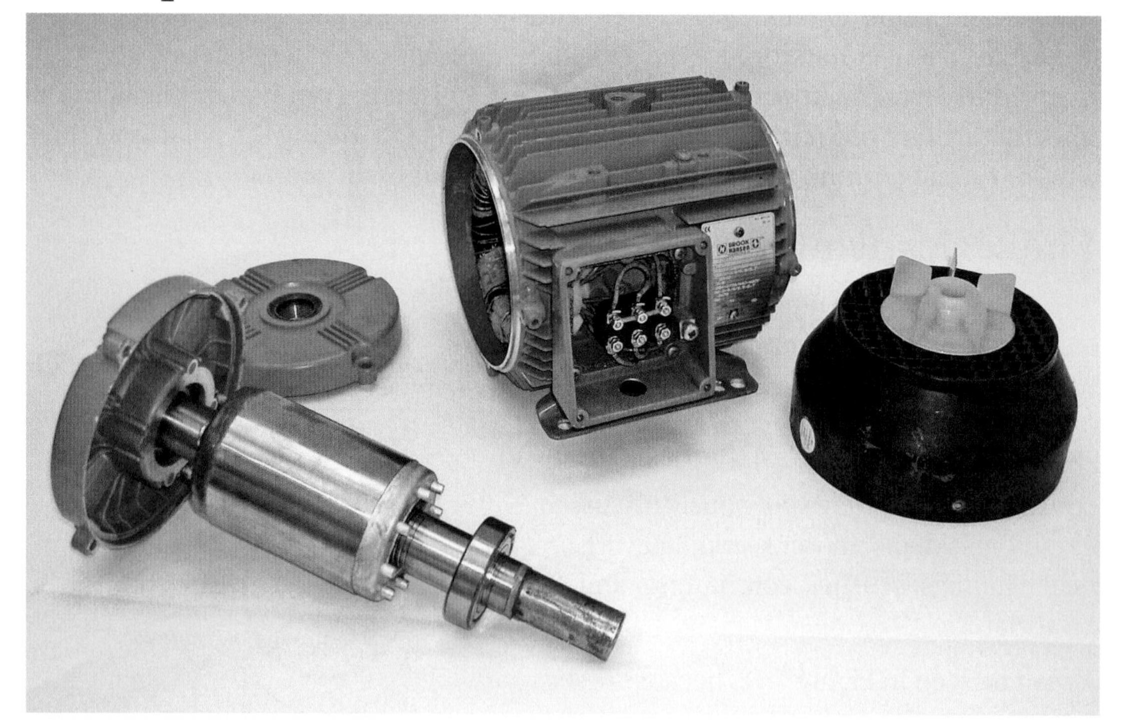

Three-phase induction motor showing component parts

Induction motors operate because a moving magnetic field induces a current to flow in the rotor. This current in the rotor then creates a second magnetic field, which combines with the field from the stator windings to exert a force on the rotor conductors, thus turning the rotor.

Production of the rotating field

Figure 7.13 shows the stator of a three-phase motor to which a three-phase supply is connected. The windings in the diagram are in star formation and two windings of each phase are wound in the same direction.

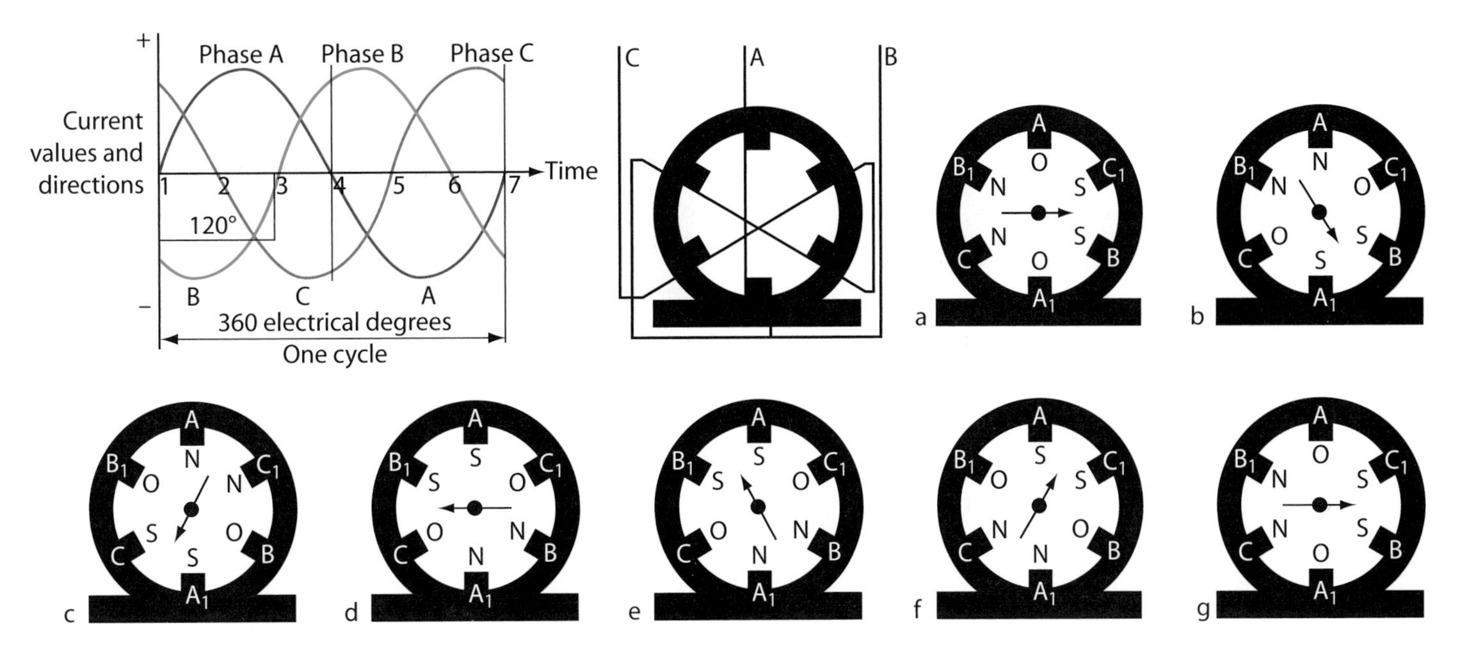

Figure 7.13 Stator of three-phase motor with three-phase supply connection

Each pair of windings will produce a magnetic field, the strength of which will depend upon the current in that particular phase at any instant of time. When the current is zero, the magnetic field will be zero. Maximum current will produce the maximum magnetic field.

As the currents in the three phases are 120° out of phase (see graph in Figure 7.13) the magnetic fields produced will also be 120° out of phase. The magnetic field set up by the three phase currents will therefore give the appearance of rotating clockwise around the stator.

The resultant magnetic field produced by the three phases is at any instant of time in the direction shown by the arrow in the diagram, where diagrams (1) to (7) in Figure 7.13 show how the direction of the magnetic field changes at intervals of 60° through one complete cycle. The speed of rotation of the magnetic field depends upon the supply frequency and the number of 'pole pairs', and is referred to as the **synchronous** speed. We will discuss synchronous speed later in this chapter.

The direction in which the magnetic field rotates is dependent upon the sequence in which the phases are connected to the windings. Reversing the connection of any two incoming phases can therefore reverse rotation of the magnetic field.

Stator construction

Field winding
Squirrel cage rotor
Rotor shaft
Steel frame or yoke

Figure 7.14 Stator construction

Stator field winding

As shown in Figure 7.14, the stator (stationary component) comprises the field windings, which are many turns of very fine copper wire wound on to formers, which are then fixed to the inside of the stator steel frame (sometimes called the yoke).

The formers have two roles:

1. to contain the conductors of the winding
2. to concentrate the magnetic lines of flux to improve the flux linkage.

The formers are made of laminated silicon steel sections to reduce eddy currents, thereby reducing the I^2R losses and reducing heat. The number of poles fitted will determine the speed of the motor.

Rotor construction and principle of operation

Essentially there are two main types of rotor:

1. squirrel-cage rotor
2. wound rotor.

Squirrel-cage rotor

In the squirrel cage (see Figure 7.15), the bars of the rotor are shorted out at each end by 'end rings' to form the shape of a cage. This shape creates numerous circuits within it for the induced e.m.f. and resultant current to flow and thus produce the required magnetic field.

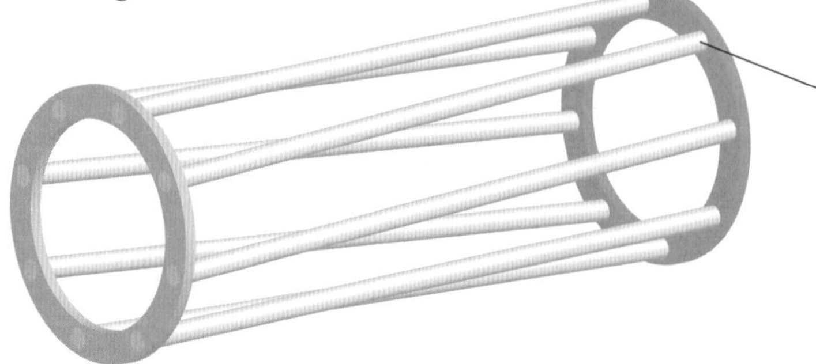

Tinned bars shorted out at each end by a tinned copper end ring

Figure 7.15 Squirrel-cage rotor

Figure 7.16 shows the cage fitted to the shaft of the motor. The rotor bars are encased within many hundreds of very thin laminated (insulated) segments of silicon steel and are skewed to increase the rotor resistance.

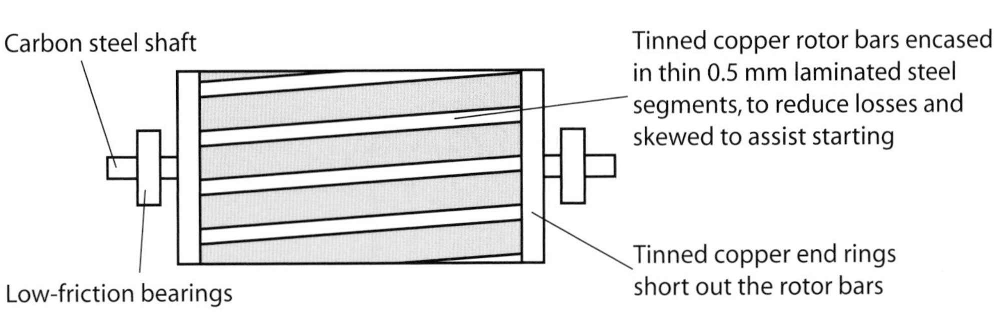

Figure 7.16 Cage fitted to shaft and motor

On the shaft will be two low-friction bearings which enable the rotor to spin freely. The bearings and the rotor will be held in place within the yoke of the stator by two end caps which are normally secured in place by long nuts and bolts that pass completely through the stator.

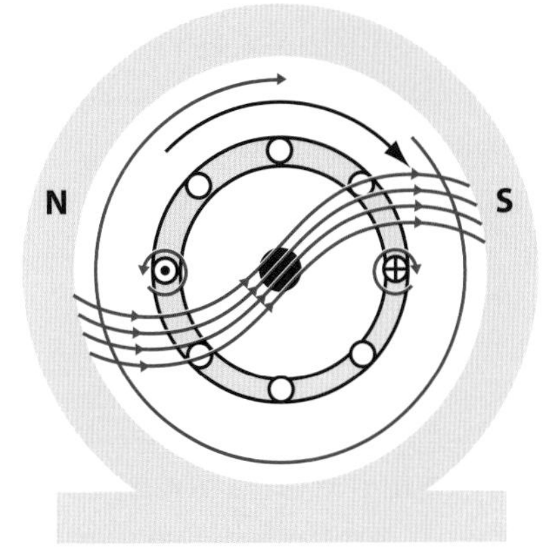

Figure 7.17 Rotating field

When a three-phase supply is connected to the field windings, the lines of magnetic flux produced in the stator, rotating at 50 revolutions per second, cut through the bars of the rotor, inducing an e.m.f. into the bars.

Faraday's Law states that 'when a conductor cuts, or is cut, by a magnetic field, then an e.m.f. is induced in that conductor the magnitude of which is proportional to the rate at which the conductor cuts or is cut by the magnetic flux.'

The e.m.f. produces circulatory currents within the rotor bars, which in turn result in the production of a magnetic field around the bars. This leads to the distortion of the magnetic field as shown in Figure 7.17. The interaction of these two magnetic fields results in a force being applied to the rotor bars, and the rotor begins to turn. This turning force is known as a torque, the direction of which is always as Fleming's left-hand rule indicates.

Wound rotor

In the wound rotor type of motor, the rotor conductors form a three-phase winding, which is starred internally. The other three ends of the windings are brought out to slip rings mounted on the shaft. Thus it is possible through brush connections to introduce resistance into the rotor circuit, albeit this is normally done on starting only to increase the starting torque. This type of motor is commonly referred to as a slip-ring motor.

Squirrel cage induction motors are sometimes referred to as the 'workhorse' of the industry, as they are inexpensive and reliable and suited to most applications

Figure 7.18 shows a completed wound-rotor motor assembly. Although it looks like a squirrel-cage motor, the difference is that the rotor bars are exchanged for heavy conductors that run through the laminated steel rotor, the ends then being brought out through the shaft to the slip rings on the end.

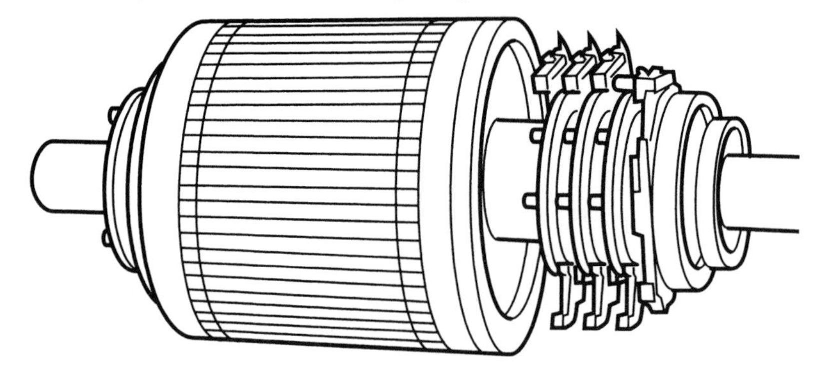

Figure 7.18 Wound-rotor motor assembly

The wound-rotor motor is particularly effective in applications where using a squirrel-cage motor may result in a starting current that's too high for the capacity of the power system. The wound-rotor motor is also appropriate for high-inertia loads having a long acceleration time. This is because you can control the speed, torque and resulting heating of the wound-rotor motor. This control can be automatic or manual. It's also effective with high-slip loads as well as adjustable-speed installations that do not require precise speed control or regulation. Typical applications include conveyor belts, hoists and elevators.

Single-phase induction motors

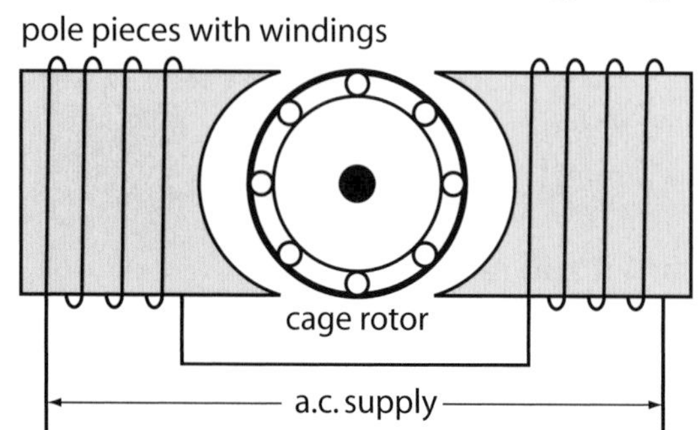

Figure 7.19 Single-phase induction motor

If we were to construct an induction motor as shown in Figure 7.19, we would find that, on connecting a supply to it, it would not run. However, if we were then to spin the shaft with our fingers, we would find that the motor *would* continue to run. Why is this?

When an a.c. supply is connected to the motor, the resulting current flow, and therefore the magnetic fields produced in the field windings, changes polarity, backwards and forwards, 100 times per second. Therefore no lines of flux cut through the rotor bars. If no lines of flux cut through the rotor bars then there is no e.m.f. being produced in them and there is therefore no magnetic field for the stator winding to interact with.

However, if we spin the rotor, we create the effect of the rotor bars cutting through the lines of force, hence the process starts and the motor runs up to speed. If we stop the motor and then connect the supply again, the motor will still not run. If this time we spin the rotor in the opposite direction, the motor will run up to speed in the new direction. So how can we get the rotor to turn on its own?

If we think back to the three-phase motor we discussed previously, we did not have this problem, because the connection of a three-phase supply to the stator automatically produced a rotating magnetic field. This is what is missing from the motor in Figure 7.19: we have no rotating field.

The split-phase motor (induction start/induction run)

We can overcome this problem if we add another set of poles, positioned 90° around the stator from our original wiring, as shown in Figure 7.20.

Now when the supply is connected, both sets of windings are energised, both windings having resistive and reactive components to them – resistance as every conductor has and also inductive reactance because the conductors form a coil. (These concepts were discussed more fully in Chapter 1 Electrical science.) These are known as the 'start' and 'run' windings.

The start winding is wound with fewer turns of smaller wire than the main winding, so it has a higher resistance. This extra resistance creates a small phase shift that results in the current in the start winding lagging that in the run winding by approximately 30°, as shown in Figure 7.21.

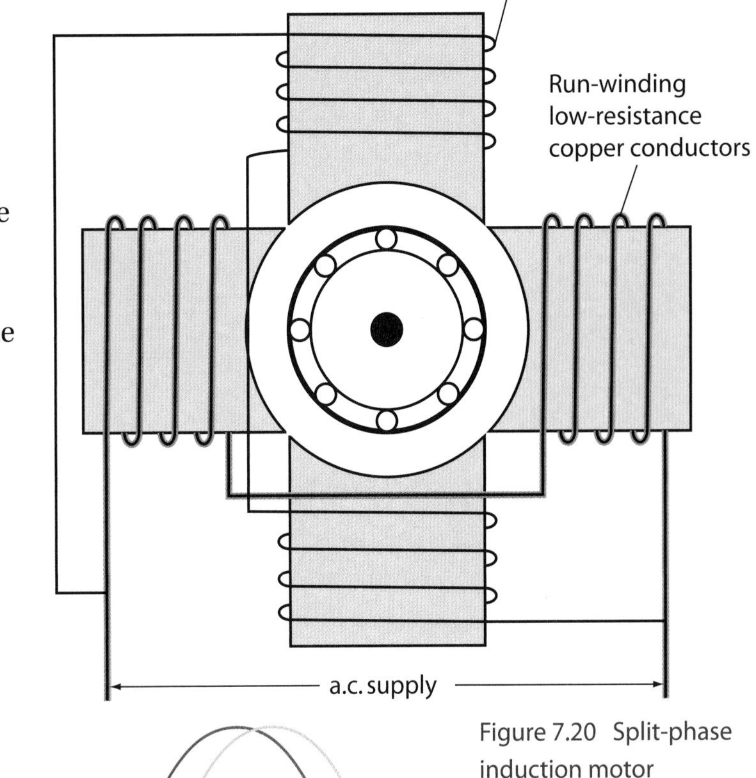

Figure 7.20 Split-phase induction motor

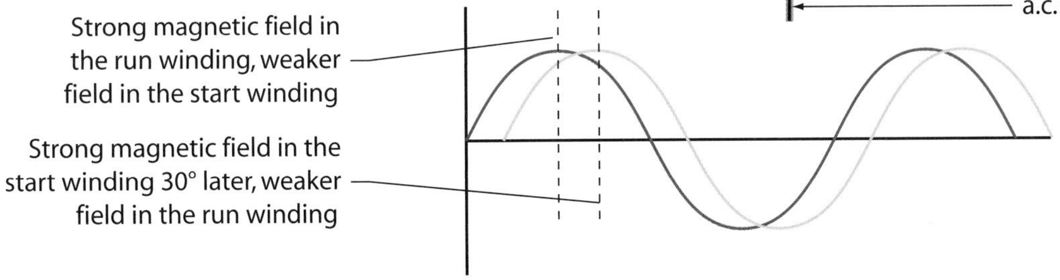

Strong magnetic field in the run winding, weaker field in the start winding

Strong magnetic field in the start winding 30° later, weaker field in the run winding

Figure 7.21 Run and start winding phases

Therefore the magnetic flux in each of the windings is growing and collapsing at different periods in time, so that for example as the run winding is having a strong north/south on the face of its pole pieces, the start winding will only have a weak magnetic field.

In the next instance, the run winding's magnetic field has started to fade, but the start winding's magnetic field is now strong, presenting to the rotor an apparent shift in the lines of magnetic flux.

In the next half of the supply cycle, the polarity is reversed and the process repeated, so there now appears to be a rotating magnetic field. The lines of force cutting through the rotor bars induce an e.m.f. into them, and the resulting current flow now produces a magnetic field around the rotor bars: the interaction between the magnetic fields of the rotor and stator takes place and the motor begins to turn.

It is because the start and run windings carry currents that are out of phase with each other that this type of motor is called the 'split-phase'.

Once the motor is rotating at about 75 per cent of its full load speed, the start winding is disconnected by the use of a device called a centrifugal switch, which is attached to the shaft; see Figure 7.22.

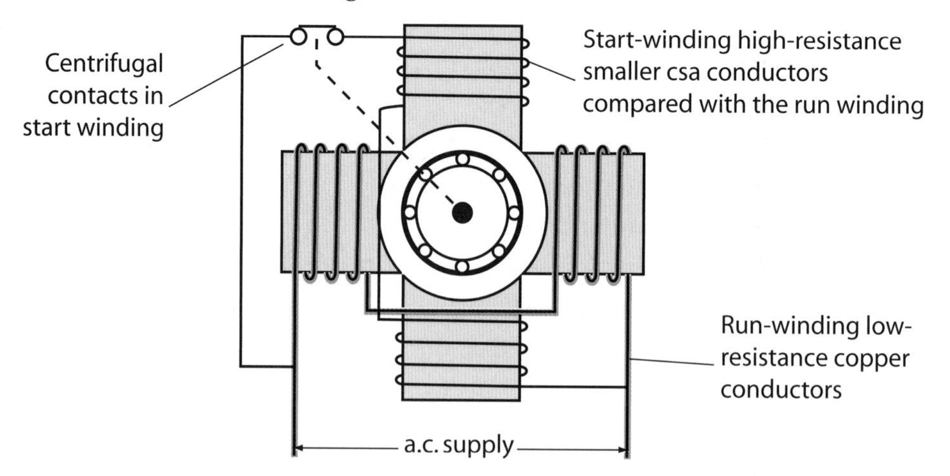

Figure 7.22 Split-phase induction motor with centrifugal switch

This switch works by centrifugal action, in that sets of contacts are held closed by a spring, and this completes the circuit to the start winding. When the motor starts to turn, a little weight gets progressively thrown away from the shaft, forcing the contacts to open and thus disconnecting the start winding. It's a bit like the fairground ride known as 'the Rotor', in which you are eventually held against the sides of the ride by the increasing speed of the spinning wheel. Once the machine has disconnected the start winding, the machine continues to operate from the run winding.

The split-phase motor's simple design makes it typically less expensive than other single-phase motors. However, it also limits performance. Starting torque is low at about 150 per cent to 175 per cent of the rated load. Also, the motor develops high starting currents of about six to nine times the full load current. A lengthy starting time can cause the start winding to overheat and fail, and therefore this type of motor shouldn't be used when a high starting torque is needed. Consequently it is used on light-load applications such as small hand tools, small grinders and fans, where there are frequent stop/starts and the full load is applied after the motor has reached its operating speed.

Reversal of direction

If you think back to the start of this section, we talked about starting the motor by spinning the shaft. We also said that we were able to spin it in either direction and the motor would run in that direction. It therefore seems logical that in order to change the direction of the motor, all we have to concern ourselves with is the start winding. We therefore need only to reverse the connections to the start winding to change its polarity, although you may choose to reverse the polarity of the run winding instead. The important thing to remember is that if you change the polarity through both the run and start windings, the motor will continue to revolve in the same direction.

The capacitor-start motor (capacitor start/induction run)

Normally perceived as being a wide-ranging industrial motor, the capacitor-start motor is very similar to the split-phase motor discussed previously. Indeed, it probably helps to think of this motor as being a split-phase motor but with an enhanced start winding that includes a capacitor in the circuit to help out with the start process. If we look at the photograph we can see the capacitor mounted on top of the motor case.

Capacitor-start motor

In this motor the start winding has a capacitor connected in series with it, and since this gives a phase difference of nearly 90° between the two currents in the windings, the starting performance is improved. We can see this represented in the sine waves shown in Figure 7.23.

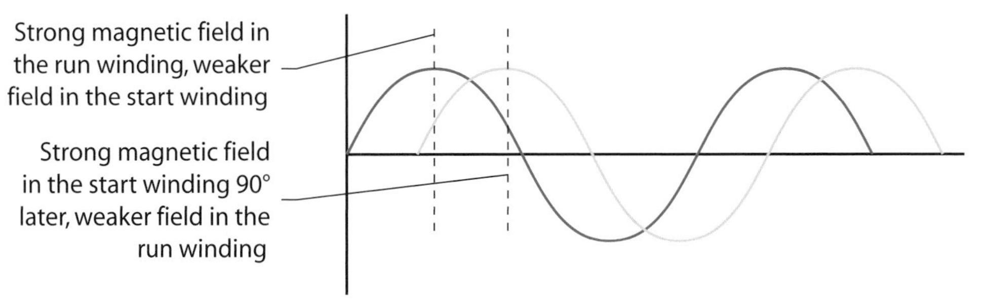

Figure 7.23 Magnetic field in capacitor-start motor

In this motor the current through the run winding lags the supply voltage due to the high inductive reactance of this winding, and the current through the start winding leads the supply voltage due to the capacitive reactance of the capacitor. The phase displacement in the currents of the two windings is now approximately 90°. Figure 7.24 shows the winding connections for a capacitor-start motor.

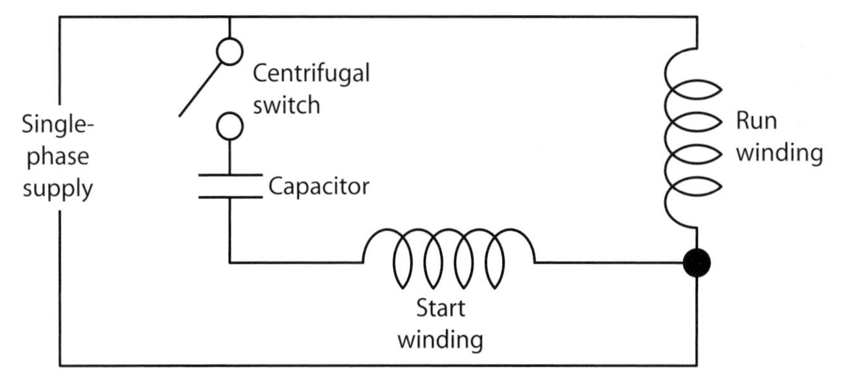

Figure 7.24 Winding connections for capacitor-start split-phase motor

The magnetic flux set up by the two windings is much greater at starting than in the standard split-phase motor, and this produces a greater starting torque. The typical starting torque for this type of motor is about 300 per cent of full-load torque, and a typical starting current is about five to nine times the full-load current.

The capacitor-start motor is more expensive than a comparable split-phase design because of the additional cost of the capacitor. But the application range is much wider because of higher starting torque and lower starting current. Therefore, because of its improved starting ability, this type of motor is recommended for loads that are hard to start, so we see this type of motor used to drive equipment such as lathes, compressors and small conveyor systems.

As with the standard split-phase motor, the start windings and the capacitor are disconnected from the circuit by an automatic centrifugal switch when the motor reaches about 75 per cent of its rated full-load speed.

Reversal of direction

Reversing the connections to the start winding will only change its polarity, although we may choose to reverse the polarity of the run winding instead.

Permanent split capacitor (PSC) motors

Permanent split capacitor (PSC) motors look exactly the same as capacitor-start motors. However, a PSC motor doesn't have either a starting switch or a capacitor that is strictly used for starting. Instead, it has a run-type capacitor permanently connected in series with the start winding, and the second winding is permanently connected to the power source. This makes the start winding an auxiliary winding once the motor reaches running speed. However, because the run capacitor must be designed for continuous use, it cannot provide the starting boost of a starting capacitor.

Typical starting torques for thus type of motor are low, from 30 to 150 per cent of rated load, so these motors are not used in difficult starting applications. However, unlike the split-phase motor, PSC motors have low starting currents, usually less than 200 per cent of rated load current, making them excellent for applications with high cycle rates.

PSC motors have several advantages. They need no starting mechanism and so can be reversed easily, and designs can easily be altered for use with speed controllers. They can also be designed for optimum efficiency and high power factor at rated load.

Permanent split capacitor motors have a wide variety of applications depending on the design. These include fans, blowers with low starting torque and intermittent cycling uses such as adjusting mechanisms, gate operators and garage-door openers, many of which also need instant reversing.

Capacitor start–capacitor run motors

Capacitor start–capacitor run motor

In appearance we can distinguish this motor because of the two capacitors that are mounted on the motor case.

This type of motor is widely held to be the most efficient single-phase induction motor, as it combines the best of the capacitor-start and the permanent split capacitor designs and is able to handle applications too demanding for any other kind of single-phase motor.

As shown in Figure 7.25, it has a start capacitor in series with the auxiliary winding like the capacitor-start motor, and this allows for high starting torque. However, like the PSC motor, it also has a run capacitor that remains in series with the auxiliary winding after the start capacitor is switched out of the circuit.

Another advantage of this type of motor is that it can be designed for lower full-load currents and higher efficiency, which means that it operates at a lower temperature than other single-phase motor types of comparable horsepower. Typical uses include woodworking machinery, air compressors, high-pressure water pumps, vacuum pumps and other high-torque applications.

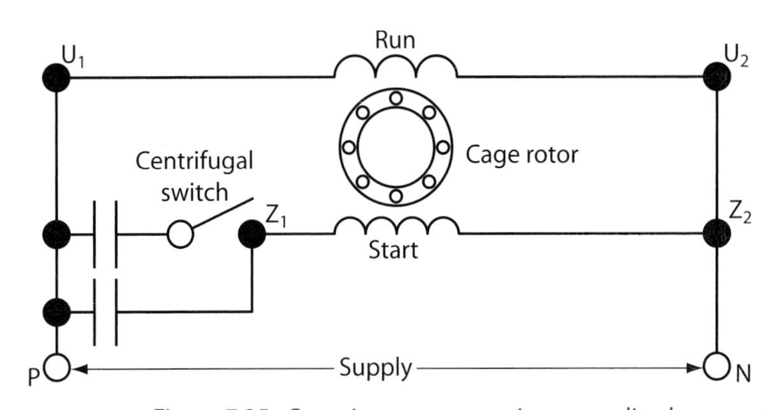

Figure 7.25 Capacitor start–capacitor run split-phase motor

Shaded pole motors

One final type of single-phase induction motor that is worthy of mention is the shaded-pole type. We cover this last as, unlike all of the previous single-phase motors we have discussed, shaded-pole motors have only one main winding and no start winding.

Starting is by means of a continuous copper loop wound around a small section of each motor pole. This 'shades' that portion of the pole, causing the magnetic field in the ringed area to lag the field in the non-ringed section. The reaction of the two fields then starts the shaft rotating.

Because the shaded pole motor lacks a start winding, starting switch or capacitor, it is electrically simple and inexpensive. In addition, speed can be controlled merely by varying voltage or through a multi-tap winding.

The shaded pole motor has many positive features, but it also has several disadvantages. As the phase displacement is very small, it has a low starting torque, typically in the region of 25 to 75 per cent of full-load torque. Also, it is very inefficient, usually below 20 per cent.

Low initial costs make shaded pole motors suitable for light-duty applications such as multi-speed fans for household use and record turntables.

Motor construction

To help you understand how motors are put together, this section includes photographs and diagrams of the components used in motor construction.

The three-phase squirrel-cage induction motor

Components of a three-phase squirrel-cage induction motor

Field winding

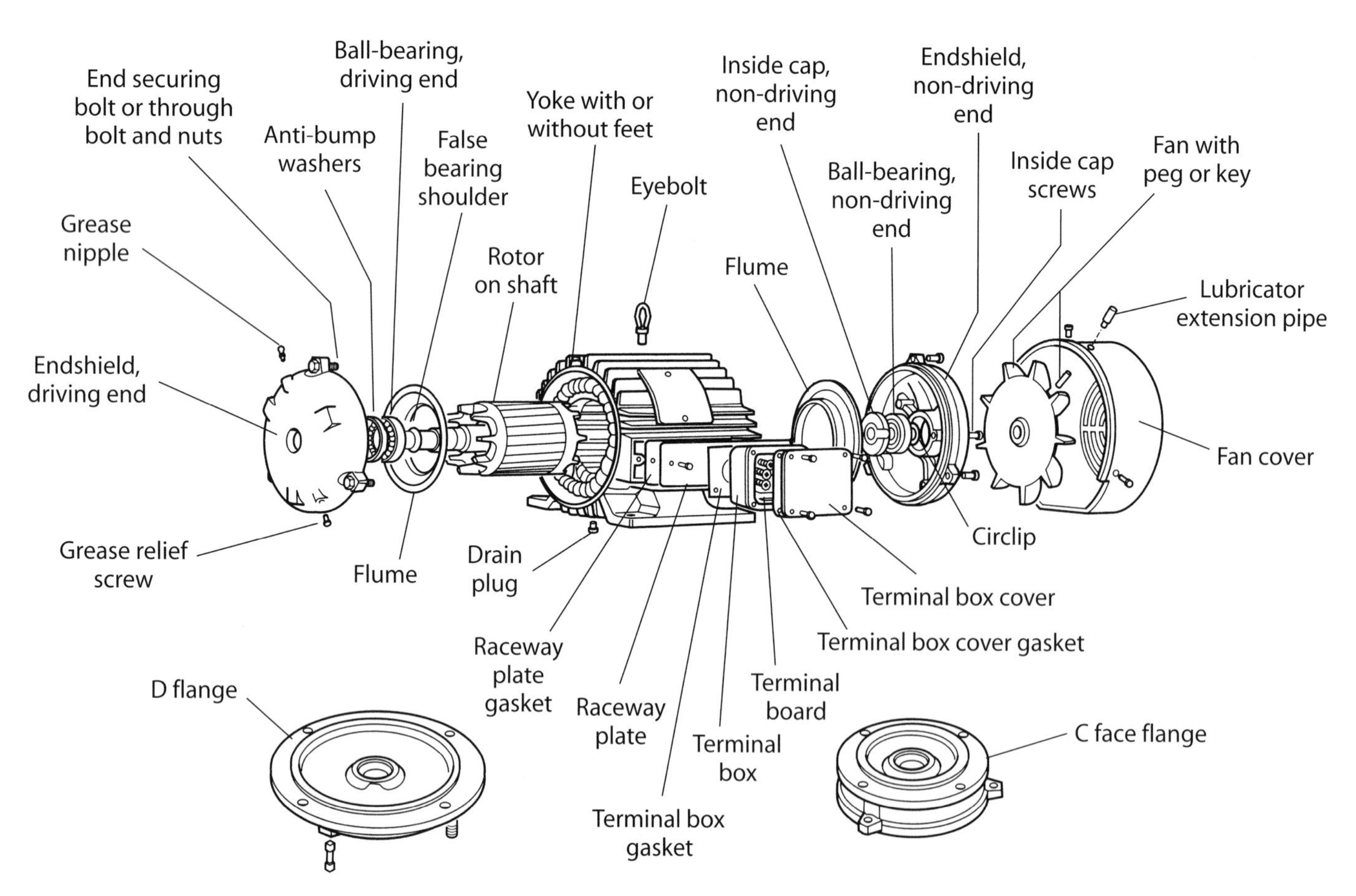

Figure 7.26 Construction of a three-phase squirrel-cage induction motor

The single-phase induction motor

This motor consists of a laminated stator wound with single-phase windings arranged for split-phase starting, and a cage rotor. The cage rotor is a laminated framework with lightly insulated longitudinal slots into which copper or aluminium bars are fitted.

The bars are then connected together at their ends by metal end-rings. No electrical connections are made to the rotor.

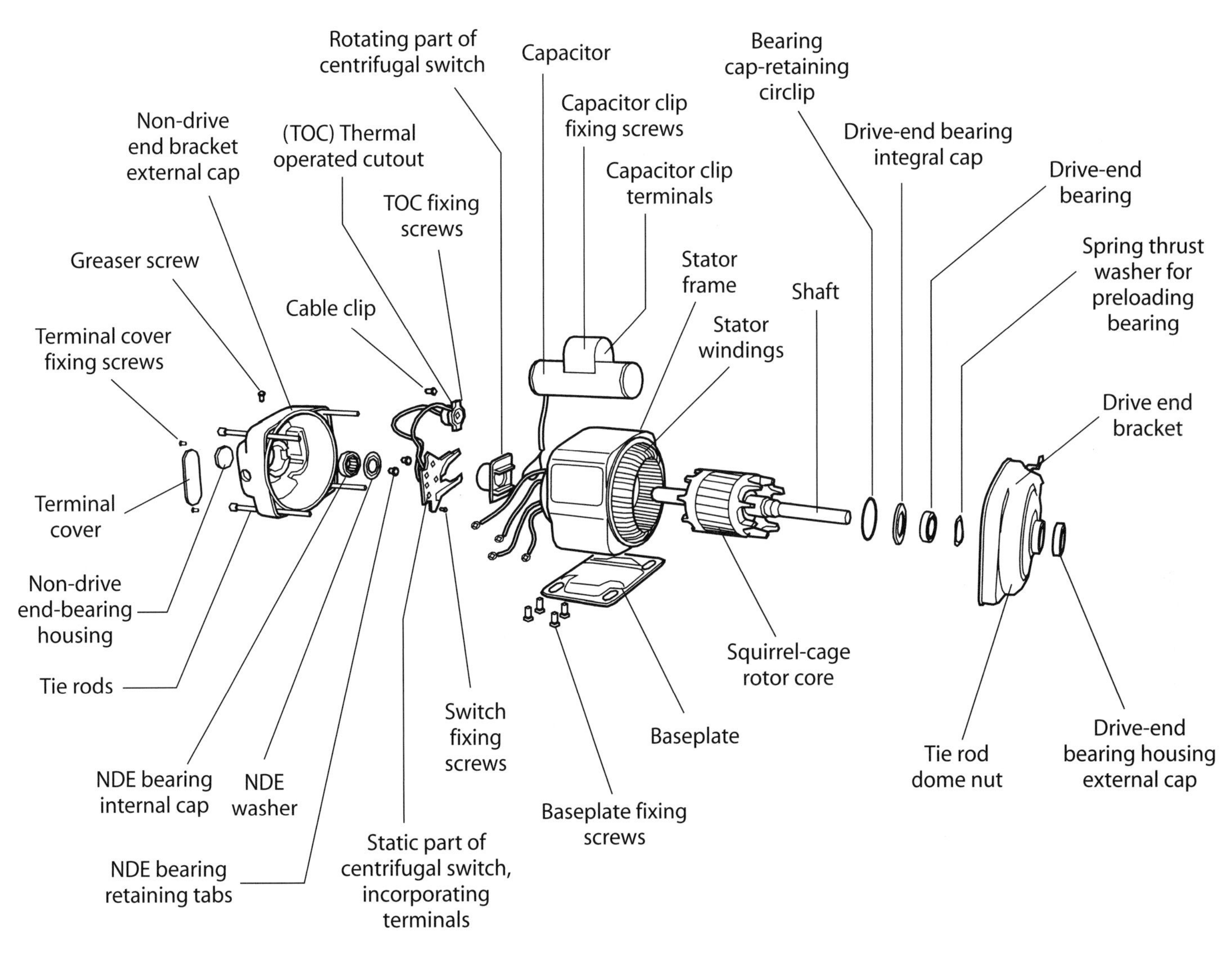

Figure 7.27 Construction of a single-phase induction motor

Motor speed and slip calculation

Speed of a motor

There are essentially two ways to express the speed of a motor.

- **Synchronous speed**. For an a.c. motor this is the speed of rotation of the stator's magnetic field. Consequently, this is really only a theoretical speed, as the rotor will always turn at a slightly slower rate.
- **Actual speed**. This is the speed at which the shaft rotates. The nameplate on most a.c. motors will give the actual motor speed rather than the synchronous speed.

The difference between the speed of the rotor and the synchronous speed of the rotating magnetic field is known as the **slip**, which can be expressed either as a unit or in percentage terms. Because of this, we often refer to the induction motor as being an asynchronous motor.

Remember, the speed of the rotating magnetic field is known as the synchronous speed, and this will be determined by the frequency of the supply and the number of pairs of poles within the machine. The speed at which the rotor turns will be between two per cent and five per cent slower, with an average of four per cent being common.

The reduced speed of the rotor is due to it having to overcome friction, as during the turning movement there is friction from the bearings in addition to any friction deriving from the load that the motor is connected to. Another factor that comes into play in determining the speed of a motor is **windage**. This means that within the enclosure there is a certain amount of air, and as the rotor rotates it has to move the air, which in turn contributes to the slowing down of the rotor. Of course there are no moving parts involved with the rotating magnetic field, so the rotor will never catch up. However, were some miracle to happen and the rotor reached the synchronous speed, we would have a different problem.

Definition

Windage – the air resistance of a moving object or the force of the wind on a stationary object

We already know that when a conductor passes at right angles through a magnetic field current is induced into the conductor. The direction of the induced current is dependant on the direction of movement of the conductor, and the strength of the current is determined by the speed at which the conductor moves. If the rotating magnetic field and the rotor are now revolving at the same speed, there will be no lines of magnetic flux cutting through the rotor bars, no induced e.m.f. and consequently no resultant magnetic field around the rotor bars to interact with the rotating magnetic field of the stator. The motor will immediately slow down and, having slowed down,will then start to speed up as the lines of magnetic flux start to cut through the rotor bars again – and so the process would continue.

Standard a.c. induction motors therefore depend on the rotor trying, but never quite managing, to catch up with the stator's magnetic field. The rotor speed is just slow enough to cause the proper amount of rotor current to flow, so that the resulting torque is sufficient to overcome windage and friction losses and drive the load.

Determining synchronous speed and slip

All a.c. motors are designed with various numbers of magnetic poles. Standard motors have two, four, six or eight poles, and these poles play an important role in determining the synchronous speed of an a.c. motor.

As we said before, the synchronous speed can be determined by the frequency of the supply and the number of pairs of poles within the machine. We can express this relationship with the following formula:

$$\text{Synchronous speed } (n_s) \text{ in revolutions per second} = \frac{\text{Frequency (f) in Hz}}{\text{The number of pole pairs (p)}}$$

Remember

When a conductor passes at right angles through a magnetic field, current is induced into the conductor. The direction of the induced current will depend on the direction of movement of the conductor, and the strength of the current will be determined by the speed at which the conductor moves

Example 1

Calculate the synchronous speed of a four-pole machine connected to a 50 Hz supply.

$$n_s = \frac{f}{P}$$

As we know the motor has four poles, this means it has two pole pairs. We can therefore complete the calculation as:

$$n_s = \frac{50}{2} \qquad \text{therefore } n_s = 25 \text{ revolutions per second (rps)}$$

To convert revolutions per second into the more commonly used revolutions per minute (rpm), simply multiply n_s by 60. This new value is referred to as N_s, which in this example will become 25×60, giving 1500 rpm.

We also said that we refer to the difference between the speed of the rotor and the synchronous speed of the rotating magnetic field as the slip, which can be expressed either as a unit (S) or in percentage terms (S per cent). We express this relationship with the following formula:

$$\text{per cent slip} = \frac{\text{Synchronous speed } (n_s) - \text{Rotor speed } (n_r)}{\text{Synchronous speed } (n_s)} \times 100$$

Example 2

In this example numbers have been rounded up for ease.

A six-pole cage-induction motor runs at 4 per cent slip.

Calculate the motor speed if the supply frequency is 50 Hz.

$$S\text{ (per cent)} = \frac{(n_S - n_r)}{n_S} \times 100$$

We therefore need first to establish the synchronous speed; as the motor has six poles it will have three pole pairs. Consequently:

Synchronous speed $n_s = \frac{f}{P}$ giving us $\frac{50}{3}$ and therefore $n_s = 16.7$ revs/sec

We can now put this value into our formula and then, by transposition, rearrange the formula to make n_r the subject. Consequently:

$$S\text{ (per cent)} = \frac{(n_S - n_r)}{n_S} \times 100 \quad \text{giving us} \quad 4 = \frac{(16.7 - n_r)}{16.7} \times 100$$

Therefore by transposition:

$$4 = \frac{(16.7 - n_r)}{16.7} \times 100 \quad \text{gives us} \quad (16.7 - n_r) = \frac{4 \times 16.7}{100}$$

When calculated :

$$(16.7 - n_r) = \frac{4 \times 16.7}{100} \quad \text{becomes} \quad (16.7 - n_r) = 0.668$$

Therefore by further transposition:

$$(16.7 - n_r) = 0.668 \quad \text{becomes} \quad 16.7 - 0.668 = n_r$$

Therefore n_r = **16.032 rps** or N_r = **962 rpm**

Synchronous a.c. induction motors

A synchronous motor, as the name suggests, runs at synchronous speed. Because of the problems discussed earlier, this type of motor is not self-starting and instead must be brought up to almost synchronous speed by some other means.

Three-phase a.c. synchronous motors

To understand how the synchronous motor works, assume that we have supplied three-phase a.c. power to the stator, which in turn causes a rotating magnetic field to be set up around the rotor. The rotor is then supplied via a field winding with d.c. and consequently acts a bit like a bar magnet, having north and south poles. The rotating magnetic field now attracts the rotor field that was activated by the d.c. This results in a strong turning force on the rotor shaft, and the rotor is therefore able to turn a load as it rotates in step with the rotating magnetic field.

It works this way once it's started. However, one of the disadvantages of a synchronous motor is that it cannot be started from a standstill by just applying a three-phase a.c. supply to the stator. When a.c. is applied to the stator, a high-speed rotating magnetic field appears immediately. This rotating field rushes past the rotor poles so quickly that the rotor does not have a chance to get started. In effect, the rotor is repelled first in one direction and then the other.

An induction winding (squirrel-cage type) is therefore added to the rotor of a synchronous motor to cause it to start, effectively meaning that the motor is started as an induction motor. Once the motor reaches synchronous speed, no current is induced in the squirrel-cage winding, so it has little effect on the synchronous operation of the motor.

Synchronous motors are commonly driven by transistorised variable-frequency drives.

Single-phase a.c. synchronous motors

Small single-phase a.c. motors can be designed with magnetised rotors. The rotors in these motors do not require any induced current so they do not slip backwards against the mains frequency. Instead, they rotate synchronously with the mains frequency. Because of their highly accurate speed, such motors are usually used to power mechanical clocks, audio turntables and tape drives; formerly they were also widely used in accurate timing instruments such as strip-chart recorders or telescope drive mechanisms. The shaded-pole synchronous motor is one version.

As with the three-phase version, inertia makes it difficult to accelerate the rotor instantly from stopped to synchronous speed, and the motors normally require some sort of special feature to get started. Various designs use a small induction motor (which may share the same field coils and rotor as the synchronous motor) or a very light rotor with a one-way mechanism (to ensure that the rotor starts in the 'forward' direction).

Motor windings

A motor can be manufactured with the windings internally connected. If this is the case and there are three terminal connections in the terminal block labelled U, V and W, you would expect the motor windings to be connected in a delta configuration. This is shown in Figure 7.28.

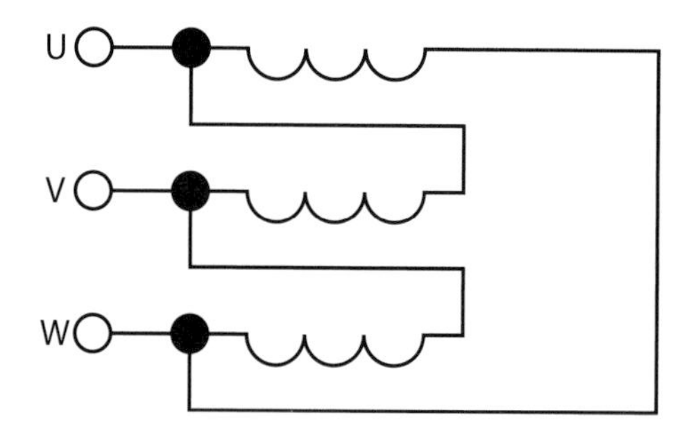

Figure 7.28 Motor windings with delta connection

However, there may be four connections in the terminal box labelled U, V, W and N. If this is the case, the windings would be arranged to give a star configuration, as shown in Figure 7.29.

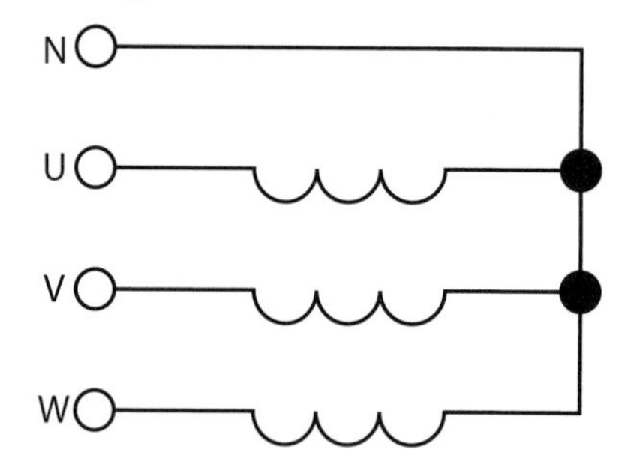

Figure 7.29 Motor windings with four connections

Alternatively the terminal block may contain six connections: U1, U2, V1, V2, W1 and W2. This is used where both star and delta configurations are required. The terminal connections can then be reconfigured for either star or delta, starting within the terminal block. Figure 7.30 illustrates the connections that would come out to the terminal block.

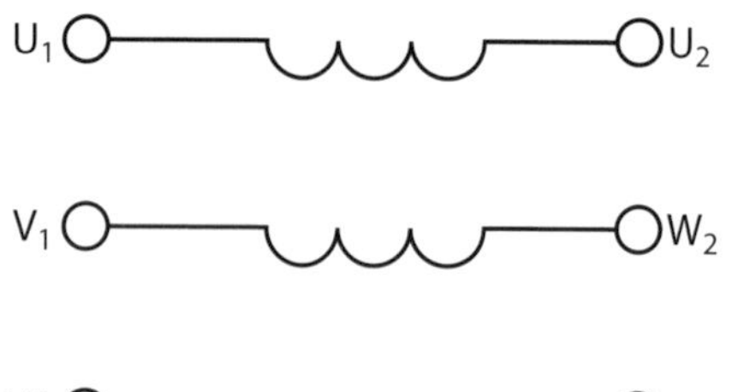

Figure 7.30 Motor windings with six connections

Remember

Some older motors may have different markings in the terminal block, and you should therefore refer to the manufacturer's data

Motor starters

A practical motor starter has to do more than just switch on the supply. It also has to have provision for automatically disconnecting the motor in the event of overloads or other faults. The overload protective device monitors the current consumption of the motor and is set to a predetermined value that the motor can safely handle. When a condition occurs that exceeds the set value, the overload device opens the motor starter control circuit and the motor is turned off. The overload protection can come in a variety of types, including solid-state electronic devices.

The starter should also prevent automatic restarting should the motor stop because of supply failure. This is called **no-volts protection** and will be discussed later in this section.

The starter must also provide for the efficient stopping of the motor by the user. Provision for this is made by the connection of remote stop buttons and safety interlock switches where required.

The Direct-On-Line (DOL) starter

This is the simplest and cheapest method of starting squirrel-cage (induction) motors.

The expression 'Direct-On-Line' starting means that the full supply voltage is directly connected to the stator of the motor by means of a contactor-starter, as shown in Figure 7.31.

DOL starter

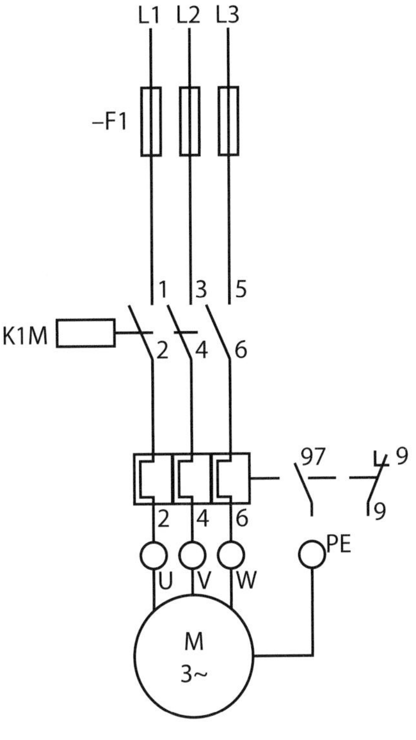

Figure 7.31 DOL starter

Since the motor is at a standstill when the supply is first switched on, the initial starting current is heavy. This 'inrush' of current can be as high as 6 to 10 times the full load current, i.e. a motor rated at 10A could have a starting current inrush as high as 60A, and the initial starting torque can be about 150 per cent of the full-load torque. Thus you may observe motors 'jumping' on starting if their mountings are not secure. As a result, Direct-On-Line starting is usually restricted to comparatively small motors with outputs of up to about 5 kW.

DOL starters should also incorporate a means of overload protection, which can be operated by either magnetic or thermal overload trips. These activate when there is a sustained increase in current flow.

To reverse the motor you need to interchange any two of the incoming three-phase supply leads. If a further two leads are interchanged then the motor will rotate in the original direction.

Operating principle of a DOL starter

A three-pole switch controls the three-phase supply to the starter. This switch normally includes fuses, which provide a means of electrical isolation and also short-circuit protection. We shall look at the operation of the DOL starter in stages.

Let's start by looking at the one-way switch again. In this circuit (Figure 7.32), the switch is operated by your finger and the contacts are then held in place mechanically.

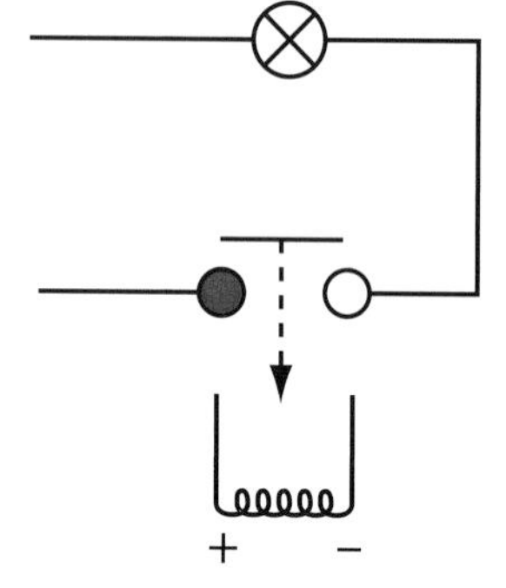

Figure 7.33 Switch relay

We could decide that we don't want to operate the switch this way and instead use a relay. In this system (Figure 7.33), when the coil is energised it creates a magnetic field. Everything in the magnetic field will be pulled in the direction of the arrow and the metal strip will be pulled onto the contacts. As long as the coil remains energised and is a magnet, the light will stay on.

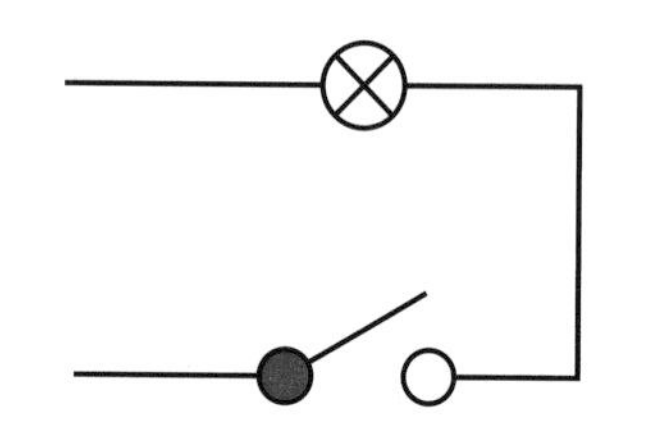
Figure 7.32 One-way switch

Looking at Figures 7.32 and 7.33, we can see that the first option works well enough for its intended purpose. However, it couldn't be used in a three-phase system as we would need one for each phase and would have to trust to luck each time we tried to hit all three switches at the same time. However, the second option does provide us with an effective method of controlling more than one thing from one switch, as long as they are all in the same magnetic field.

Let's apply this knowledge to a DOL starter. We know that we can't have three one-way switches in the starter. But it helps to try and think of the contacts: where the switches aren't operated by your fingers, but are pulled in the direction of the arrow by the magnetic effect of the coil (see Figure 7.34), items affected by the same magnetic effect are normally shown linked by a dotted line. For ease of explanation we'll use a 230V coil.

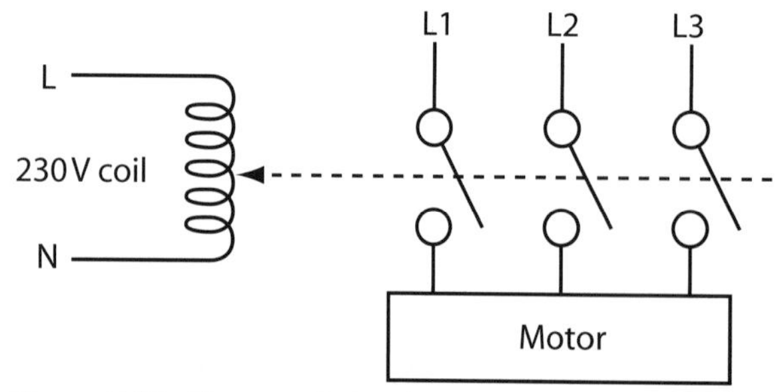

Figure 7.34 Three-switch relay

From this we can see that as soon as we put a supply onto the coil it energises, becomes a magnet and pulls the contacts in. Obviously, this would be no good for our starter, as every time the power is put on, the starter will become active and start operating whatever is connected to it. In the case of machinery this could be very dangerous. The starter design therefore goes one step further to include no-volts protection. The simple addition of a 'normally open' Start button gives this facility, as shown in Figure 7.35.

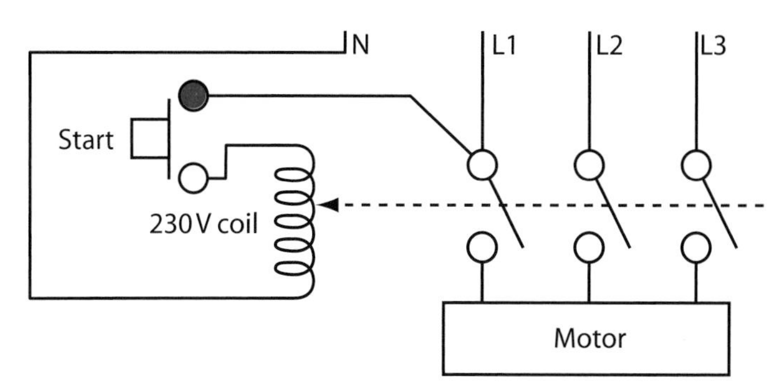

Figure 7.35 Three-switch relay with 'normally open' Start button

So far so good. We now have to make a conscious decision to start the motor.

Our next problem though, is that every time we take our finger off the Start button, the button springs back out, the supply to the coil is lost and the motor stops. This is the 'no volt' protection element in operation.

What we need for normal operation is a device called the 'hold in' or 'retaining' contact, as shown in Figure 7.36. This is a 'normally closed' contact (position 1) which is also placed in the magnetic field of the coil. Consequently, when the coil is energised, it is also pulled in the direction shown, and in this case, across and onto the terminals of the Start button (position 2).

We can now take our finger off the Start button, as the supply will continue to feed the coil by running through the 'hold in' contact that is linking out the Start button terminals (position 2).

This now means that we can only break the coil supply and stop the motor by fitting a Stop button. In the case of starters fitted with a 230 V coil, this will be a normally closed button placed in the neutral wire.

Now, for the fraction of a second that we push the Stop button, the supply to the coil is broken, the coil ceases to be a magnet and the 'hold in' contact returns to its original position (position 1). Since the Start button had already returned to its original open position when we released it, when we take our finger off the Stop button everything will be as we first started. Therefore any loss of supply will immediately break the supply to the coil and stop the motor: if a supply failure was restored, the equipment could not restart itself – someone would have to take the conscious decision to do so.

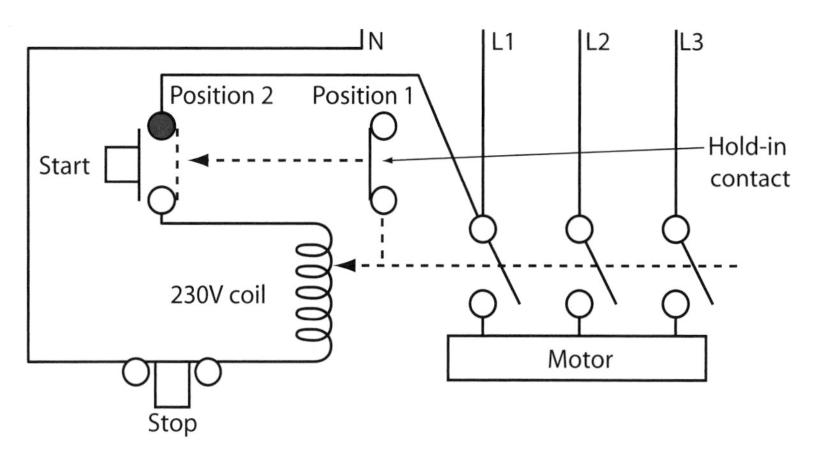

Figure 7.36 Restarting the motor

This system is basically the same as that of a contactor. In fact, many people refer to this item as the contactor starter. Such starters are also available with a 400 V coil, which is therefore connected across two of the phases.

Remote stop/start control

In the DOL starter as described, we have the means of stopping and starting the motor from the buttons provided on the starter enclosure. However, there are situations where the control of the motor needs to take place from some remote location. This could be, for example, in a college workshop where, in the case of an emergency, emergency stop buttons located throughout the workshop can be activated to break the supply to a motor. Equally, because of the immediate environment around the motor, it may be necessary to operate it from a different location.

Did you know?

Inch control gives the ability to partially start a motor without it commencing its normal operation

Commonly known as a remote stop/start station, the enclosure usually houses a start and a stop button connected in series. However, depending on the circumstances it is also possible to have an additional button included to give 'inch' control of a motor.

If we use the example of our DOL starter as described in the previous diagrams, but now include the remote stop/start station, the circuit would look as in Figure 7.37 below, where for ease the additional circuitry has been shown in red.

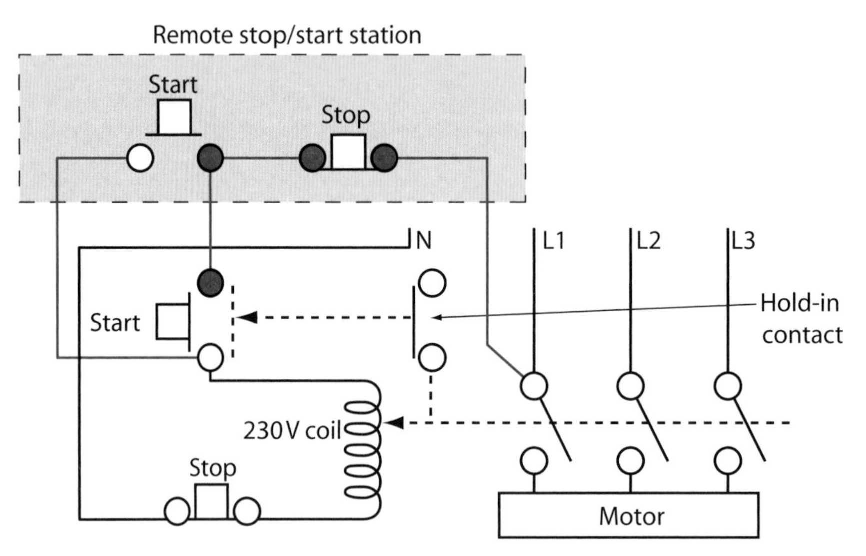

Figure 7.37 Remote stop/start control

As can be seen, the remote start button is effectively in parallel with the start button on the main enclosure with the supply to both of these buttons being routed via the stop button of the remote station.

If the intention is to provide only emergency stops, omit the remote station shown so that these are all connected in series with the stop button on the main enclosure.

Hand-operated star-delta starter

This is a two-position method of starting a three-phase squirrel-cage motor, in which the windings are connected firstly in star for acceleration of the rotor from standstill, and then secondly in delta for normal running.

The connections of the motor must be arranged for star-delta, starting with both ends of each phase winding – six in all – brought out to the motor terminal block. The starter itself, in its simplest form, is in effect a changeover switch. Figure 7.38 gives the elementary connections for both star and delta.

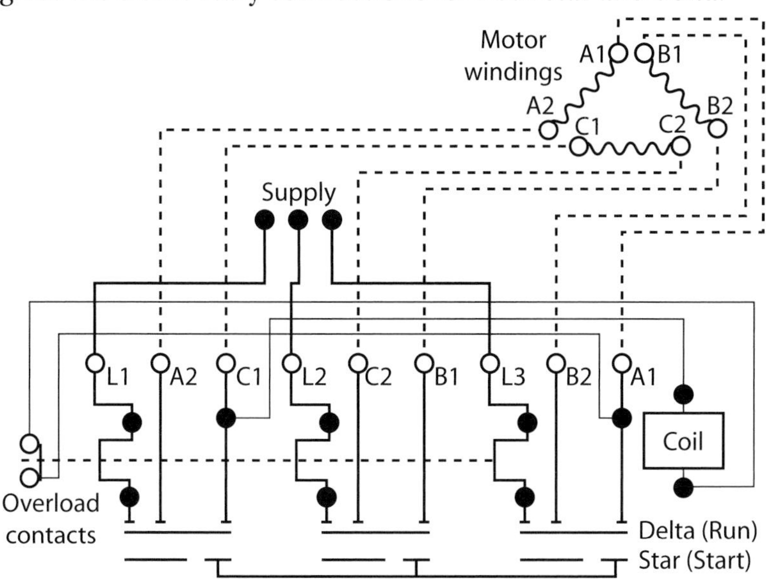

Figure 7.38 Hand-operated star-delta connections

When the motor windings are connected in star, the voltage applied to each phase winding is reduced to 58 per cent of the line voltage, thus the current in the winding is correspondingly reduced to 58 per cent of the normal starting value.

Applying these reduced values to the typical three-phase squirrel-cage induction motor, we would have: initial starting current from two to three-and-a-half times full-load current and initial starting torque of about 50 per cent of the full-load value.

The changeover from star to delta should be made when the motor has reached a steady speed on star connection, at which point the windings will now receive full line voltage and draw full-rated current.

If the operator does not move the handle quickly from the start to run position, the motor may be disconnected from the supply long enough for the motor speed to fall considerably. When the handle is eventually put into the run position, the motor will therefore take a large current before accelerating up to speed again. This surge current could be large enough to cause a noticeable voltage dip. To prevent this, a mechanical interlock is fitted to the operating handle. However, in reality the handle must be moved quickly from start to run position, otherwise the interlock jams the handle in the start position.

The advantage of this type of starter is that it is relatively cheap. It is best suited for motors against no load or light loads, and it also incorporates no-volts protection and overload protection.

Remember

Although hand-operated star-delta starters are now rare, you may still come across them in your career as an electrician

Star-delta starter

Automatic star-delta starter

Bearing in mind the user actions required of the previous hand-operated starter, the fully automatic star-delta contactor starter (as shown in Figure 7.39) is the most satisfactory method of starting a three-phase cage-induction motor. The starter consists of three triple-pole contactors, one employing thermal overload protection, the second having a built-in time-delay device, and the third providing the star points.

The changeover from star to delta is carried out automatically by the timing device, which can be adjusted to achieve the best results for a particular situation.

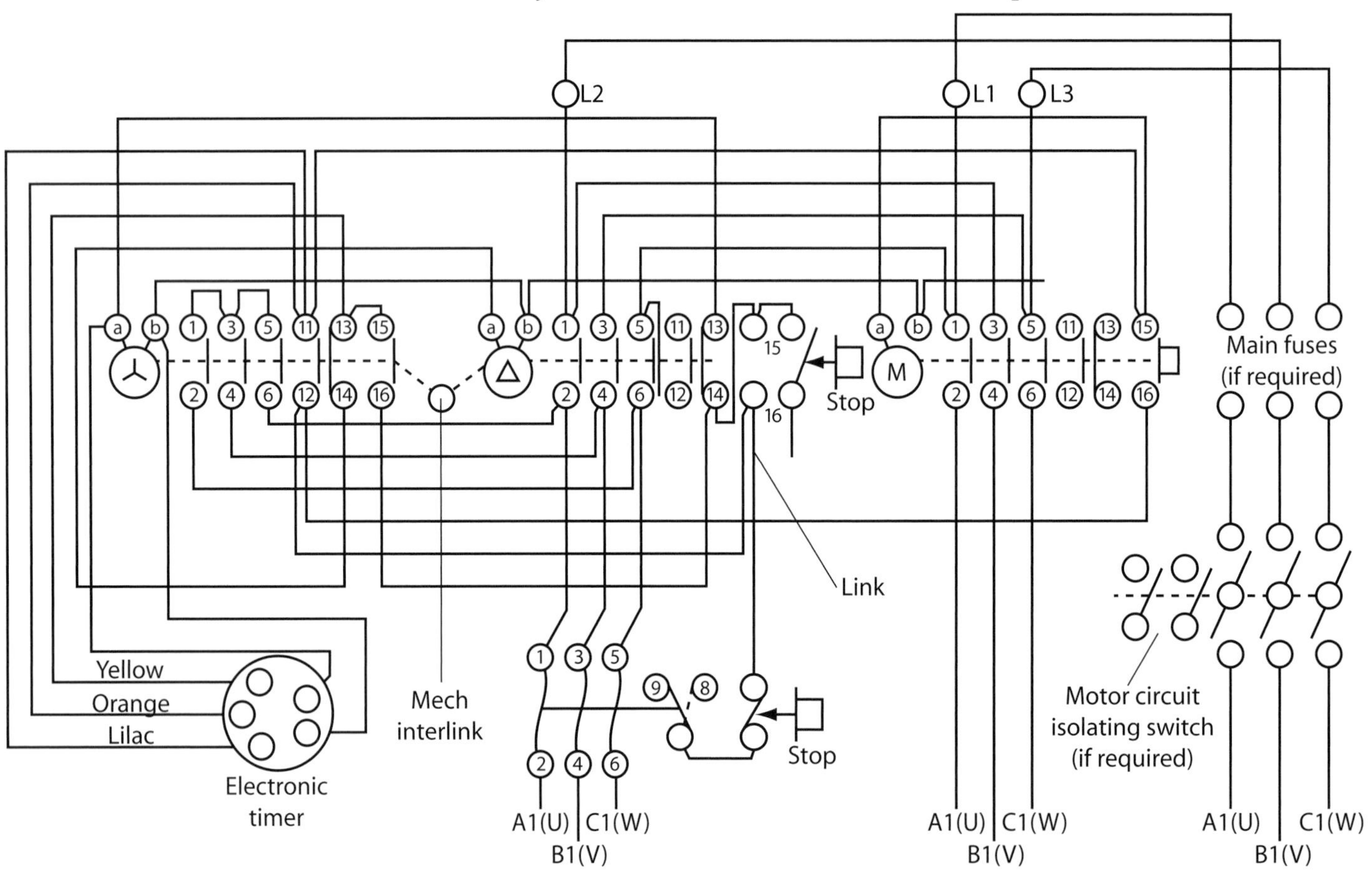

Figure 7.39 Automatic star-delta contactor starter

Soft starters

A soft starter is a type of reduced-voltage starter that reduces the starting torque for a.c. induction motors. The soft starter is in series with the supply to the motor, and uses solid-state devices to control the current flow and therefore the voltage applied to the motor. In theory, soft starters can be connected in series with the line voltage applied to the motor, or can be connected inside the delta loop of a delta-connected motor, controlling the voltage applied to each winding. Soft starters can also have a soft-stop function, which is the exact opposite to soft start, and sees the voltage gradually reduced and thus a reduction in the torque capacity of the motor.

The auto-transformer starter

This method of starting is used when star-delta starting is unsuitable, either because the starting torque would not be sufficiently high using that type of starter, or because only three terminals have been provided at the motor terminal box – a practice commonly found within the UK water industry.

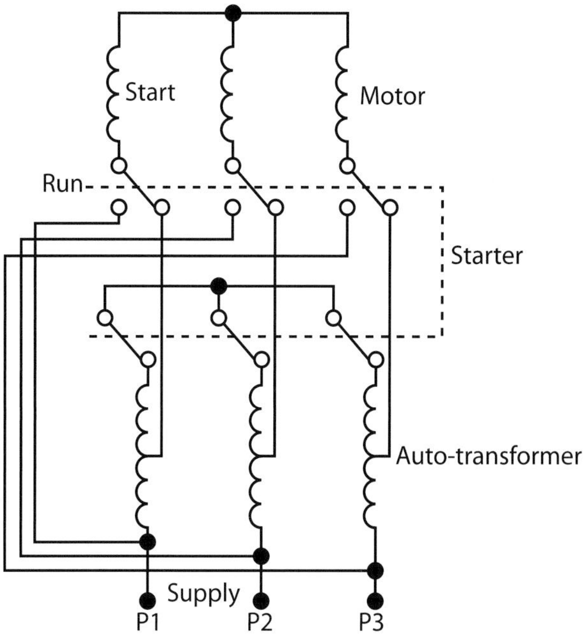

This again is a two-stage method of starting three-phase squirrel-cage induction motors, in which a reduced voltage is applied to the stator windings to give a reduced current at start. The reduced voltage is obtained by means of a three-phase auto transformer, the tapped windings of which are designed to give 40 per cent, 60 per cent and 75 per cent of the line voltage respectively. Although there are a number of tappings, only one tapping is used for the initial starting, as the reduced voltage will also result in reduced torque. Once this has been selected for the particular situation in which the motor is operating, it is left at that position and the motor is started in stages – much like the star-delta starter in that once the motor has reached sufficient speed the changeover switch moves onto the run connections, thus connecting the motor directly to the three-phase supply. Figure 7.40 illustrates the connections for an auto-transformer starter.

Figure 7.40 Connections for an auto-transformer starter

The rotor-resistance starter

This type of starter is used with a slip-ring wound-rotor motor. These motors and starters are primarily used where the motor will start against full load, as an external resistance is connected to the rotor windings through slip rings and brushes, which serves to increase the starting torque.

When the motor is first switched on, the external rotor resistance is at a maximum. As the motor speed increases, the resistance is reduced until at full speed, when the external resistance is completely eliminated and the machine runs as a squirrel-cage induction motor.

The starter is provided with no-volts and overload protection and an interlock to prevent the machine being switched on with no rotor resistance connected. (For clarity these are not shown in Figure 7.41, since the purpose of the diagram is to show the principle of operation.)

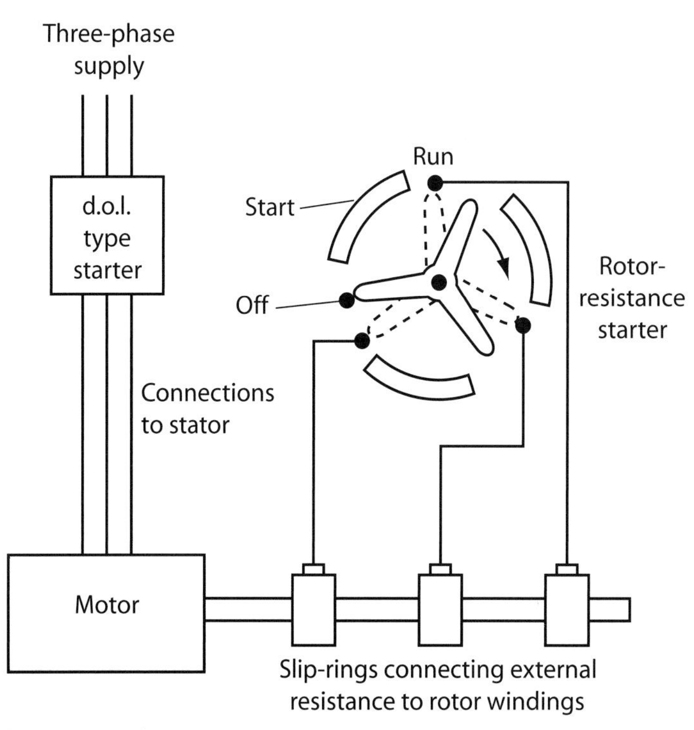

Figure 7.41 Rotor-resistance starter

Motor speed control

Speed control of d.c. machines

We said at the beginning of this chapter that there are three types of d.c. motor: series, shunt and compound, and that one of the advantages of the d.c. machine is the ease with which the speed may be controlled.

Some of the more common methods used to achieve speed control on a d.c. machine are described below.

Armature resistance control

With this system of control we are effectively reducing the voltage across the armature terminals by inserting a variable resistor into the armature circuit of the motor. In effect we are creating the illusion of applying a lower-than-rated voltage across the armature terminals.

The disadvantages of this method of control are that we see much of the input energy dissipated in the variable resistor, a loss of efficiency in the motor and poor speed regulation in the shunt and compound motors.

Although not discussed in this book, the principle of applying a lower-than-rated voltage across the armature terminals forms the basis of the Ward-Leonard system of speed control, which in essence provides a variable voltage to the armature terminals by controlling the field winding of a separate generator. Although expensive, this method gives excellent speed control and therefore finds use in situations such as passenger lifts.

Field control

This method works on the principle of controlling the magnetic flux in the field winding. This can be controlled by the field current and as a result controls the motor speed. As the field current is small, the power dissipated by the variable resistor is reasonably small. We can control the field current in the various types of motor as follows:

- **Series motor** – Place a variable resistor in parallel with the series field winding.
- **Shunt motor** – Place a variable resistor in series with the field shunt winding.

This method of speed control is not felt to be suitable for compound machines, as any reduction in the flux of the shunt is offset by an increase in flux from the series field because of an increase in armature current.

Pulse width modulation (PWM)

We know all about the problem with the variable resistor: although it works well, it generates heat and hence wastes power. PWM d.c. motor control uses electronics to eliminate this problem. It controls the motor speed by driving the motor with short pulses. These vary in duration to change the speed of the motor. The longer the pulses, the faster the motor turns, and vice versa. The main disadvantages of PWM circuits are their added complexity and the possibility of generating radio frequency interference (RFI), although this can be minimised by the use of short leads or additional filtering on the power supply leads.

Speed control of a.c. induction motors

We have already established that synchronous speed is directly proportional to the frequency of the supply and inversely proportional to the number of pole pairs. Therefore the speed of an induction motor can be changed by varying the frequency and/or the number of poles. We can also control the speed by changing the applied voltage and the armature resistance.

For adjustable speed applications, variable speed drives (VSD) use these principles by controlling the voltage and frequency delivered to the motor. This gives control over motor torque and reduces the current level during starting. Such drives can control the speed of the motor at any time during operation.

The phrase 'a.c. drive' has different meanings to different people. To some it means a collection of mechanical and electro-mechanical components (the variable frequency inverter and motor combination), which, when connected together, will move a load. More commonly – and also for our purposes – an a.c. drive should be considered as being a variable-frequency inverter unit (the drive) with the motor as a separate component. Manufacturers of variable-frequency inverters will normally refer to these units as being a variable-frequency drive (VFD).

A variable-frequency drive is a piece of equipment that has an input section – the converter – which contains diodes arranged in a bridge configuration. This converts the a.c. supply to d.c. The next section of the VFD, known as the constant-voltage d.c. bus, takes the d.c. voltage and filters and smoothes out the wave form. The smoother the d.c. wave form is, the cleaner the output wave form from the drive.

The d.c. bus then feeds the final section of the drive, the inverter, and as we know from reading earlier chapters of this book, this section of the drive will invert the d.c. back to a.c. using Insulated Gate Bipolar Transistors (IGBT), which create a variable a.c. voltage and frequency output.

On the job: Motor noise

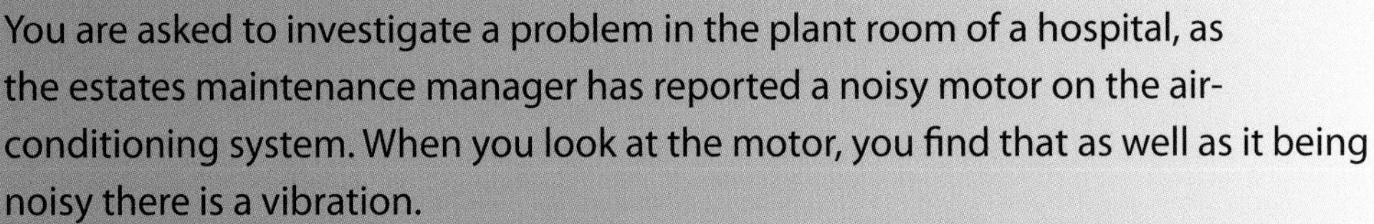

You are asked to investigate a problem in the plant room of a hospital, as the estates maintenance manager has reported a noisy motor on the air-conditioning system. When you look at the motor, you find that as well as it being noisy there is a vibration.

1. Do you think this is a real problem? If so, what actions would you recommend?

(A suggested answer can be found on page 334)

Knowledge check

1. What will be the percentage slip of an induction motor having a synchronous speed of 1,500 rpm and a rotor speed of 1,425 rpm?
2. What is the type of motor most likely to be used in a domestic vacuum cleaner?
3. Describe the principle of operation of a series-wound (universal) motor.
4. How would you change the direction of rotation of a three-phase induction motor?
5. What are the component parts of a squirrel-cage rotor?
6. Describe how to reverse the direction of a single-phase split-phase motor.
7. What is the purpose of the centrifugal switch in a capacitor start motor?
8. What type of motor uses the rotor-resistance starter?
9. In Fleming's left-hand (motor) rule, the second finger is used to represent which quantity?

Chapter 8

Electronics

OVERVIEW

In today's modern technical world electricians can no longer say they do not need to learn about electronics because the use of devices such as security alarms, telephones, dimmers, boiler controls and speed controllers have now brought electronics into general electrical installation work. This chapter will cover:

- **Resistors**
- **Capacitors**
- **Semiconductor devices**
- **Rectification**
- **Transistors**
- **Integrated circuits (ICs)**
- **Thyristors**
- **Field effect transistors (FETs)**

Resistors

There are two basic types of resistor: fixed and variable. The resistance value of a fixed resistor cannot be changed by mechanical means (though its normal value can be affected by temperature or other effects). Variable resistors have some means of adjustment (usually a spindle or slider). The method of construction, specifications and features of both fixed and variable resistor types vary, depending on what they are to be used for.

Fixed resistors

Making a resistor simply consists of taking some material of a known resistivity and making the dimensions (csa and length) of a piece of that material such that the resistance between the two points at which leads are attached (for connecting into a circuit) is the value required.

Most of the very earliest resistors were made by taking a length of resistance wire (wire made from a metal with a relatively high resistivity, such as brass) and winding this on to a support rod of insulating material. The resistance value of the resulting resistor depended on the length of the wire used and its cross sectional area.

Wire wound resistor

This method is still used today, though it has been somewhat refined. For example, the resistance wire is usually covered with some form of enamel glazing or ceramic material to protect it from the atmosphere and mechanical damage. The external and internal view of a typical wire wound resistor is shown in Figure 8.1.

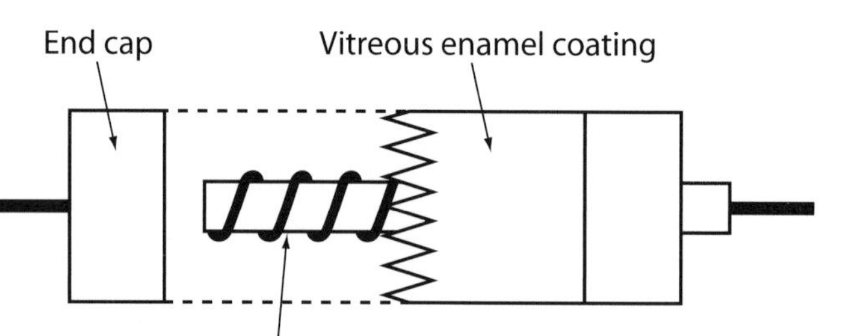

Figure 8.1 Typical wire wound resistor

Most wire wound resistors can operate at fairly high temperatures without suffering damage, so they are useful in applications where some power may be dissipated. They are, however, relatively difficult to mass-produce, which makes them expensive. Techniques for making resistors from materials other than wire have now been developed for low power applications.

Metal oxide and carbon-composition light-dependent resistors

Resistor manufacture advanced considerably when techniques were developed for coating an insulating rod (usually ceramic or glass) with a thin film of resistive material (see Figure 8.2). The resistive materials in common use today are carbon and metal oxides. Metal end caps fitted with leads are pushed over the ends of the coated rod and the whole assembly is coated with several layers of very tough varnish or similar material to protect the film from the atmosphere and from knocks during handling. These resistors can be mass-produced with great precision at very low cost.

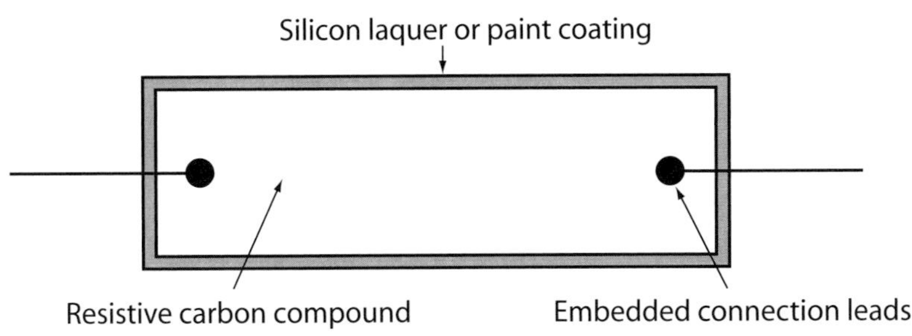

Figure 8.2 Resistor construction

Variable resistors

The development of the techniques for manufacturing variable resistors followed fairly closely that of fixed resistors, though they required some sort of sliding contact together with a fixed resistor element.

Wire wound variable resistors are often made by winding resistance wire onto a flat strip of insulating material, which is then wrapped into a nearly complete circle. A sliding contact arm is made to run in contact with the turns of wire as they wrap over the edge of the wire strip as in Figure 8.3 below. Straight versions are also possible. A straight former is used and the wiper travels in a straight line along it as shown in Figure 8.4.

While wire wound resistors are ideal for certain applications, there are many others where their size, cost and other disadvantages make them unattractive, and as a consequence alternative types have been developed.

The early alternative to the wire wound construction was to make the resistive element (on which the wiper rubs) out of a carbon composition, deposited or moulded as a track and shaped as a nearly completed circle on an insulating support plate. Alternative materials for the track are carbon films, or metal alloys of a metal oxide and a ceramic (cermet) and again straight versions are possible.

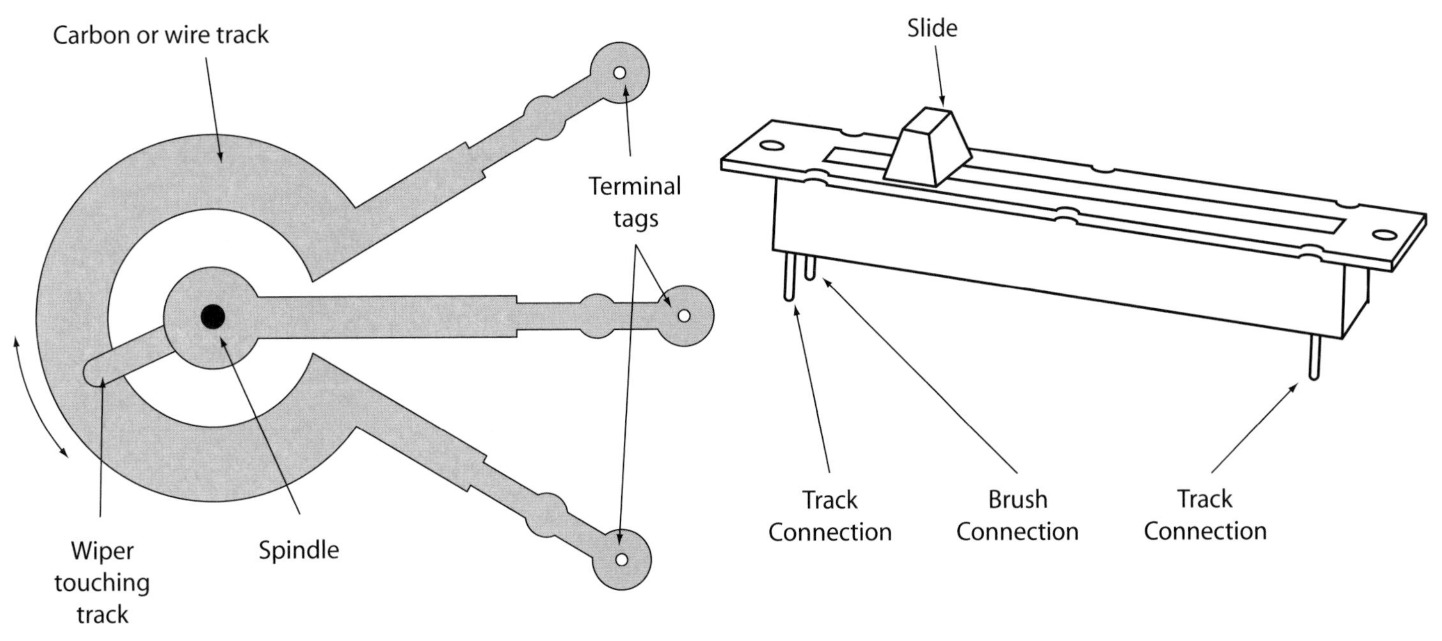

Figure 8.3 Layout of internal track of rotary variable resistor

Figure 8.4 Linear variable resistor

Preferred values

In theory, there's no reason why you couldn't have resistors in every imaginable resistance value; from zero to, say, tens or hundreds of megohms. In reality, however, such an enormous range would be totally impractical to manufacture and store. And from the point of view of the circuit designer it's not usually necessary, either.

So, rather than an overwhelming number of individual resistance values, what manufacturers do is make a limited range of preferred resistance values. In electronics, we use the preferred value closest to the actual value we need.

A resistor with a preferred value of 1000 Ω and a 10 per cent tolerance can have any value between 900 Ω and 1100 Ω. The next largest preferred value, which would give the maximum possible range of resistance values without too much overlap, is 1200 Ω. This can have a value between 1080 Ω and 1320 Ω.

Together, these two preferred value resistors cover all possible resistance values between 900 Ω and 1320 Ω. The next preferred values would be 1460 Ω, 1785 Ω etc.

There is a series of preferred values for each tolerance level as shown in Table 8.1, so that every possible numerical value is covered.

E6 series 20% Tol	E12 series 10% Tol	E24 series 5% Tol
10	10	10
		11
	12	12
		13
15	15	15
		16
	18	18
		20
22	22	22
		24
	27	27
		30

E6 series 20% Tol	E12 series 10% Tol	E24 series 5% Tol
33	33	33
		36
	39	39
		43
47	47	47
		51
	56	56
		62
68	68	68
		75
	82	82
		91

Table 8.1 Table of preferred values

Resistance markings

There is obviously the need for the resistor manufacturer to provide some sort of markings on each resistor so that it can be identified.

The user should be able to tell, by looking at the resistor, what its nominal resistance value is and its tolerance. Various methods of marking this information onto each resistor have been used and sometimes a resistor code will use numbers and letters rather than colours.

Where physical size permits, putting the actual value on the resistor in figures and letters has an obvious advantage in terms of easy interpretation. However, again because of size restrictions, we don't use the actual words and instead use a code system. This code is necessary because when using small text on a small object, certain symbols and the decimal point become very hard to see. This code system is also commonly used to represent resistance values on circuit diagrams for the same reason.

In reality resistance values are generally given in either Ω, kΩ or MΩ using numbers from 1–999 as a prefix (e.g. 10 Ω, 567 kΩ etc.). In the code system we replace Ω, kΩ and MΩ and represent them instead by using the following letters:

- Ω = R
- kΩ = K
- MΩ = M.

These letters are now inserted wherever the decimal point would have been in the value. So for example a resistor of value 10 Ω resistor would now be shown as 10R, and a resistor of value 567 kΩ resistor would become 567 K.

Table 8.2 gives some more examples of this code system. Table 8.3 shows the letters that are then commonly used to represent the tolerance values. These letters are added at the end of the resistor marking so that, for example, a resistor of value 2.7 MΩ with a tolerance of ±10% would be shown as 2M7K.

Remember

Whole numbers could have a decimal point at the end (e.g. 10.0 or 567.0), but we normally miss them out when we write the numbers down (e.g. 10 or 567)

0.1 Ω	is coded	R10
0.22 Ω	is coded	R22
1.0 Ω	is coded	1R0
3.3 Ω	is coded	3R3
15 Ω	is coded	15R
390 Ω	is coded	390R
1.8 Ω	is coded	1R8
47 Ω	is coded	47R
820 kΩ	is coded	820K
2.7 MΩ	is coded	2M7

Table 8.2 Examples of resistance value codes

F	=	± 1%
G	=	± 2%
J	=	± 5%
K	=	± 10%
M	=	± 20%
N	=	± 30%

Table 8.3 Codes for common tolerance values

Remember

Before you read a resistor, turn it so that the end with bands is on the left hand side. Now you read the bands from left to right (as shown in Figures 8.5 to 8.8)

Resistor coding

Standard colour code

Many resistors are so small that it is impractical to print their value on them. Instead, they are marked with a code that uses bands of colour. Located at one end of the component, it is these bands that identify the resistor's value and tolerance.

Most general resistors have four bands of colour, but high precision resistors are often marked with a five-colour band system. No matter which system is being used, the value of the colours is the same.

Resistor colour code

Band colour	Value
Black	0
Brown	1
Red	2
Orange	3
Yellow	4
Green	5
Blue	6
Violet	7
Grey	8
White	9
Gold	0.1
Silver	0.01

Figure 8.5 Resistor colour code

Tolerance colour code

Band colour	±%
Brown	1
Red	2
Gold	5
Silver	10
None	20

Figure 8.6 Tolerance colour code

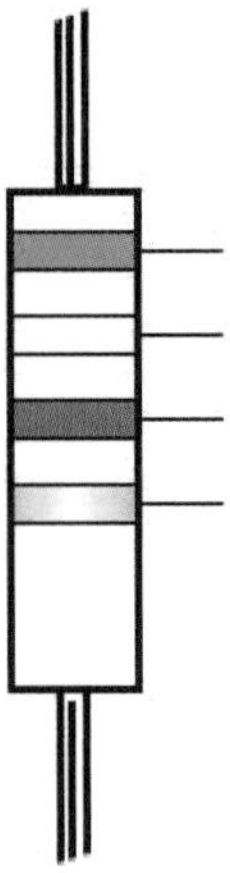

What this means

Band 1 First figure of value
Band 2 Second figure of value
Band 3 Number of zeros/multiplier
Band 4 Tolerance (±%) See below

Note that the bands are closer to one end than the other

Figure 8.7 What this means

Brown 1, Green 5, Orange 000, Gold 5%
Resistor is 15000Ω or 15K ± 5%

Yellow 4, Violet 7, Silver ×0.01, Gold 5%
Resistor is 47 × 0.01Ω or 0.47K ± 2%

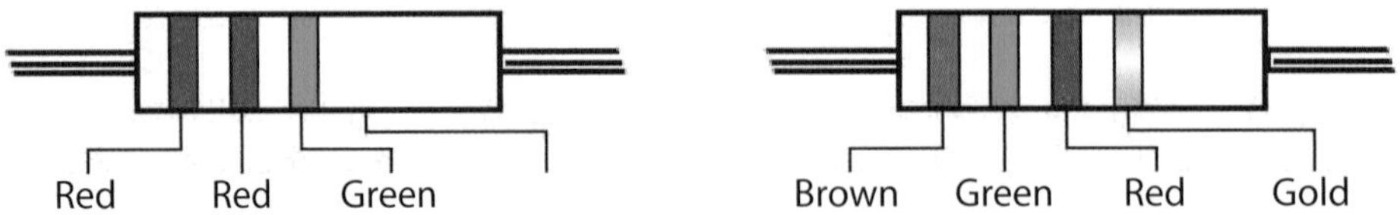

Red 2, Red 2, Green 00000, 20%
Resistor is 2200000Ω or 2.2M ± 20%

Brown 1, Green 5, Red 00, Gold 5%
Resistor is 1500Ω or 1.5K ± 5%

Figure 8.8 Examples of colour coding

Example 1

A resistor is colour coded red, yellow, orange, gold. Determine the value of the resistor.

- first band red (First digit) 2
- second band yellow (Second digit) 4
- third band orange (No of zeros) 3
- fourth band gold (Tolerance) 5%.

The value is 24,000 Ω ±5%.

Example 2

A resistor is colour coded yellow, yellow, blue, silver. Determine the value of the resistor.

- first band yellow (First digit) 4
- second band yellow (Second digit) 4
- third band blue (No of zeros) 6
- fourth band silver (Tolerance) 10%.

The value is 44,000,000 Ω ±10%.

Example 3

A resistor is colour coded violet, orange, brown, gold. Determine the value of the resistor.

- first band violet (First digit) 7
- second band orange (Second digit) 3
- third band brown (No of zeros) 1
- fourth band gold (Tolerance) 5%.

The value is 730 Ω ± 5%.

Example 4

A resistor is colour coded green, red, yellow, silver. Determine the value of the resistor.

- first band green (First digit) 5
- second band red (Second digit) 2
- third band yellow (No of zeros) 4
- fourth band silver (Tolerance) 10%.

The value is 520,000 Ω ±10%.

Remember

To help you remember the resistor colour, learn this rhyme:

Barbara	Black	0
Brown	Brown	1
Runs	Red	2
Over	Orange	3
Your	Yellow	4
Garden	Green	5
But	Blue	6
Violet	Violet	7
Grey	Grey	8
Won't	White	9

Testing resistors

Resistors must be removed from a circuit before testing, otherwise readings will be false. To measure the resistance, the leads of a suitable ohmmeter should be connected to each resistor connection lead and a reading obtained which should be close to the preferred value and within the tolerance stated.

Resistors as current limiters

A resistor is often provided in a circuit to limit, restrict or reduce the current flowing in the circuit to some level that better suits the ratings of some other component in the circuit. For example, consider the problem of operating a solenoid valve from a 36 V d.c. supply, given the information that the energising current of the coil fitted to the valve is 100 mA and its resistance is 240 Ω.

Note that the coil, being a wound component, is actually an inductor. However, we are concerned here with the steady d.c. current through the coil and not the variation in coil current at the instant the supply is connected, so we can ignore the effects of its inductance and consider only the effects of its resistance.

If the solenoid valve were connected directly across the 36 V supply, as shown in Figure 8.9, then from Ohm's law the steady current through its coil would be:

$$I = \frac{V}{R}$$

$$= \frac{36}{240}$$

$$= 0.15 \text{ A or } 150 \text{ mA}$$

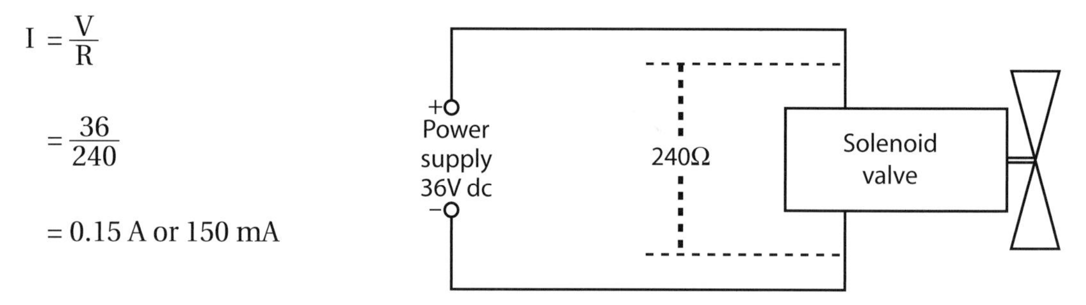

Figure 8.9 Solenoid valve connected across 36 V supply

As the coil was designed to produce an adequate magnetic 'pull' when energised at 100 mA, any increase in the energising current is unnecessary and may in fact be highly undesirable due to the resulting increase in the power, which would be dissipated (as heat) within the coils.

Note: The power dissipated in the coil of the solenoid valve, when energised at the recommended current of 100 mA is:

$$P = I^2 \times R$$

$$= 0.1^2 \text{ A} \times 240\,\Omega = 2.4 \text{ watts}$$

Now if the current were to be 0.15 A on the 36 V supply it would be:

$$P = V \times I$$

$$= 36 \times 0.15$$

$$= 5.4 \text{ watts}$$

Thus connecting the coil directly across a 36 volt supply would result in the power dissipation in it being more than doubled. If the valve is required to be energised for more than very brief periods of time, the coil could be damaged by overheating.

Some extra resistance must therefore be introduced into the circuit so that the current through the coil is limited to 100 mA even though the supply is 36 V.

For a current of 100 mA to flow from a 36 V supply, the total resistance R_t connected across the 36 V must, from Ohm's law, be:

$$R_t = \frac{36}{0.1}$$

$$= 360\ \Omega$$

of which there is already 240 Ω in the coil.

A resistor of value 120 Ω must therefore be fitted in series with the coil to bring the value of R_t to 360 Ω. This limits the current through the coil to 100 mA when the series combination of coil and resistor is connected across the 36 V supply as shown in Figure 8.10.

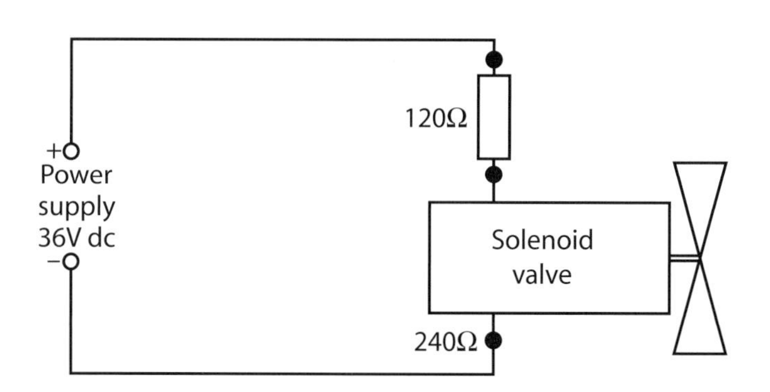

Figure 8.10 Series combination of coil and resistor connected across 36 V supply

Resistors for voltage control

Within a circuit it is often necessary to have different voltages at different stages and we can achieve this by using resistors.

For example, if we physically opened up a resistor and connected its ends across a supply, we would find that, if we then measured the voltage at different points along the resistor, the values would vary along its length. In doing so, we are effectively imitating the 'tapping' technique that we used in the transformers section of this book.

However, reality will stop us from doing this as resistors are sealed components. But we can create the same tapping effect by combining two resistors in series as shown in Figure 8.11, and then our tapping becomes a connection point made between the two resistors.

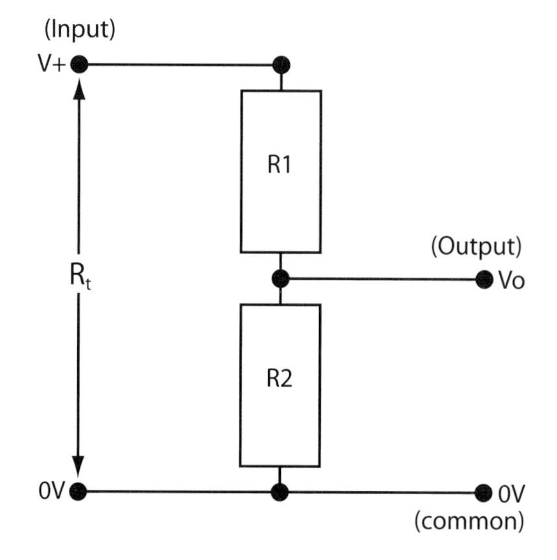

Figure 8.11 Series circuit for voltage control

If we look at Figure 8.11, we can see that the series combination of resistors R_1 and R_2 is connected across a supply that is provided by two rails. One is shown as V+ (the positive supply rail or in other words our input) and the other as 0V (or common rail of the circuit).

The total resistance of our network (R_t) will be:

$$R_t = R_1 + R_2$$

We know, using Ohm's law, that $V = I \times R$ and the same current flows through both resistors. Therefore for this network, we can see that $V+ = I \times R_t$, and the voltage dropped across resistor R_2, $V_o = I \times R_2$.

We now have two expressions, one for V+ and one for Vo. We can find out what fraction Vo is of V+ by putting Vo over V+ on the left hand side of an equation and then putting what we said each one is equal to in the corresponding positions on the right hand side. This gives us the following formula:

$$\frac{Vo}{V+} = \frac{I \times R_2}{I \times R_t}$$

As current is common on the right hand side of our formula, they cancel each other out. This leaves us with:

$$\frac{Vo}{V+} = \frac{R_2}{R_t}$$

To establish what Vo (our output voltage) actually is, we can transpose again, which would give us:

$$Vo = \frac{R_2 \times V+}{R_t}$$

Finally, we can replace R_t by what it is actually equal to, and this will give us the means of establishing the value of the individual resistors needed to give a desired output voltage (Vo). By transposition, our final formula now becomes:

$$Vo = \frac{R_2}{R_1 + R_2} \times V+$$

This equation is normally referred to as the potential divider rule.

In reality R_1 and R_2 could each be a combination (series or parallel) of many resistors.

However, as long as each combination is replaced by its equivalent resistance so that the simplified circuit looks like Figure 8.11, then the potential divider rule can be applied.

The potential divider circuit is very useful where the full voltage available is not required at some point in a circuit and, as we have seen, by a suitable choice of resistors in the potential divider, the desired fraction of the input voltage can be produced.

In applications where the fraction produced needs to be varied from time to time, the two resistors are replaced by a variable resistor (also known as a potentiometer, which is often abbreviated to the word pot), which would be connected as shown in Figure 8.12.

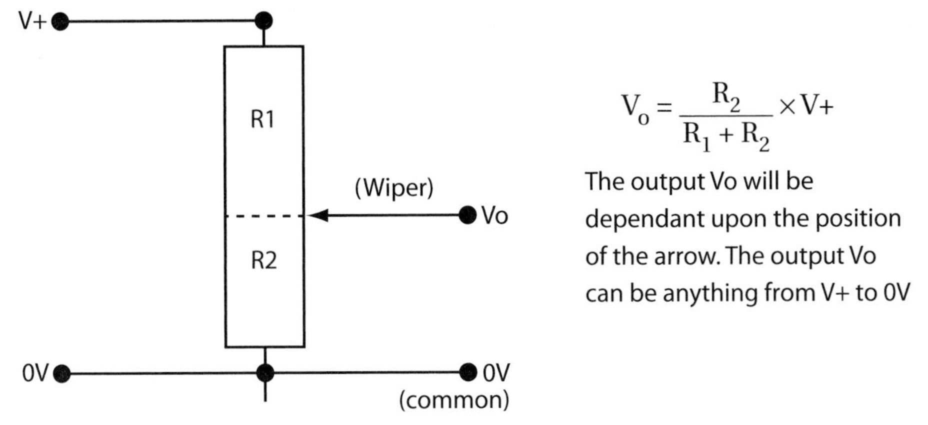

Figure 8.12 Circuit diagram for voltage applied across potentiometer

The potentiometer has a resistor manufactured in the form of a track, the ends of which effectively form our V+ and 0V connections. Our output voltage (V_o) is achieved by means of a movable contact that can touch the track anywhere on its length and this is called the wiper. We have therefore effectively created a variable tapping point.

To compare this with our potential divider, we can say that the part of the track above the wiper can be regarded as R_1 and that part below the wiper as R_2. The fraction of the input voltage appearing at the output can therefore be calculated for any setting of the wiper position by using our potential divider equation:

$$V_o = \frac{R_2}{R_1 + R_2} \times V+$$

Obviously, when the wiper is at the top of the track, R_1 becomes zero and the equation would give the result that Vo is equal to V+. Equally, with the wiper right at the bottom of the track, R_2 now becomes zero and therefore Vo also becomes zero, which is not too surprising as the wiper is now more or less directly connected to the 0V rail.

This sort of circuit finds practical application in a wide variety of control functions such as volume or tone controls on audio equipment, brightness and contrast controls on televisions and shift controls on oscilloscopes.

Power ratings

Resistors often have to carry comparatively large values of current so they must be capable of doing this without overheating and causing damage. As the current has to be related to the voltage, it is the power rating of the resistor that needs to be identified.

The power rating of a resistor is thus really a convenient way of stating the maximum temperature at which the resistor is designed to operate without damage to itself. In general the more power a resistor is designed to be capable of dissipating, the larger physically the resistor is; the resulting larger surface area aids heat dissipation.

Resistors with high power ratings may even be jacketed in a metal casing provided with cooling ribs and designed to be bolted flat to a metal surface – all to improve the radiation and conduction of heat away from the resistance element.

Power is calculated by:

$$P = V \times I$$

Instead of V we can substitute $I \times R$ for V and $\frac{V}{R}$ for I. We can then use the following equations to calculate power:

$$P = I^2 \times R$$

Or:

$$P = \frac{V^2}{R}$$

What would the power rating of the 50 Ω resistor in Figure 8.13 be?

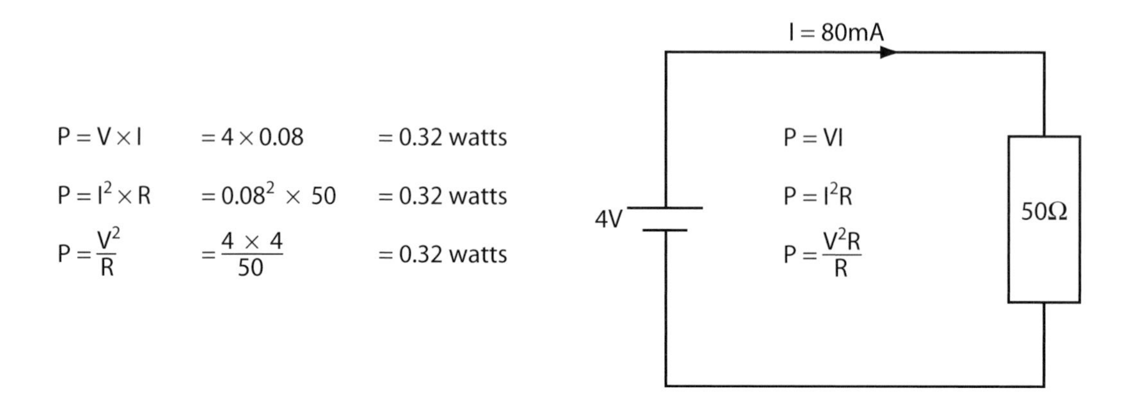

Figure 8.13 Typical power ratings for resistors

Normally only one calculation is required. Typical power ratings for resistors are shown in Table 8.4.

Carbon resistors	0 to 0.5 watts
Ceramic resistors	0 to 6 watts
Wire wound resistors	0 to 25 watts

Table 8.4 Typical power ratings for resistors

Manufacturers also always quote a maximum voltage rating for their resistors on their data sheets. The maximum voltage rating is basically a statement about the electrical insulation properties of those parts of the resistor, which are supposed to be insulators (e.g. the ceramic or glass rod which supports the resistance element or the surface coating over the resistance element).

If the maximum voltage rating is exceeded there is a danger that a flashover may occur from one end of the resistor to the other. This flashover usually has disastrous results. If it occurs down the outside of the resistor it can destroy not only the protective coating but, on film resistors, the resistor film as well.

If it occurs down the inside of the resistor the ceramic or glass rod is frequently cracked (if not shattered) and, of course, this mechanical damage to the support for the resistance element results in the element itself being damaged as well.

Thermistors

A thermistor is a resistor which is temperature sensitive. The general appearance is shown in the photograph. They can be supplied in various shapes and are used for the measurement and control of temperature up to their maximum useful temperature limit of about 300°C. They are very sensitive and because of their small construction they are useful for measuring temperatures in inaccessible places.

Thermistors are used for measuring the temperature of motor windings and sensing overloads. The thermistor can be wired into the control circuit so that it automatically cuts the supply to the motor when the motor windings overheat, thus preventing damage to the windings.

Did you know?

Thermistors are used for monitoring the temperature of the water in a motor car

Thermistors can have a temperature coefficient that may be positive (PTC) or negative (NTC). With a PTC thermistor the resistance of the thermistor increases as the surrounding temperature increases. With the NTC thermistor the resistance decreases as the temperature increases.

The rated resistance of a thermistor may be identified by a standard colour code or by a single body colour used only for thermistors. Typical values are shown in Table 8.5.

Colour	Resistance
Red	3,000 Ω
Orange	5,000 Ω
Yellow	10,000 Ω
Green	30,000 Ω
Violet	100,000 Ω

Table 8.5 Colour coding for rated resistance of thermistor

Thermistors

Light-dependent resistors

These resistors are sensitive to light. They consist of a clear window with a cadmium sulphide film under it. When light shines onto the film its resistance varies, with the resistance reducing as the light increases.

These resistors are commonly found in street lighting. You may sometimes observe street lights switching on during a thunderstorm in the daytime. This is because the sunlight is obscured by the dark thunderclouds, thus increasing the resistance, which in turn controls the light 'on' circuit.

Light-dependent resistor

Capacitors

Just as resistors enable us to introduce known amounts of resistance into a circuit to serve our purposes, so we can use components known as capacitors to introduce capacitance into a circuit. Like resistance, capacitance always exists in circuits – though, as you'll see when we have discussed the subject in more detail, capacitance exists between conductors whereas resistance exists in conductors.

Safety tip

Never pick a capacitor up by the terminals as it may still be charged and you will receive a shock. Always ensure the capacitor has been discharged before handling. Some capacitors have a discharge resistor connected in the circuit for this reason

Basic principles

A capacitor is basically two metallic surfaces usually referred to as plates, separated by an insulator commonly known as the dielectric. The plates are usually, though not necessarily, metal and the dielectric is any insulating material. Air, glass, ceramic, mica, paper, oils and waxes are some of the many materials commonly used. The common symbols used for capacitors are identified in Figure 8.14.

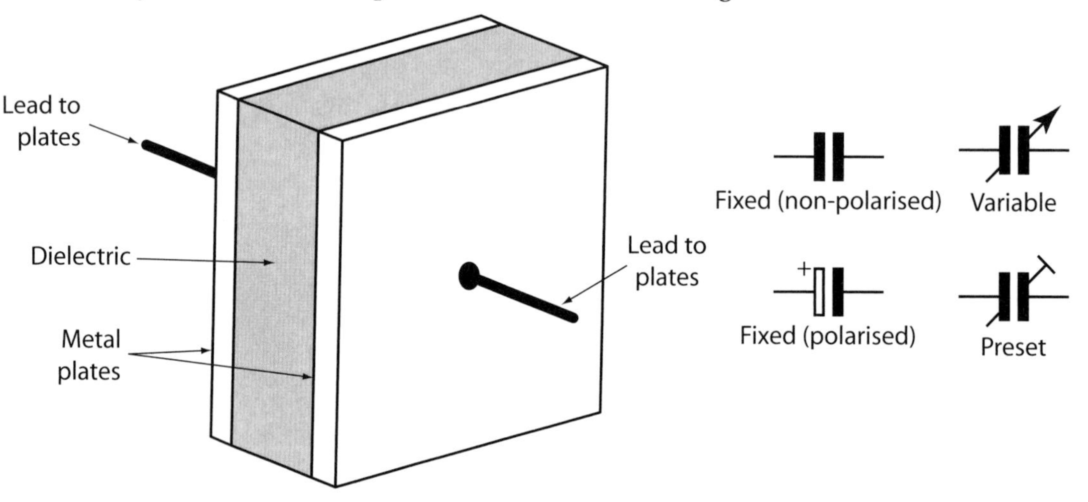

Figure 8.14 Basic construction of a capacitor and circuit symbols

Remember

In any application the capacitor to be used should meet or preferably exceed the capacitor voltage rating. The voltage rating is often called 'the working voltage' and refers to d.c. voltage values. When applied to a.c. circuits the peak voltage value must be used as a comparison to the d.c. working voltage of a capacitor

These two plates are not in contact with each other and as such they do not form a circuit in the same way that conductors with resistors do. However, the capacitor stores a small amount of electric charge and as such it can be thought of as a small rechargeable battery, which can be quickly recharged.

The capacitance of any capacitor depends on three factors:

1. It depends on the working area of the plates i.e. the area of the conducting surfaces facing each other. We can think of the degree of crowding of excess electrons near the surface of one plate of a capacitor (and the corresponding sparseness of electrons near the surface of the other) as being directly related to the potential difference (p.d.) applied across the capacitor, for example connecting it directly across a battery. If we increase the area of the plates, then more electrons can flow onto one of the plates before the same degree of crowding is reached. The battery voltage determines this level of crowdedness. There is of course a similar increased loss of electrons from the other plate. The working area of the plates is directly proportional to the capacitance. If we double the area of the plates we double the capacitance of the capacitor.

2. Capacitance depends on the thickness of the **dielectric** between the plates. As was mentioned earlier, the capacitance effect depends on the forces of repulsion or attraction caused by an electron surplus or shortage on the plates on either side of the dielectric. The further apart the plates are, the weaker these factors become. As a result, the degree of crowding of electrons on one plate (and the shortage of electrons on the other) produced by a given p.d across the capacitor decreases.
3. The capacitance depends on the nature of the dielectric or spacing material used. This fundamental principle of capacitors and the time constant of capacitor resistor circuits will be looked at later under the heading of electrostatics.

Capacitor types

There are two major types of capacitor, fixed and variable, both of which are used in a wide range of electronic devices. Fixed capacitors can be further subdivided into electrolytic and non-electrolytic types and together they represent the majority of the market.

All capacitors possess some resistance and inductance because of the nature of their construction. These undesirable properties result in limitations, which often determine their applications.

Fixed capacitors

Electrolytic capacitors

These capacitors have a much higher capacitance, volume for volume, than any other type. This is achieved by making the plate separation extremely small by using a very thin dielectric (insulator). The dielectric is often mica or paper.

They are constructed on the Swiss roll principle as are the paper dielectric capacitors used for Power Factor correction in electrical installation circuits, for example fluorescent lighting circuits.

The main disadvantage of an electrolytic capacitor is that it is polarised and must be connected to the correct polarity in a circuit, otherwise a short circuit and destruction of the capacitor will result.

Figure 8.15 illustrates a newer type of electrolytic capacitor using tantalum and tantalum oxide to give a further capacitance/size advantage. It looks like a raindrop with two leads protruding from the bottom. The polarity and values may be marked on the capacitor (see photograph in Polarity section on page 280) or the colour code (see Figure 8.22 on page 296) can be used.

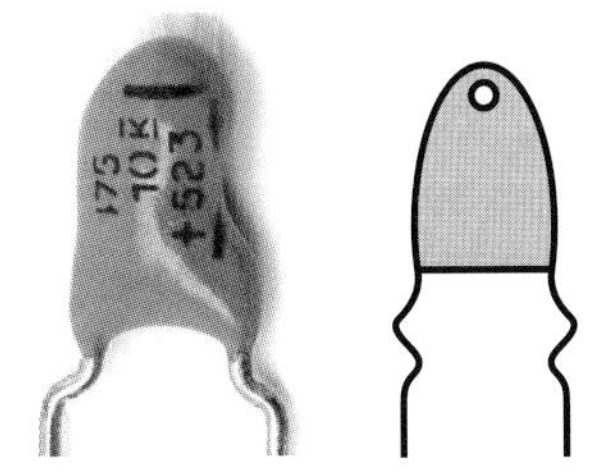

Figure 8.15 Tantalum capacitor

Non-electrolytic capacitors

There are many different types of non-electrolytic capacitor. However, only mica, ceramic and polyester are of any significance. Older types using glass and vitreous enamel are expected to disappear over the next few years and even mica will be replaced by film types.

Did you know?

Mica is a common rock-forming mineral. You find it in rocks such as granite and some sandstones and mudstones

Mica

Mica is a naturally occurring dielectric and has a very high resistance; this gives excellent stability and allows the capacitors to be accurate within a value of ±1 per cent of the marked value. Since costs usually increase with increased accuracy, they tend to be more expensive than plastic film capacitors. They are used where high stability is required, for example in tuned circuits and filters required in radio transmission. Figure 8.16 illustrates a typical mica capacitor.

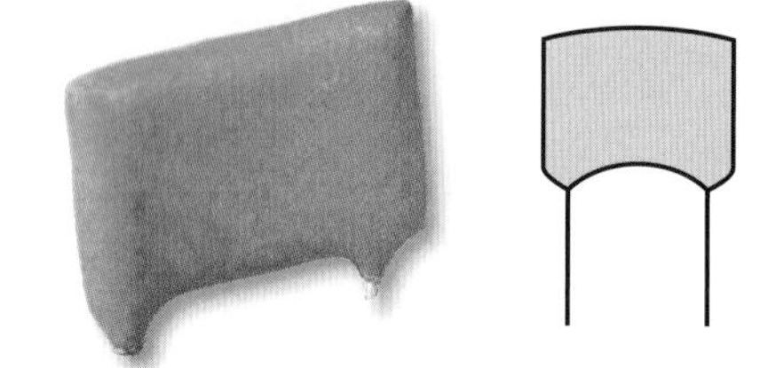

Figure 8.16 Mica capacitor

Ceramic capacitors

These consist of small rectangular pieces of ceramic with metal electrodes on opposite surfaces. Figure 8.17 illustrates a typical ceramic capacitor. These capacitors are mainly used in high-frequency circuits subjected to wide temperature variations. They have high stability and low loss.

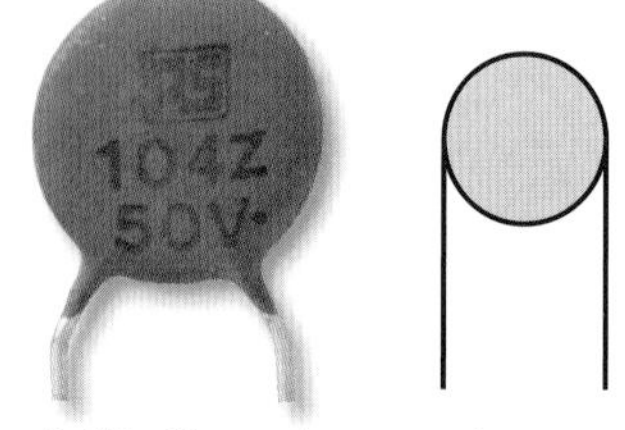

Figure 8.17 Ceramic capacitor

Polyester capacitors

These are an example of a plastic film capacitor. Polypropylene, polycarbonate and polystyrene capacitors are other types of plastic film capacitors. They are widely used in the electronics industry due to their good reliability and relative low cost but are not suitable for high-frequency circuits. Figure 8.18 illustrates a typical polyester capacitor; however, they can also be a tubular shape (see Figure 8.19).

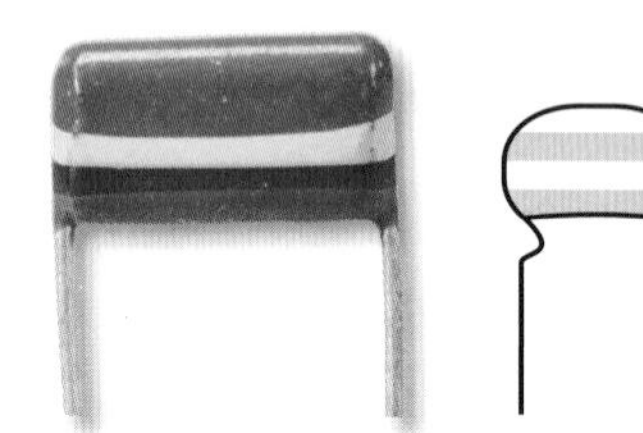

Figure 8.18 Polyester capacitor

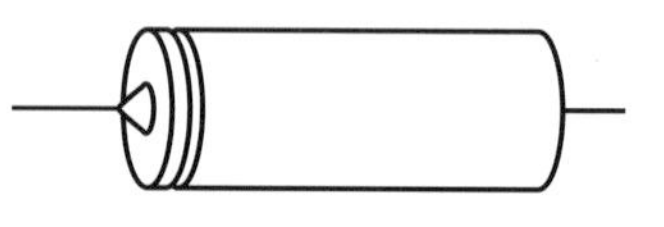

Figure 8.19 Tubular capacitor

Variable capacitors

Variable capacitors generally have air or a vacuum as the dielectric, although ceramics are sometimes used. The two main sub-groups are tuning and trimmer capacitors.

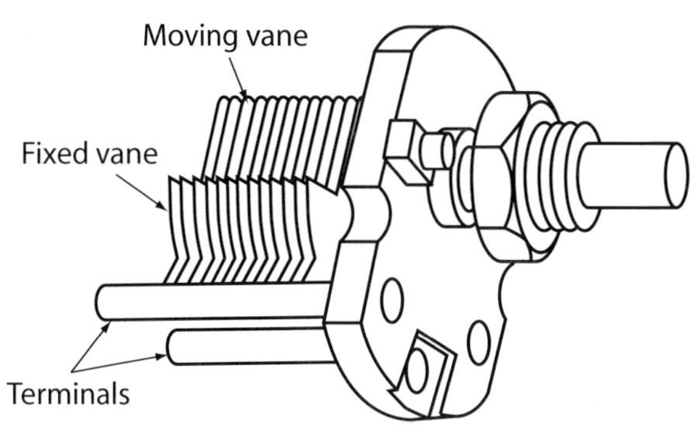

Figure 8.20 A typical variable capacitor of the tuning type

Tuning capacitors

These are so called because they are used in radio tuning circuits and consist of two sets of parallel metal plates, one isolated from the mounting frame by ceramic supports while the other is fixed to a shaft which allows one set to be rotated into or out of the first set. The rows of plates interlock like fingers, but do not quite touch each other.

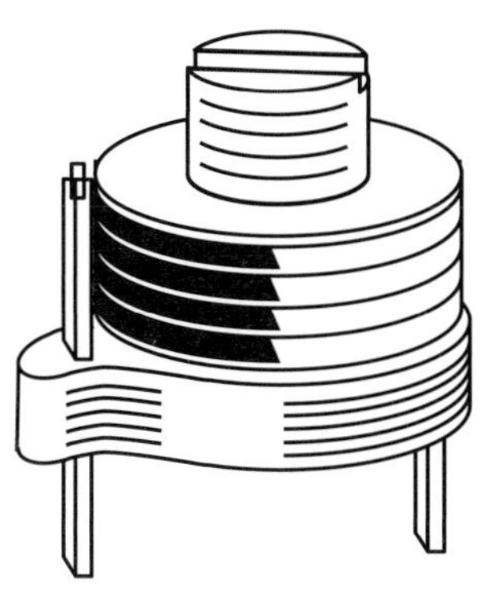
Figure 8.21 A typical capacitor used as a trimmer

Trimmer capacitors

These are constructed of flat metal leaves separated by a plastic film; these can be screwed towards each other. They have a smaller range of variation than tuning capacitors, and so are only used where a slight change in value is needed.

Capacitor coding

To identify a capacitor the following details must be known: the capacitance, working voltage, type of construction and polarity (if any). The identification of capacitors is not easy because of the wide variation in shapes and sizes. In the majority of cases the capacitance will be printed on the body of the capacitor, which often gives a positive identification that the component is a capacitor.

The capacitance value is the farad (symbol F); this was named after the English scientist Michael Faraday. However, for practical purposes the farad is much too large and in electrical installation work and electronics we use fractions of a farad as follows:

- 1 microfarad = 1 μF = 1×10^{-6} F
- 1 nanofarad = 1 nF = 1×10^{-9} F
- 1 picofarad = 1 pF = 1×10^{-12} F.

The power factor correction capacitor found in fluorescent luminaries would have a value typically of 8 μF at a working voltage of 400 V. One microfarad is one million times greater than one picofarad.

The working voltage of a capacitor is the maximum voltage that can be applied between the plates of the capacitor without breaking down the dielectric insulating material.

An ideal capacitor, which is isolated, will remain charged forever. But in practice no dielectric insulating material is perfect, and therefore the charge will slowly leak between the plates, gradually discharging the capacitor. The loss of charge by leakage should be very small for a practical capacitor

It was quite common for capacitors to be marked with colour codes but today relatively few capacitors are colour coded. At one time nearly all plastic foil type capacitors were colour coded, as in Figure 8.22, but this method of marking is rarely encountered. However, it is a useful skill to know and be able to use the colour-coding method as shown in Table 8.6.

This method is based on the standard four-band resistor colour coding. The first three bands indicate the value in normal resistor fashion, but the value is in picofarads. To convert this into a value in nanofarads it is merely necessary to divide by 1000. Divide the marked value by 1,000,000 if a value in microfarads is required. The fourth band indicates the tolerance, but the colour coding is different to the resistor equivalent. The fifth band shows the maximum working voltage of the component. Details of this colour coding are shown in Figure 8.22 and Table 8.6.

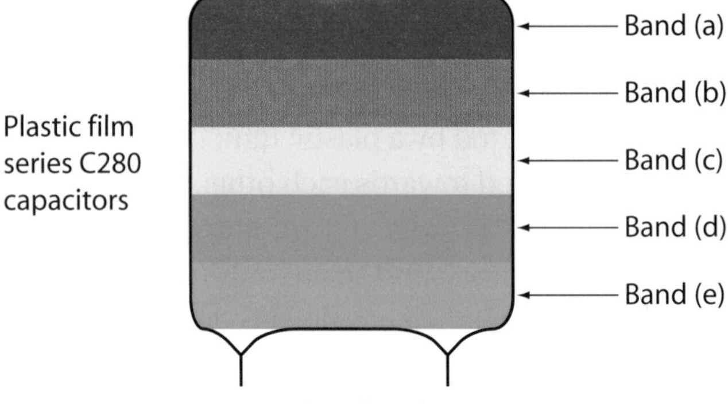

Figure 8.22 Capacitor colour bands

Standard capacitor colour coding

Colour	1st Digit	2nd Digit	3rd Digit	Tol. Band	Max. Voltage
Black		0	None	20%	
Brown	1	1	1		100 V
Red	2	2	2		250 V
Orange	3	3	3		
Yellow	4	4	4		400 V
Green	5	5	5	5%	
Blue	6	6	6		630 V
Violet	7	7	7		
Grey	8	8	8		
White	9	9	9	10%	

Table 8.6 Standard capacitor colour coding

Bands are then read from top to bottom. Digit 1 gives the first number of the component value; the second digit gives the second number. The third band gives the number of zeros to be added after the first two numbers and the fourth band indicates the capacitor tolerance, which is normally black 20%, white 10% and green 5%.

Example 1

A plastic film capacitor is colour coded from top to bottom brown, red, yellow, black, red. Determine the value of the capacitor, its tolerance and working voltage.

- band (a) – brown = 1
- band (b) – red = 2
- band (c) – yellow = 4 multiply by 10,000
- band (d) – black = 20% tolerance
- band (e) – red = 250 volts.

The capacitor has a value of 120,000 pF or 0.12 μF with a tolerance of 20% and a maximum working voltage of 250 volts.

Example 2

A plastic film capacitor is colour coded from top to bottom orange, orange, yellow, green, yellow. Determine the value of the capacitor, its tolerance and working voltage.

- band (a) – orange = 3
- band (b) – orange = 3
- band (c) – yellow = 4 multiply by 10,000
- band (d) – green = 5%
- band (e) – yellow = 400 volts.

The capacitor has a value of 330,000 pF or 0.33 μF with a tolerance of 5% and a maximum working voltage of 400 volts.

Example 3

A plastic film capacitor is colour coded from top to bottom violet, blue, orange, black, and brown. Determine the value of the capacitor, its tolerance and working voltage.

- band (a) – violet = 7
- band (b) – blue = 6
- band (c) – orange = 3 multiply by 1,000
- band (d) – black = 20%
- band (e) – brown = 100 volts.

The capacitor has a value of 76,000 pF or 0.076 μF with a tolerance of 20% and a maximum working voltage of 100 volts.

Often the value of a capacitor is simply written on its body, possibly together with the tolerance and/or its maximum operating voltage. The tolerance rating may be omitted, and it is generally higher for capacitors than resistors. Most modern resistors have tolerances of 5% or better, but for capacitors the tolerance rating is generally 10% or 20%. The tolerance figure is more likely to be marked on a close tolerance capacitor than a normal 10% or 20% type.

The most popular form of value marking on modern capacitors is for the value to be written on the components in some slightly cryptic form. Small ceramic capacitors generally have the value marked in much the same way that the value is written on a circuit diagram.

Where the value includes a decimal point, it is standard practice to use the prefix for the multiplication factor in place of the decimal point. This is the same practice as was used for resistors.

The abbreviation μ means microfarad; n means nanofarad; p means picofarad. Therefore:

- a 3.5 pF capacitor would be abbreviated to 3p5
- a 12 pF capacitor would be abbreviated to 12p
- a 300 pF capacitor would be abbreviated to 300p or n30
- a 4,500 pF capacitor would be abbreviated to 4n5
- 1,000 pF = 1 nF = 0.001 μF.

Polarity

Once the size, type and d.c. voltage rating of a capacitor have been determined it now remains to ensure that its polarity is known. Some capacitors are constructed in such a way that if the component is operated with the wrong polarity its properties as a capacitor will be destroyed, especially electrolytic capacitors. Polarity may be indicated by a + or – as appropriate. Electrolytic capacitors that are contained within metal cans will have the can casing as the negative connection. If there are no markings a slight indentation in the case will indicate the positive end.

Capacitor showing polarity

Tantalum capacitors have a spot on one side as shown in Figure 8.23.

When this spot is facing you the right hand lead will indicate the positive connection.

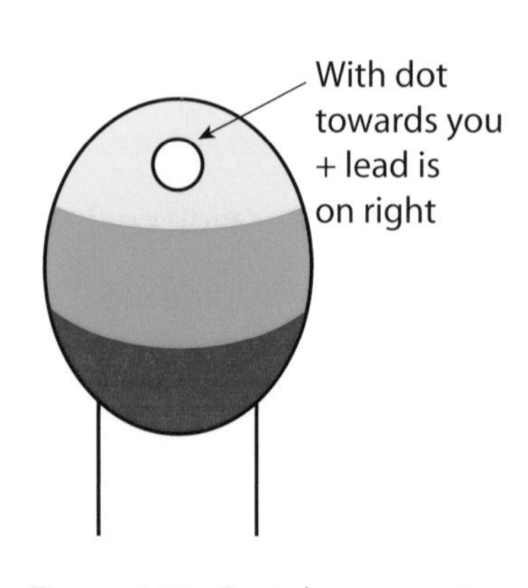

Figure 8.23 Tantalum capacitor

Electrostatics and calculations with capacitors

The charge stored on a capacitor is dependent on three main factors: the area of the facing plates; the distance between the plates; and the nature of the dielectric. The charge stored by a capacitor is measured in coulombs (Q) and is related to the value of capacitance and the voltage applied to the capacitor:

Charge (coulombs) = Capacitance (farads) × Voltage (volts)

$$Q = C \times V$$

The formula for energy stored in a capacitor can be calculated by using the formula:

$$W = \frac{1}{2}CV^2$$

Capacitors in combination

Capacitors, like resistors, may be joined together in various combinations of series or parallel connections. Figures 8.24 and 8.25 illustrate the equivalent capacitance C_t of a number of capacitors. C_t can be found by applying similar formulae as for resistors. However, these formulae are the opposite way round to series and parallel resistors.

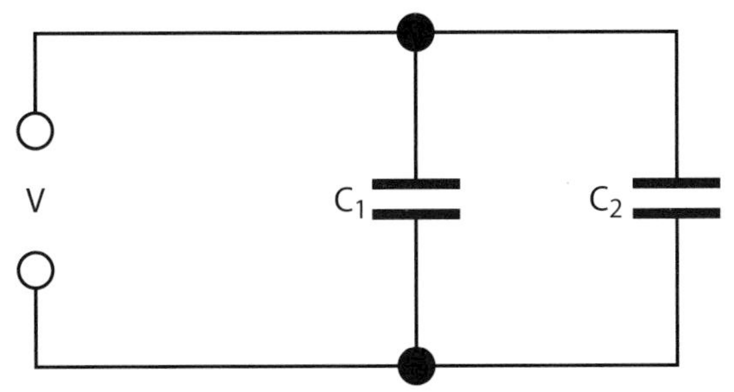

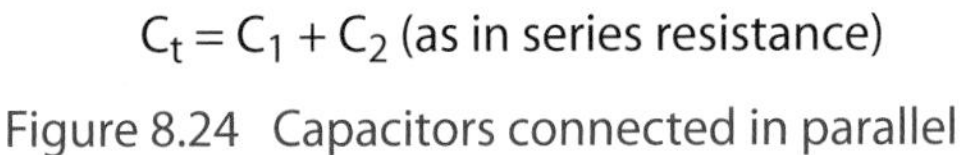

$C_t = C_1 + C_2$ (as in series resistance)

Figure 8.24 Capacitors connected in parallel

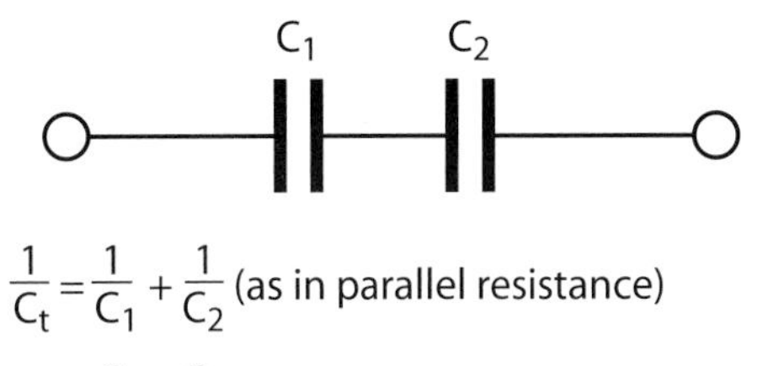

$$\frac{1}{C_t} = \frac{1}{C_1} + \frac{1}{C_2} \text{ (as in parallel resistance)}$$

or

$$C_t = \frac{C_1 \times C_2}{C_1 + C_2} \text{ (when there are two capacitors in series)}$$

Figure 8.25 Capacitors connected in series

Example 1

Capacitors of 10 μF and 40 μF are connected in series and then in parallel. Calculate the effective capacitance for each connection.

Series:

$$\frac{1}{C_t} = \frac{1}{C_1} + \frac{1}{C_2}$$

$$\frac{1}{C_t} = \frac{1}{10\,\mu F} + \frac{1}{40\,\mu F}$$

$$\frac{1}{C_t} = \frac{4\,\mu F + 1\,\mu F}{40\,\mu F}$$

$$\frac{1}{C_t} = \frac{5\,\mu F}{40\,\mu F}$$

Therefore:

$$\frac{C_t}{1\,\mu F} = \frac{40\,\mu F}{5\,\mu F}$$

$$C_t = 8\mu F$$

Parallel:

$$C_t = C_1 + C_2$$

$$C_t = 10\,\mu F + 40\,\mu F$$

$$C_t = 50\,\mu F$$

Example 2

Three capacitors of 30 μF, 20 μF and 15 μF are connected in series across a 400 V d.c. supply. Calculate the total capacitance and the charge on each capacitor.

$$\frac{1}{C_t} = \frac{1}{C_1} + \frac{1}{C_2} + \frac{1}{C_3}$$

$$= \frac{1}{30\ \mu F} + \frac{1}{20\ \mu F} + \frac{1}{15\ \mu F}$$

$$= \frac{9\ \mu F}{60\ \mu F}$$

$$\therefore C_t = 6.66\ \mu F$$

Q, the charge, is common to each capacitor. Therefore:

$$Q = C \times V$$

$$Q = 6.66 \times 10^{-6} \times 400$$

$$Q = 2.664 \text{ m coulombs}$$

Example 3

Three capacitors of 30 μF, 20 μF and 15 μF are connected in parallel across a 400 V d.c. supply. Calculate the total capacitance, the total charge and the charge on each capacitor.

$$C_t = C_1 + C_2 + C_3$$

$$C_t = 30 + 20 + 15$$

$$C_t = 65\ \mu F$$

Total charge $Q = C \times V$

$$Q_t = C_1 \times V$$

$$Q_t = 65 \times 10^{-6} \times 400$$

$$Q_t = 2.6 \text{ m coulombs}$$

$$Q_1 = C_1 \times V$$

$$Q_1 = 30 \times 10^{-6} \times 400$$

$$Q_1 = 1.2 \text{ m coulombs}$$

$$Q_2 = C_2 \times V$$

$$Q_2 = 20 \times 10^{-6} \times 400$$

$$Q_2 = 0.8 \text{ m coulombs}$$

$$Q_3 = C_3 \times V$$

$$Q_3 = 15 \times 10^{-6} \times 400$$

$$Q_3 = 0.06 \text{ m coulombs}$$

Charging and discharging capacitors

When a capacitor is connected to a battery, positive and negative charges are deposited on the capacitor plates and the capacitor will charge up. These charges build up fast but not instantaneously; they follow a set pattern of charge. A typical charge and discharge circuit is illustrated in Figure 8.26.

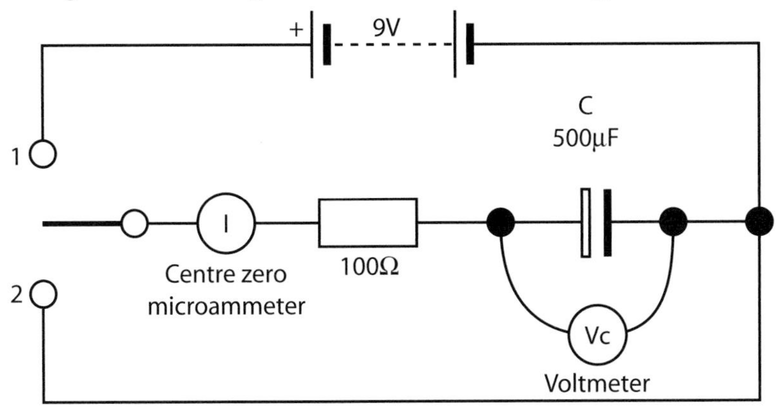

Figure 8.26 A typical charge and discharge circuit

When the switch contact is connected to 1, current flows from the battery to the capacitor plates and the electrolytic capacitor will charge up to the same voltage as the battery.

From the time the current (measured by the microammeter) is switched on a charge curve starts that is related to the charge voltage (measured by the voltmeter) and time. The maximum voltage that the capacitor can take is that of the supply; this then sets the voltage scale from 0 volts to maximum voltage (9 V in this example).

The time scale is divided into sections called time constants. The pattern of charge is always the same. Figure 8.27 illustrates a graph of voltage against time for charging the capacitor.

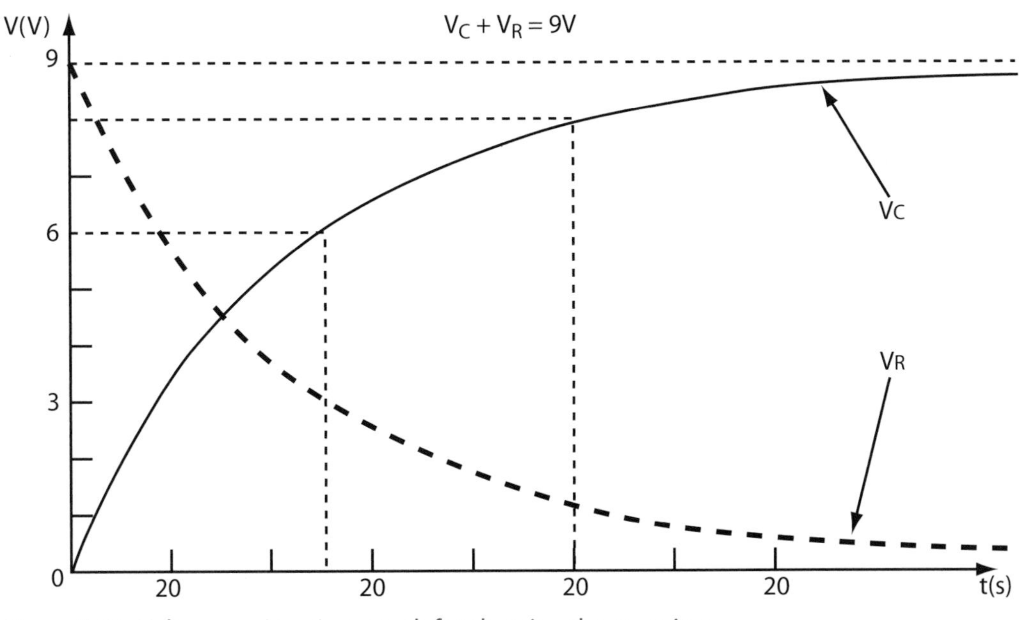

Figure 8.27 Voltage against time graph for charging the capacitor

At 'switch on' the voltage V_C across the capacitor is zero and the voltage V_R across the resistor is at maximum supply voltage.

As the current flow charges up the capacitor plates, V_C increases with the charging of the capacitor plates. V_R reduces over the charge time because the rate of current flow reduces since the supply voltage and the voltage across the capacitor will be at the same potential at completion of charge, therefore no current will be flowing. The current will follow a similar line to V_R on the graph.

The time scale has been subdivided into equal time periods, each time period known as a time constant. In the first time period the voltage reaches approximately two-thirds of the maximum volts (6 V), in the next period the charge goes from the finishing point in the first time constant to finish about two-thirds of what is left (2 V).

This pattern continues until the capacitor is fully charged. The rate of charge will be exponential as shown in Figure 8.28 and it will take five time constants for the capacitor to be fully charged. This also applies if the capacitor is discharging through a series resistor; the voltage (V_C) across the capacitor will fall exponentially and it will again take five time constants for the capacitor to be fully discharged.

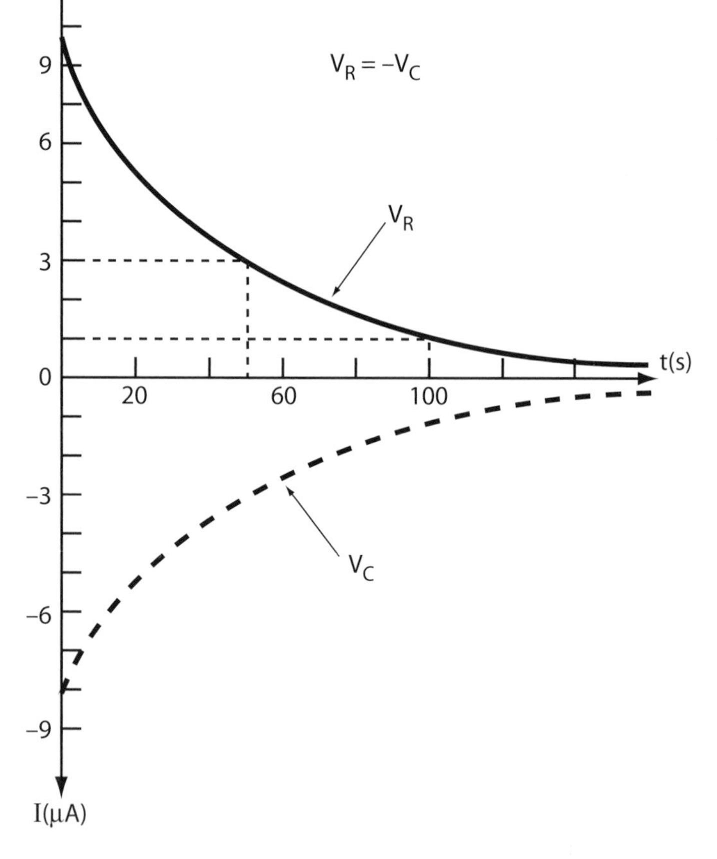

Graph of voltages V_R and V_C

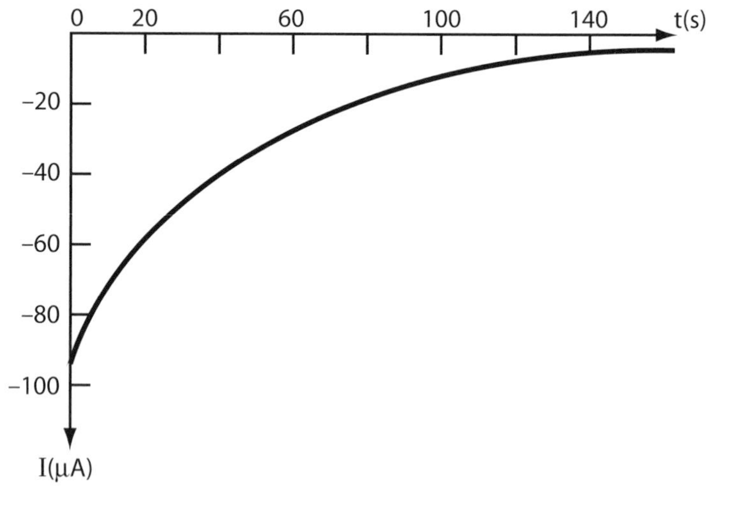

Graph of current I(μA)

Figure 8.28 Discharge curve

Semiconductor devices

The crystal radio that can be constructed from modern electronic game sets depends on the detector action produced by a 'cat's whisker' and a crystal (a cat's whisker was a piece of wire, the point of which was pressed firmly into contact with a suitably mounted piece of natural crystal).

This crystal detector was in fact a diode. It is the use of diodes, semiconductors and semiconductor devices that we will investigate in this section.

Semiconductor basics

Try to think of a semiconductor as being a material that has an electrical quality somewhere between a conductor and an insulator, in that it is neither a good conductor nor a good insulator. Typically, we use semi-conducting materials such as silicon or germanium, where the atoms of these materials are arranged in a 'lattice' structure. The lattice has atoms at regular distances from each other, with each atom 'linked' or 'bonded' to the four atoms surrounding it. Each atom then has four **valence electrons**.

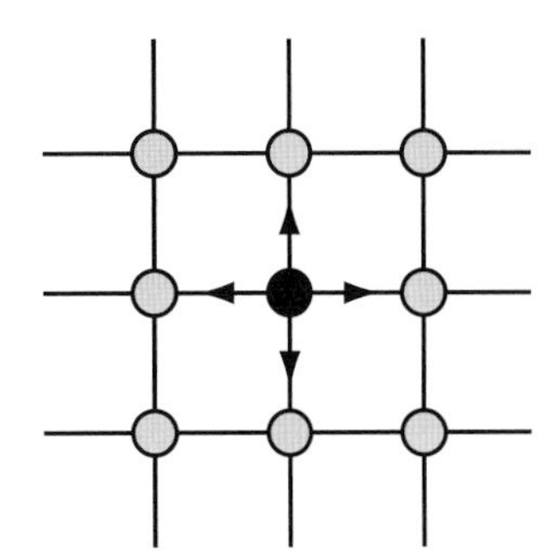

Figure 8.29 Lattice structure of semi-conducting material

Definition

Valence electrons – the electrons in an atom's outermost orbit

However, we have a problem in that, with atoms of pure silicon or germanium no conduction is possible because we have no free electrons. To allow conduction to take place we add an impurity to the material via a process known as **doping**. When we dope the material we can add two types of impurity:

- pentavalent – e.g. arsenic which contains five valence electrons
- trivalent – e.g. aluminium that contains three valence electrons.

As we can see by the number of valence electrons in each, adding a pentavalent (five) material introduces an extra electron to the semiconductor and adding a trivalent (three) material to the semiconductor 'removes' an electron (also known as creating a hole).

When we have an extra electron, we have a surplus of negative charge and call this type of material 'n-type'. When we have 'removed' an electron we have a surplus of positive charge and call this material 'p-type'. It is the use of these two materials that will allow us to introduce the component responsible for rectification, the diode.

The p–n junction

A semiconductor diode is basically created when we bring together an 'n-type' material and a 'p-type' material to form a p–n junction. The two materials form a barrier where they meet which we call the **depletion** layer. In this barrier, the coming together of unlike charges causes a small internal p.d. to exist.

We now need to connect a battery across the ends of the two materials, where we call the end of the p-type material the anode and the end of the n-type material the cathode.

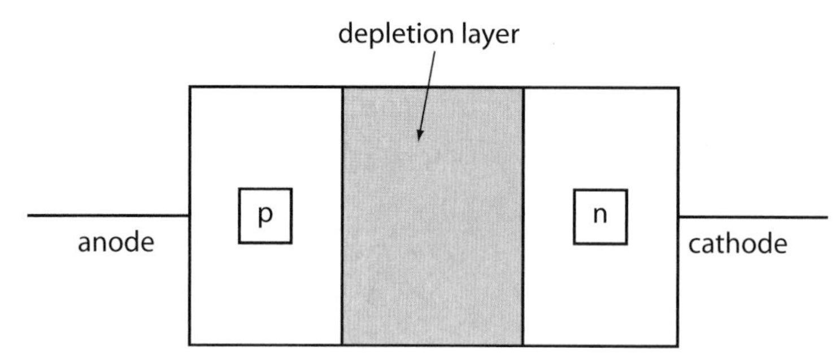

Figure 8.30 p–n junction

If the anode is positive and the battery voltage is big enough, it will overcome the effect of the internal p.d. and push charges (both positive and negative) over the junction. In other words, the junction has a low enough resistance for current to flow. This type of connection is known as being forward biased.

Diodes

Reverse the battery connections so that the anode is now negative and the junction becomes high resistance and no current can flow. This type of connection is known as being reverse biased.

When the junction is forward biased, it only takes a small voltage (0.7 V for silicon) to overcome the internal barrier p.d.

When reversed biased, it takes a large voltage (1200 V for silicon) to overcome the barrier and thus destroy the diode, effectively allowing current to flow in both directions. As a general summary of its actions, we can therefore say that a diode allows current to flow through it in one direction only.

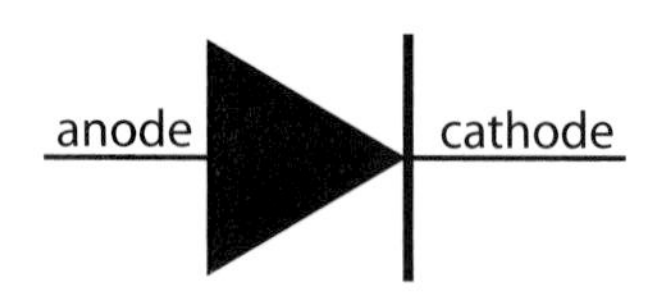

Figure 8.31 Symbol representing a diode

We normally use the symbol in Figure 8.31 to represent a diode.

In this symbol, the direction of the arrow can be taken to represent the direction of current flow.

Zener diode

We have just established that a conventional diode will not allow current flow if reverse biased and below its reverse breakdown voltage. We also said that when forward biased (in the direction of the arrow), the diode exhibits a voltage drop of roughly 0.6V for a typical silicon diode. If we were to exceed the breakdown voltage, the internal barrier of the diode would be destroyed, thus allowing current flow in both directions. However, this normally results in the total destruction of the device.

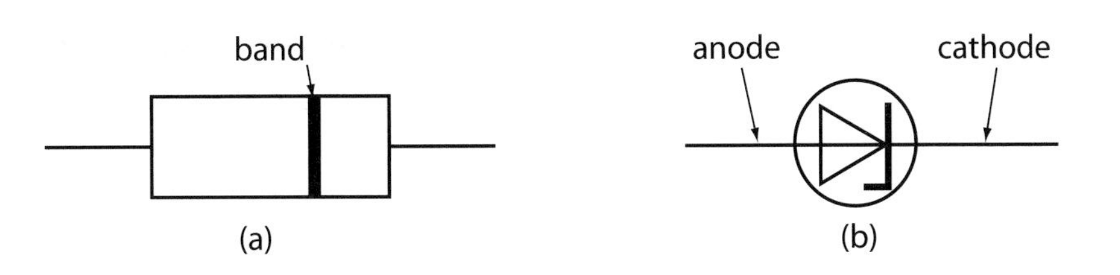

Figure 8.32 Zener diode characteristics

Zener diodes are p–n junction devices that are specifically designed to operate in the reverse breakdown region without completely destroying the device. The breakdown voltage of a zener diode V_Z, (known as the zener voltage, named after the American physicist Clarence Zener who first discovered the effect) is set by carefully controlling the doping level during manufacture. The breakdown voltage can be controlled quite accurately in the doping process and tolerances to within 0.05 per cent are available, although the most widely used tolerances are 5 per cent and 10 per cent.

Therefore, a reverse biased zener diode will exhibit a controlled breakdown, allow current to flow and thus keep the voltage across the zener diode at the predetermined zener voltage. Because of this characteristic, the zener diode is commonly used as a form of voltage limiting/regulation when connected in parallel across a load.

When connected so that it is reverse biased in parallel with a variable voltage source, a zener diode acts as a short circuit when the voltage reaches the diode's reverse breakdown voltage and therefore limits the voltage to a known value.

A zener diode used in this way is known as a shunt voltage regulator (shunt meaning connected in parallel and voltage regulator being a class of circuit that produces a fixed voltage).

For a low current power supply, a simple voltage regulator could be made with a resistor (to limit the operating current) and a reverse biased zener diode as shown in Figure 8.33. Here, Vs is the supply voltage, remembering that V_Z is our zener breakdown voltage.

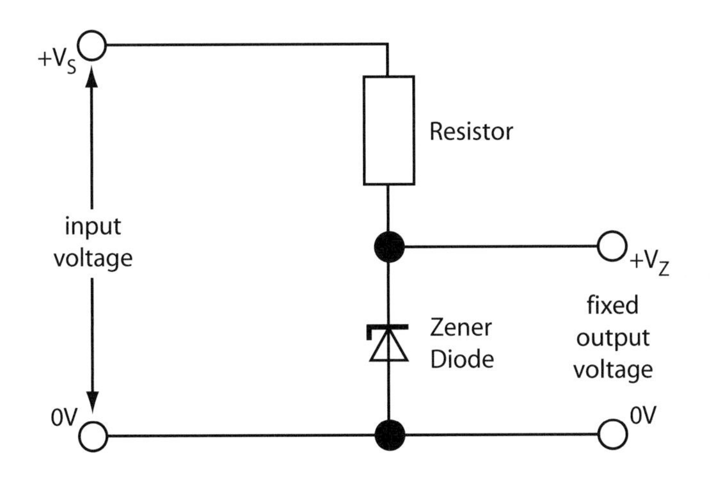

Figure 8.33 A simple voltage regulator made with a resistor and a reverse biased zener diode

As a summary, we can therefore say that a zener's properties are as follows:

- When forward biased (although not normally used for a zener) the behaviour is like an ordinary semiconductor diode.
- When reverse biased, at voltages below V_Z the device essentially doesn't conduct, and it behaves just like an ordinary diode.
- When reverse biased, any attempt to apply a voltage greater than V_Z causes the device to be prepared to conduct a very large current. This has the effect of limiting the voltage we can apply to around V_Z.

As with any characteristic curve, the voltage at any given current, or the current at any given voltage, can be found from the curve of a zener diode.

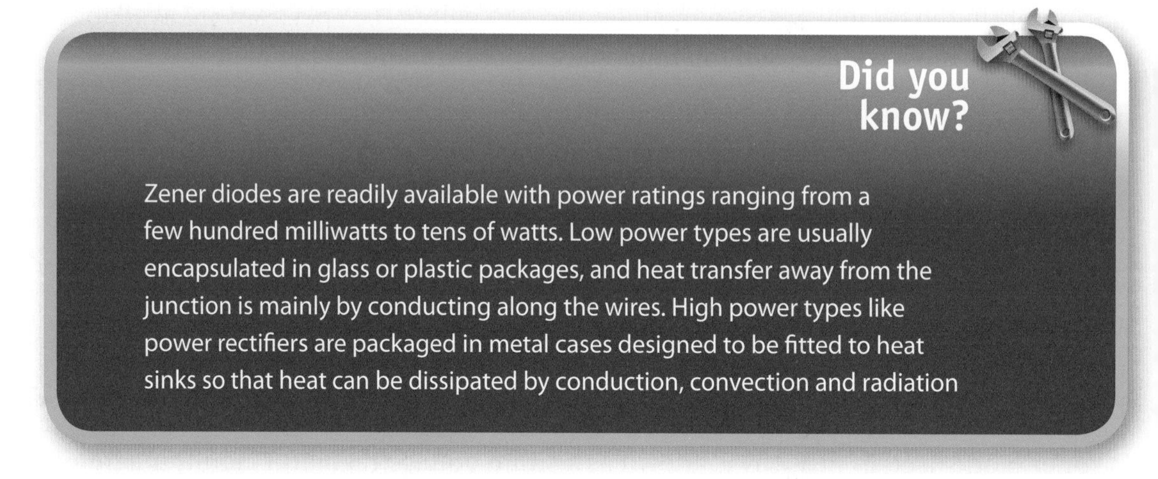

Did you know?

Zener diodes are readily available with power ratings ranging from a few hundred milliwatts to tens of watts. Low power types are usually encapsulated in glass or plastic packages, and heat transfer away from the junction is mainly by conducting along the wires. High power types like power rectifiers are packaged in metal cases designed to be fitted to heat sinks so that heat can be dissipated by conduction, convection and radiation

Light emitting diodes (LEDs)

The light emitting diode is a p–n junction especially manufactured from a semi-conducting material, which emits light when a current of about 10 mA flows through the junction. No light is emitted if the diode is reverse biased and if the voltage exceeds 5 volts then the diode may be damaged. If the voltage exceeds 2 volts then a series connected resistor may be required.

Figure 8.34 illustrates the general appearance of a LED.

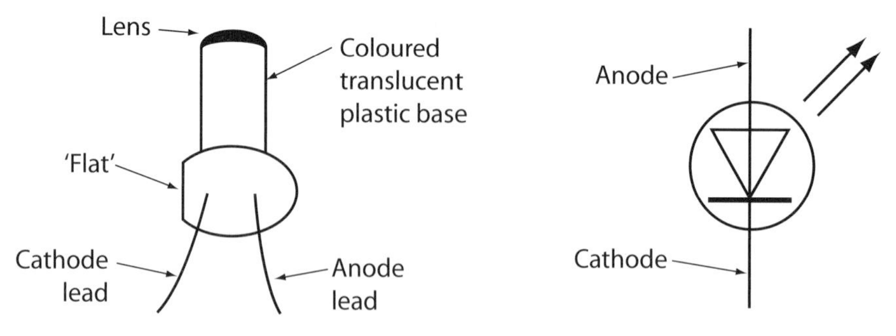

Figure 8.34 Light emitting diode

Photo cell and light-dependent resistor

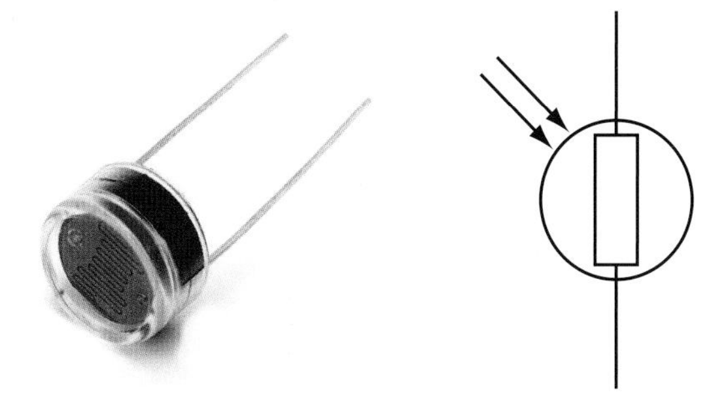

Figure 8.35 Photo cell and its circuit symbol

The photo cell shown in Figure 8.35 changes light (also infrared and ultraviolet radiation) into electrical signals and is useful in burglar and fire alarms as well as in counting and automatic control systems. Photoconductive cells or light-dependent resistors make use of the semiconductors whose resistance decreases as the intensity of light falling on them increases. The effect is due to the energy of the light setting free electrons from donor atoms in the semiconductor, making it more conductive. The main use of this type of device is for outside lights along the streets, roads and motorways. There are also smaller versions for domestic use within homes and businesses.

Photodiode

A photodiode is a p–n junction designed to be responsive to optical input. As a result, they are provided with either a window or optical fibre connection that allows light to fall on the sensitive part of the device.

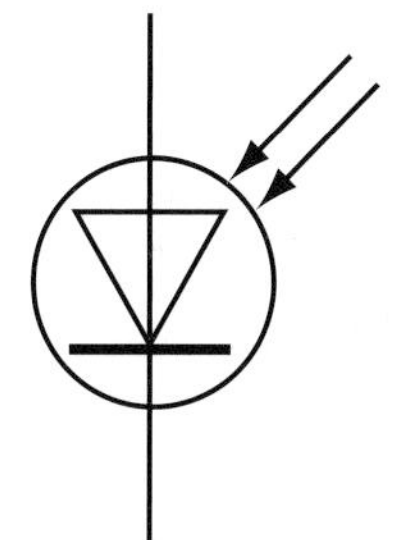

Figure 8.36 Circuit symbol for photodiode

Photodiodes can be used in either zero bias or reverse bias. When zero biased, light falling on the diode causes a voltage to develop across the device, leading to a current in the forward bias direction. This is called the photovoltaic effect and is the principle of operation of the solar cell, a solar cell being a large number of photodiodes. The use of the solar cell can then be seen in providing power to equipment such as calculators, solar panels and satellites orbiting the earth

When reverse biased, diodes usually have extremely high resistance. This resistance is reduced when light of an appropriate frequency shines on the junction. When light falls on the junction, the energy from the light breaks down bonds in the 'lattice' structure of the semiconductor material, thus producing electrons and allowing current to flow. Circuits based on this effect are more sensitive to light than ones based on the photovoltaic effect. Consequently, the photodiode is used as a fast counter or in light meters to measure light intensity, as in Figure 8.36.

Opto-coupler

The opto-coupler, also known as an opto-isolator, consists of an LED combined with a photodiode or phototransistor in the same package, as shown in Figure 8.37 and photograph.

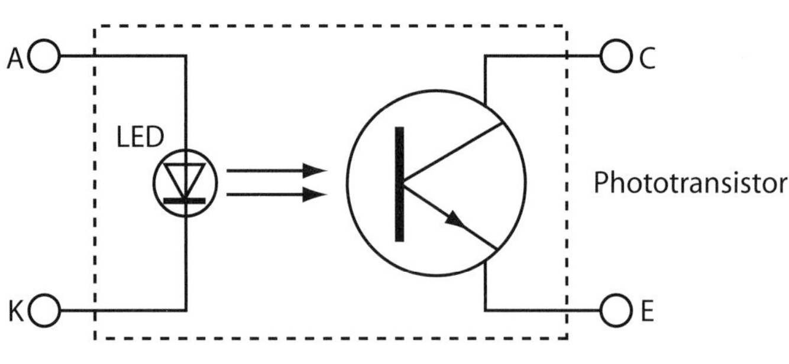

Figure 8.37 Opto-coupler circuit

Opto-coupler package

The opto-coupler package allows the transfer of signals, analogue or digital, from one circuit to another in cases where the second circuit cannot be connected electrically to the first, for example, due to different voltages.

Light (or infrared) from the LED falls on the photodiode/transistor which is shielded from outside light. A typical use for one of these is in a VCR to detect the end or start of the tape.

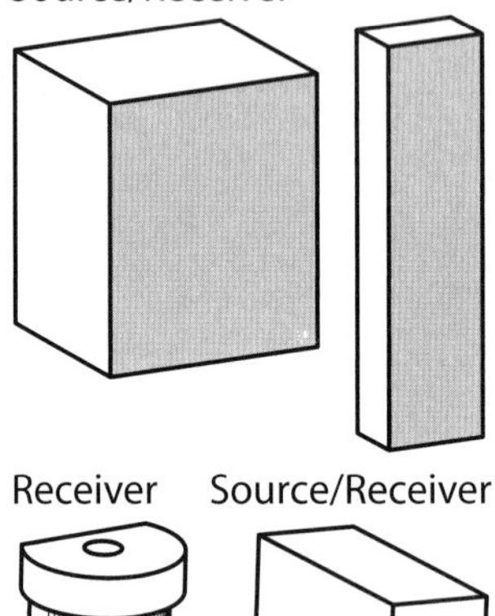

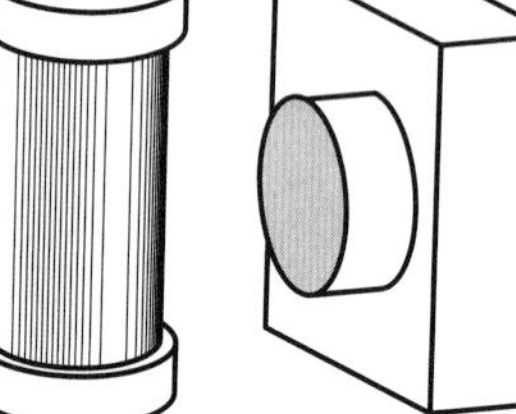

Figure 8.38 Housings for infrared source output and sensors for security alarms

Infrared source and sensor

An infrared beam of light is projected from an LED which is a semiconductor made from gallium arsenide crystal. The light emitted is not visible light, but very close to the white light spectrum. Figure 8.38 shows various housings for both the source output and the sensors within the security alarms industry.

Infrared beams have a receiver, which reacts to the beam in differing ways depending upon its use. Infrared sources/receivers are used for alarm detection and as remote control signals for many applications. The passive infrared (PIR) detector is housed in only one enclosure and uses ceramic infrared detectors. The device does not have a projector but detects the infrared heat radiated from the human body.

Fibre optic link

The simplified block diagram in Figure 8.39 shows a system of communication, which can be several thousand kilometres in length. On the far left we input information such as speech or visual pictures as electrical signals. They are then pulse code modulated in the coder and changed into equivalent digital light signals by the optical transmitter via a miniature laser or LED at the end of the fibre optic cable.

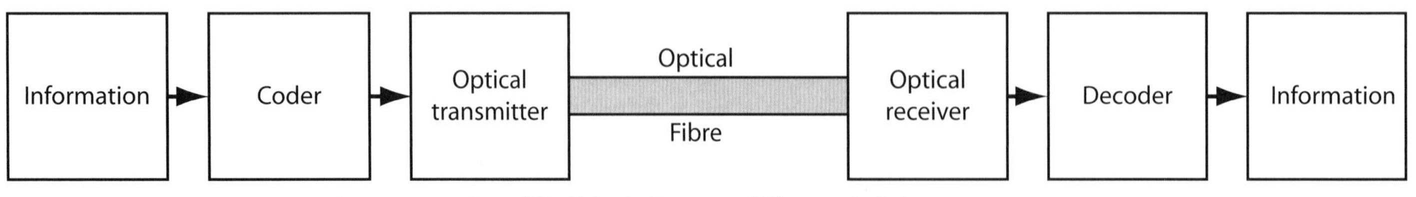

Figure 8.39 Simplified block diagram of fibre optic link

The light is transmitted down the cable to the optical receiver which uses a photodiode or phototransistor which converts the incoming signals back to electrical signals before they are decoded back into legible information.

The advantages of this type of link over a conventional communication system are:

- high information carrying capacity
- free from the noise of electrical interference
- greater distance can be covered, as there is no volt drop
- the cable is lighter, smaller and easier to handle than copper
- crosstalk between adjacent channels is negligible
- it offers greater security to the user.

The fibre optic cable

The fibre optic cable (see Figure 8.40) has a glass core of higher refractive index than the glass cladding around it. This maintains the light beam within the core by total internal reflection at the point where the core and the cladding touch.

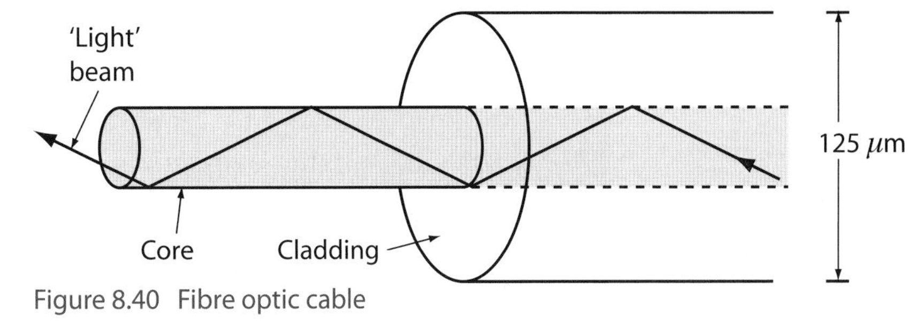

Figure 8.40 Fibre optic cable

This is similar to the insulation of the single core cable preventing the current leaking from the conductor. The beam of light bounces off the outer surface of the core in a zigzag formation along its length.

There are two main types of cable: multimode and singlemode (monomode).

Multimode

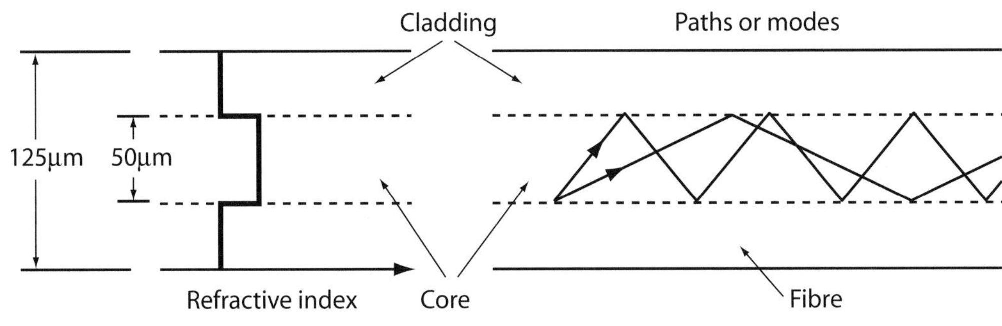

Figure 8.41 Multimode fibre optic cable

The wider core of the multimode fibre (Figure 8.41) allows the infrared to travel by different reflected paths or modes. Paths that cross the core more often are longer and take more time to travel along the fibre. This can sometimes cause errors and loss of information.

Singlemode

The core of the singlemode fibre (Figure 8.42) is about one tenth of the multimode and only straight through transmission is possible.

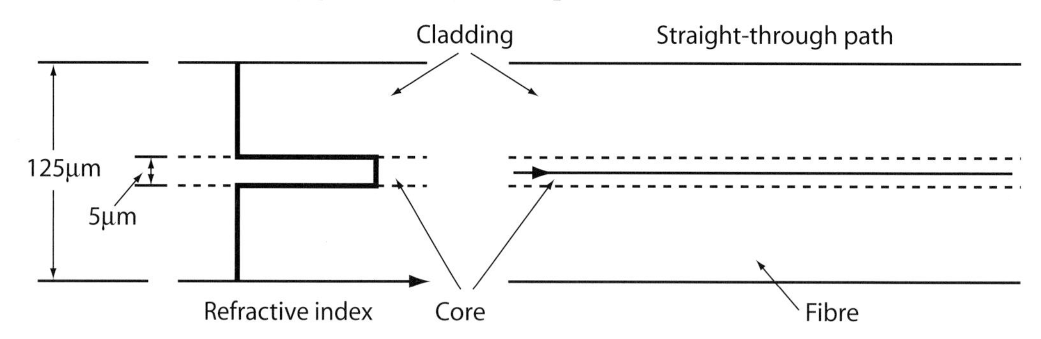

Figure 8.42 Singlemode fibre optic cable

Solid state temperature device (thermistors)

Thermistors monitor the changes in temperature of components, which could become damaged due to excessive heat. As the temperature rises there is a reaction within the semiconductor for its resistance to either rise (positive temp. coefficient) or fall (negative temp. coefficient) depending upon its make up. There are two types of thermistor:

- ptc – made from barium titanate
- ntc – made from oxides of nickel, manganese, copper and cobalt.

We also used **thermocouples**, which are two different metals bonded together and each has a lead. When the bonded metals are heated a voltage appears across the two leads. The hotter the metals become, the larger the voltage (mV). An example of their use is for measuring the temperature of furnaces within the steel industry.

Thermocouples

Diode testing

The p–n junction diode has a low resistance when a voltage is applied in the forward direction and a high resistance when applied in the reverse direction.

Connecting an ohmmeter with the red positive lead to the anode of the junction diode and the black negative lead to the cathode would give a very low reading, reversing the lead connections would give a high resistance reading in a good component.

Figures 8.43 and 8.44 illustrate these connections.

Remember

When using a multimeter you need to put it into the diode test mode, otherwise when testing on the ohms range you could get a reversed reading

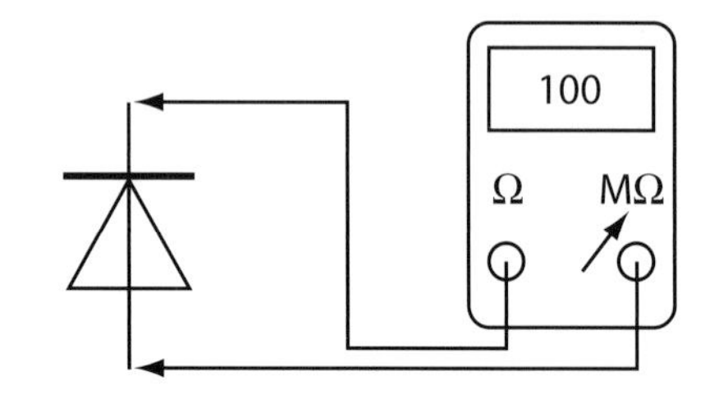

Figure 8.43 High resistance connection

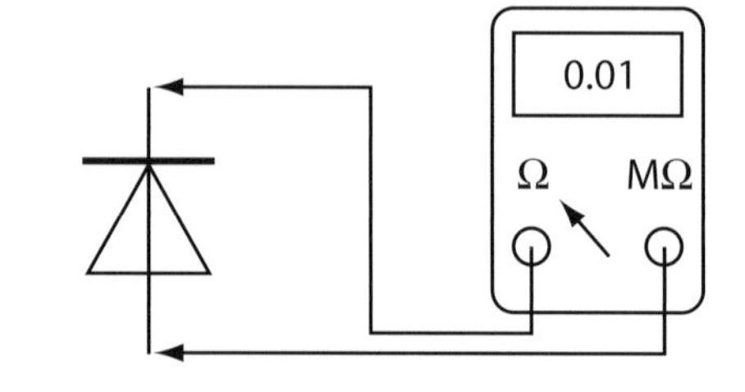

Figure 8.44 Low resistance connection

Rectification

Rectification is the conversion of an a.c. supply into a d.c. supply. Despite the common use of a.c. systems in our day-to-day work as an electrician, there are many applications (e.g. electronic circuits and equipment) that require a d.c. supply. The following section looks at the different forms this can take.

Half-wave rectification

We now know that a diode will only allow current to flow in one direction and it does this when the anode is more positive than the cathode. In the case of an a.c. circuit, this means that only the positive half cycles are allowed 'through' the diode and, as a result, we end up with a signal that resembles a series of 'pulses'. This tends to be unsuitable for most applications, but can be used in situations such as battery charging. A transformer is also commonly used at the supply side to ensure that the output voltage is to the required level. The waveform for this form of rectification would look as in Figure 8.45.

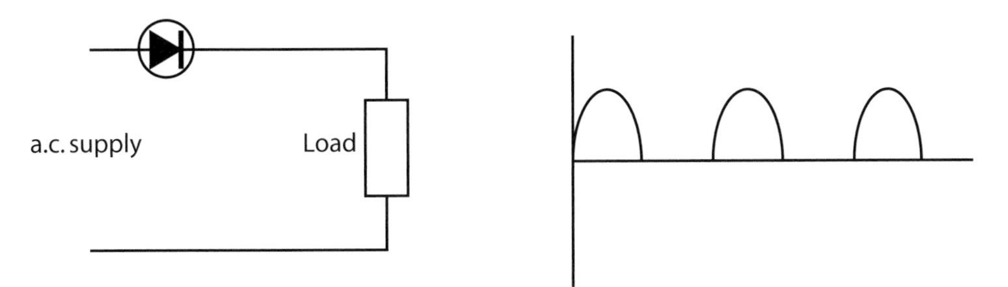

Figure 8.45 Half-wave rectification

Full-wave rectification

We have seen that half-wave rectification occurs when one diode allows the positive half cycles to pass through it. However, we can connect two diodes together to give a more even supply. We call this type of circuit **biphase**. In this method, we connect the anodes of the diodes to the opposite ends of the secondary winding of a centre-tapped transformer. As the anode voltages will be 180° out of phase with each other, one diode will effectively rectify the positive half cycle and one will rectify the negative half cycle. The output current will still appear to be a series of pulses, but they will be much closer together, with the waveform shown in Figure 8.46.

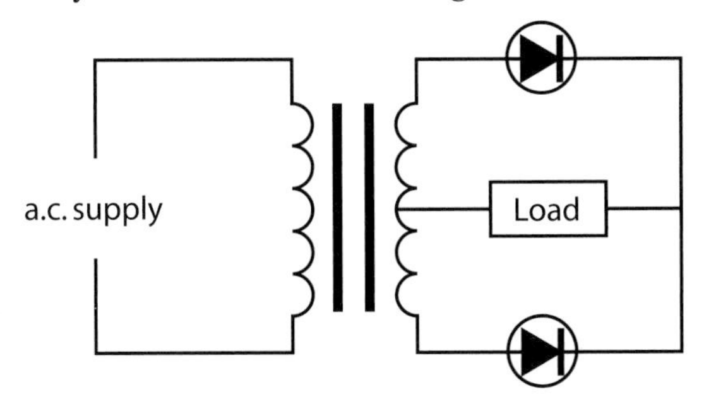

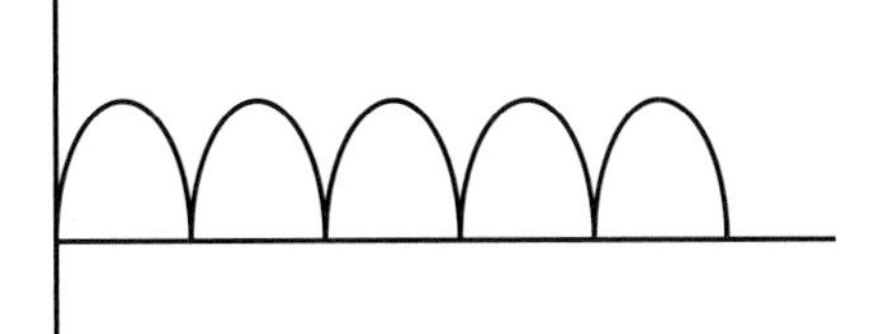

Figure 8.46 Full-wave rectification

The full-wave bridge rectifier

This method of rectification does have to rely on the use of a centre-tapped transformer, but the output waveform will be the same as that of the biphase circuit previously described. In this system, we use four diodes, connected in such a way that at any instant in time two of the four will be conducting. The connections would be as in Figure 8.47, where we have shown two drawings to represent the route through the network for each half cycle.

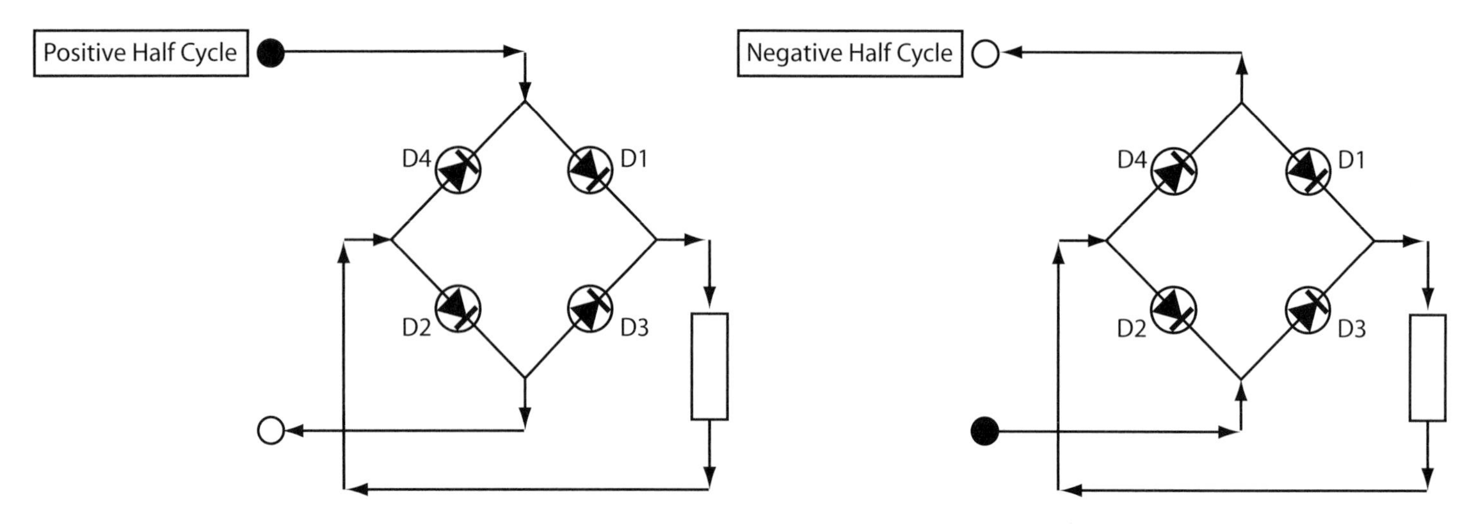

Figure 8.47 Full-wave bridge rectifier

The circuits that we have looked at so far convert a.c. into a supply which, although never going negative, is still not a true d.c. supply. This brings us to the next stage of the story – smoothing.

Smoothing

We have seen that the waveform produced by our circuits so far could best be described as having the appearance of a rough sea. The output current is not at a constant value, but constantly changing. This, as we have said before, is acceptable for battery charging, but useless for electronic circuits where a smooth supply voltage is required.

To make it useful for electronic circuits, we need to smooth out the waveform by creating a situation that is sometimes referred to as **ripple-free** and in essence there are three ways to achieve this, namely capacitor smoothing, choke smoothing and filter circuits.

Capacitor smoothing

If we connect a capacitor in parallel across the load, then the capacitor will charge up when the rectifier allows a flow of current and discharge when the rectifier voltage is less than the capacitor. However, the most effective smoothing comes under no-load conditions. The heavier the load current, the heavier the ripple. This means that the capacitor is only useful as a smoothing device for small output currents.

Choke smoothing

If we connect an inductor in series with our load, then the changing current through the inductor will induce an e.m.f. in opposition to the current that produced it. This means that the e.m.f. will try to maintain a steady current. Unlike the capacitor, this means that the heavier the ripple (rate of change of current) the more that ripple will be smoothed. This effectively means that the choke is more useful in heavy current circuits.

Filter circuits

This is the name given to a circuit that removes the ripple and is basically a combination of the two previous methods. The most effective of these is the capacitor input filter, shown in Figure 8.48.

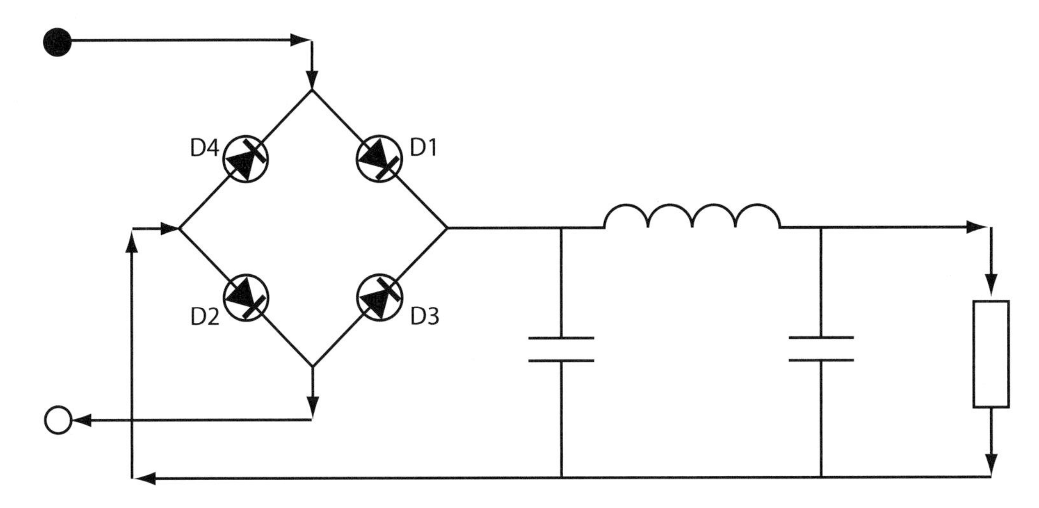

Figure 8.48 Capacitor input filter

The waveform for the filter circuit that we have just spoken about is shown in Figure 8.49, where the dotted line indicates the waveform before smoothing.

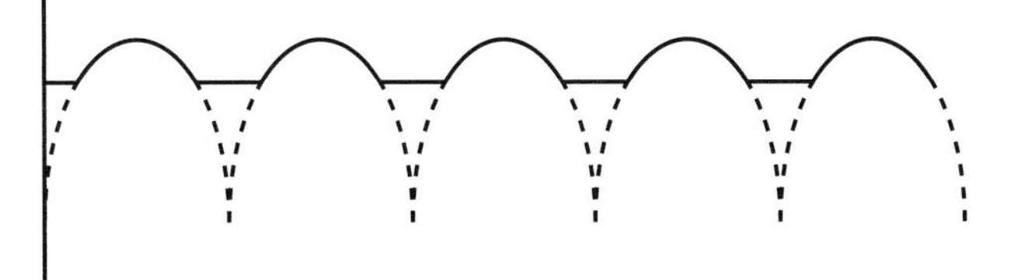

Figure 8.49 Waveform for capacitor input filter

Three-phase rectifier circuits

Whereas Figure 8.49 indicates that a reasonably smooth waveform can be obtained from a single-phase system, we can get a much smoother wave from the three-phase supply mains. To do so we use six diodes connected as a three-phase bridge circuit. These types of rectifier are used to provide high-powered d.c. supplies.

Transistors

Did you know?

The term 'transistor' is derived from the two words 'transfer-resistor'. This is because in a transistor approximately the same current is transferred from a low to a high resistance region. The term 'bipolar' means both electrons and holes are involved in the action of this type of transistor

What is usually referred to as simply a transistor, but is more accurately described as a bipolar transistor, is a semiconductor device, which has two p–n junctions. It is capable of producing current amplification and, with an added load resistor, both a load and voltage power gain can be achieved.

Transistor

Transistor basics

A bipolar transistor consists of three separate regions or areas of doped semiconductor material and, depending on the configuration of these regions, it is possible to manufacture two basic types of device.

When the construction is such that a central n-type region is sandwiched between two p-type outer regions, a pnp transistor is formed as in Figure 8.50.

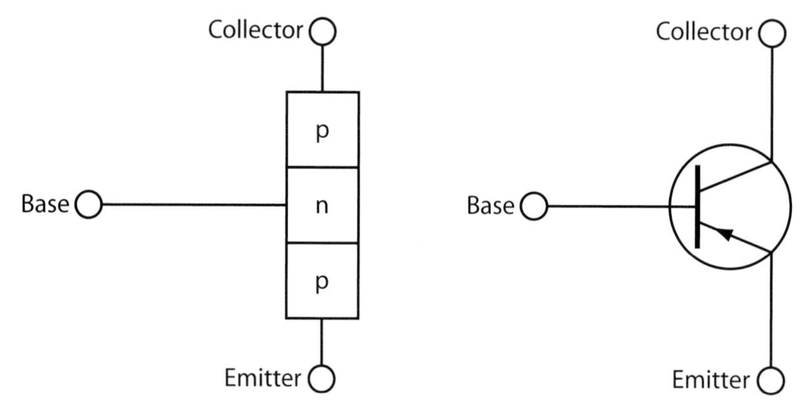

Figure 8.50 pnp transistor and its associated circuit symbols

If the regions are reversed as in Figure 8.51, an npn transistor is formed.

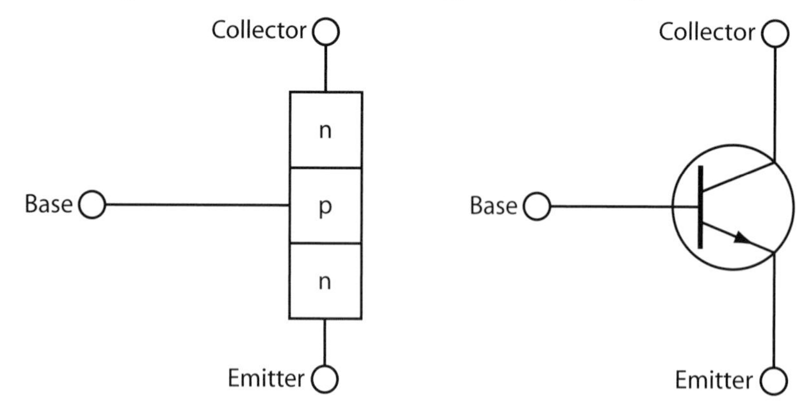

Figure 8.51 npn transistor and its associated circuit symbols

In both cases the outer regions are called the emitter and collector respectively and the central area the base.

The arrow in the circuit symbol for the pnp device points towards the base, whereas in the npn device it points away from it. The arrow indicates the direction in which conventional current would normally flow through the device.

Electron flow is of course in the opposite direction as explained in Electrical Installations Book 1. Note that in these idealised diagrams, the collector and emitter regions are shown to be the same size. This is not so in practice; the collector region is made physically larger since it normally has to dissipate the greater power during operation. Further, the base region is physically very thin, typically only a fraction of a micron (a micron is one millionth of a metre).

Hard-wire connections are made to the three regions internally; wires are then brought out through the casing to provide an external means of connection to each region. Either silicon or germanium semiconductor materials may be used in the fabrication of the transistor but silicon is preferred for reasons of temperature stability.

Transistor operation

For transistors to operate three conditions must be met:

1. The base must be very thin.
2. Majority carriers in the base must be very few.
3. The base-emitter junction must be forward biased and the base-collector junction reverse biased.

See Figure 8.52.

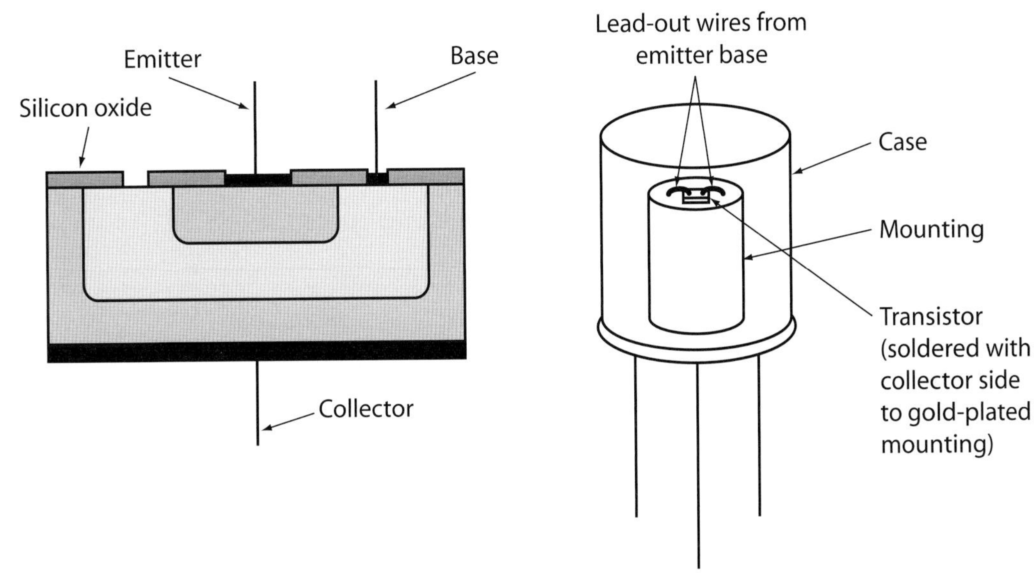

Figure 8.52 Transistor operation

Electrons from the emitter enter the base and diffuse through it. Due to the shape of the base most electrons reach the base-collector junction and are swept into the collector by the strong positive potential. A few electrons stay in the base long enough to meet the indigenous holes present and recombination takes place.

To maintain the forward bias on the base-emitter junction, holes enter the base from the base bias battery. It is this base current which maintains the base-emitter forward bias and therefore controls the size of the emitter current entering the base. The greater the forward bias on the base-emitter junction, the greater the number of emitter current carriers entering the base.

The collector current is always a fixed proportion of the emitter current set by the thinness of the base and the amount of doping. Holes from the emitter enter the base and diffuse through it – see Figure 8.53.

Due to the shape of the base most holes reach the base-collector junction and are swept into the collector by the strong negative potential.

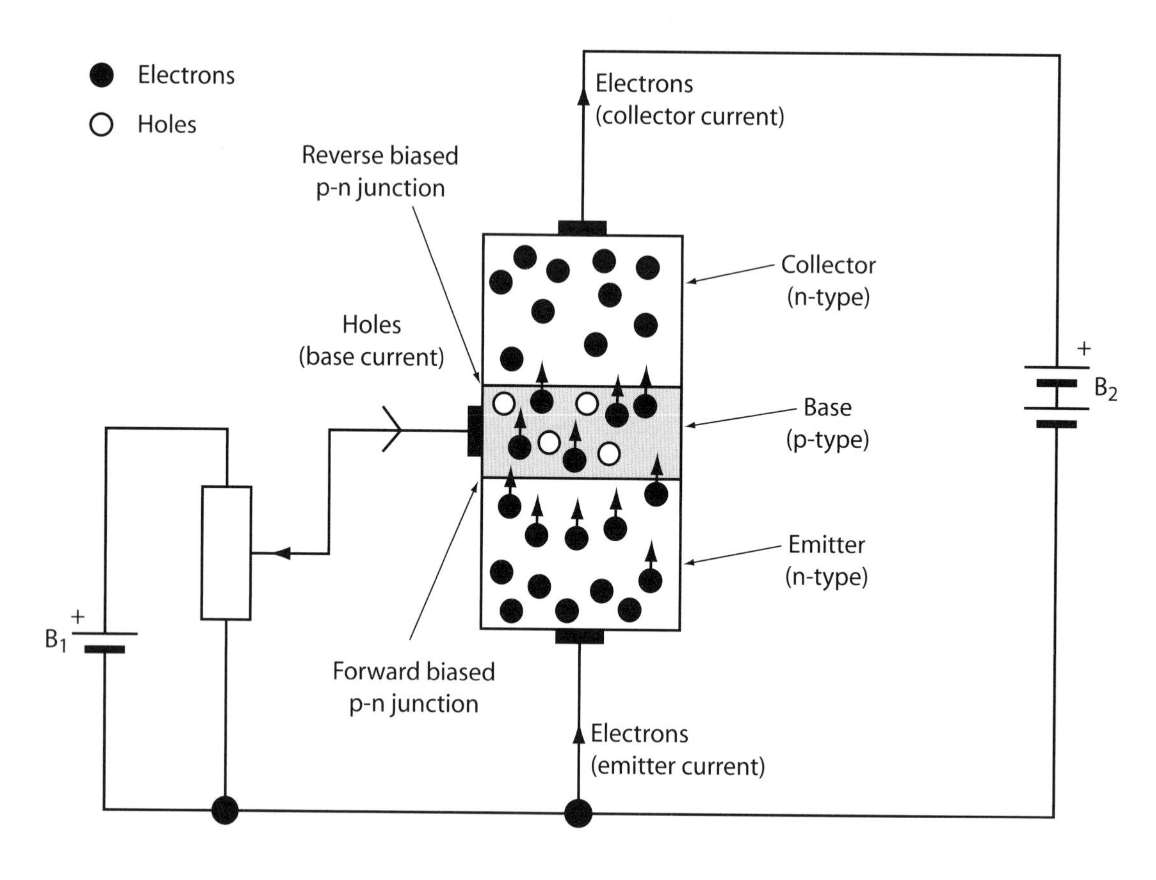

Figure 8.53 Circuit diagram for operation of transistor

Current amplification

Consider, for example, as in Figure 8.54, a base bias of some 630 mV has caused a base current of 0.5 mA to flow but more importantly has initiated a collector current of 50 mA.

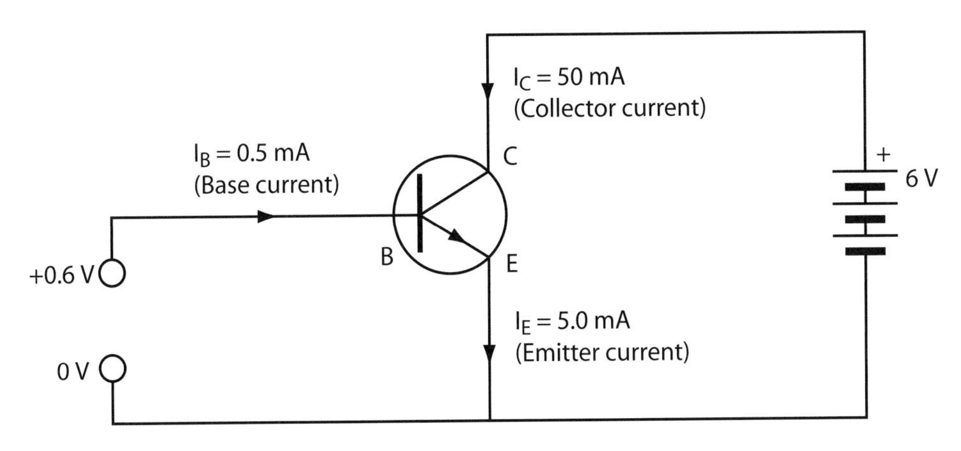

Figure 8.54 Current flow in transistor

Although these currents may seem insignificant, it's the comparison between the base and collector currents which is of interest. This relationship between I_B and I_C is termed the 'static value of the short-circuit forward current transfer' – we normally just call it the gain of the transistor, and it is simply a measure of how much amplification we would get. The symbol that we use for this is h_{FE}. This is the ratio between the continuous output current (collector current) and the continuous input current (base current). Thus when I_B is 0.5 mA and I_C is 50 mA the ratio is:

$$h_{FE} = \frac{I_C}{I_B}$$

$$= \frac{50 \text{ mA}}{0.5 \text{ mA}}$$

i.e. approximately equal to 100.

Note: There are no units since this is a ratio.

It can therefore be said that a small base current initiated by the controlling forward base-bias voltage produces a significantly higher value of collector current to flow dependent on the value of h_{FE} for the transistor. Thus current amplification has been achieved.

Voltage amplification

Mention was previously made to the derivation of the word 'transistor', and it quoted the device as transferring current from a low resistive circuit to approximately the same current in a high resistive circuit. This is an npn transistor so the low resistive reference is the emitter circuit and the high resistive reference, the collector circuit, the current in both being almost identical.

The reason the emitter circuit is classed as having a low resistance is because it contains the forward biased (pn) base-emitter junction. Conversely, the collector circuit contains the reverse biased (np) base-collector junction, which is of course in the order of tens of thousands of ohms (it varies with I_C). In order to produce a voltage output from the collector, a load resistor (R_B) is added to the collector circuit as indicated in the circuit diagram in Figure 8.55.

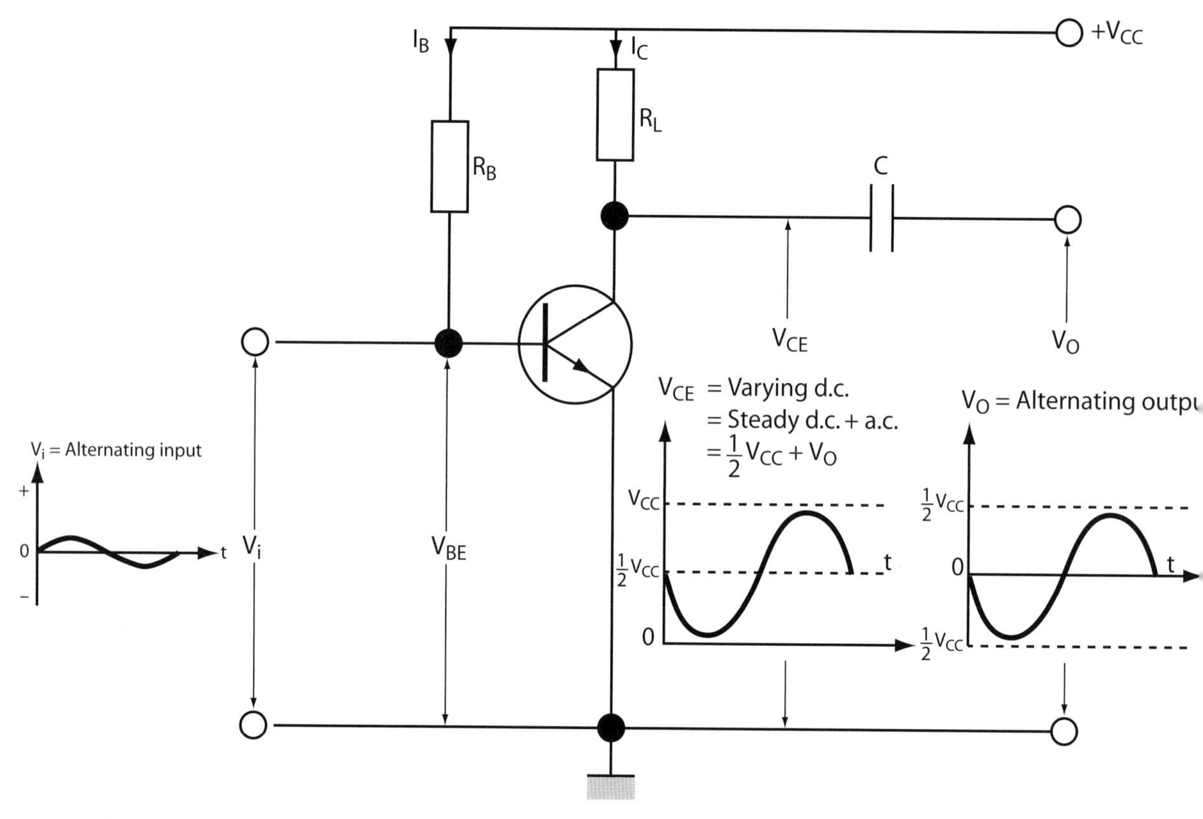

Figure 8.55 Voltage amplification

This shows the simplest circuit for a voltage amplifier. To see how voltage amplification occurs we have to consider that there is no input across V_i, which is called the quiescent (quiet) state. For transistor action to take place the base emitter junction V_{BE} must be forward biased (and has to remain so when V_i goes positive and negative due to the a.c. signal input).

By introducing resistor R_B between collector and base, a small current I_B will flow from V_{CC} through R_B into the base and down to 0V via the emitter, thus keeping the transistor running (ticking over).

Component resistor values R_B and R_L are chosen so that the steady base current I_B makes the quiescent collector-emitter voltage V_{CE} about half the power supply voltage V_{CC}. This allows V_O to replicate the input signal V_i at an amplified voltage with a 180° phase shift. When an a.c. signal is applied to the input V_i and goes positive it increases V_{BE} slightly to around 0.61 V. When V_i swings negative, V_{BE} drops slightly to 0.59 V. As a result a small alternating current is superimposed on the quiescent base current I_B, which in effect is a varying d.c. current.

The collector emitter voltage (V_{CE}) is a varying d.c. voltage, or an alternating voltage superimposed on a normal steady d.c. voltage. The capacitor C is there to block the d.c. voltage, but allow the alternating voltage to pass on to the next stage. So in summing up, a bipolar transistor will act as a voltage amplifier if:

- It has a suitable collector load R_L.
- It is biased so that so that the quiescent value V_{CE} is around half the value of V_{CC}, which is known as the class A condition.
- The transistor and load together bring about voltage amplification.
- The output is 180° out of phase with the input signal as Figure 8.55 indicates.
- The emitter is common to the input, output and power supply circuits and is usually taken as the reference point for all voltages, i.e. 0 V. It is called 'common', 'ground' or 'earth' if connected to earth.

Transistor as a switch

We have looked at the transistor as an amplifier of current and voltage. If we connect the transistor as in Figure 8.56, we can operate it as a switch. Compared with other electrically operated switches, transistors have many advantages, whether in discrete or integrated circuit (IC) form. They are small, cheap, reliable, have no moving parts and can switch millions of times per second – the perfect switch that has infinite resistance when 'off', no resistance when 'on' and changes instantaneously from one state to another, using up no power.

Figure 8.56 shows the basic circuit for an npn common emitter as in previous diagrams with a load resistor R_L connected in series with the supply (V_{CC}) and the collector.

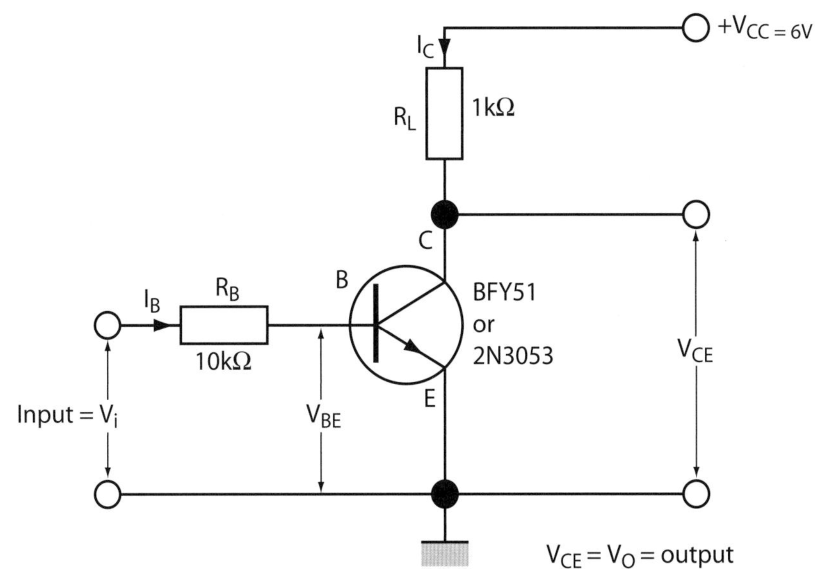

Figure 8.56 Circuit diagram for transistor used as a switch

R_B prevents excessive base currents, which would seriously damage the transistor when forward biased. With no input across V_I, the transistor is basically turned off. This means then that there will be no current (I_C) through R_L, therefore there will be no volt drop across R_L so the $+V_{CC}$ voltage (6 V) will appear across the output V_{CE}.

If we now connect a supply of between 2 V–6 V across V_i input, the transistor will switch on, current will flow through the collector load resistor R_L and down to common, making the output V_{CE} around 0 V. From this we can state that:

- when the input V_i = 0 V, the output V_{CE} = 6 V
- when the input V_i = 2 V–6 V, the output V_{CE} = 0 V.

From this we can see that the transistor is either High (6 V) or Low (0 V), or we can confirm, like a switch i that it is 'On' (6 V) or 'Off' (0 V).

This circuit can be used in alarms and switch relays for all types of processes and is the basic stage for programmable logic control (plc) which uses logic gates with either one or zero to represent what the output is from a possible input.

In Figure 8.57 are identified basic logic gate circuits with their inputs/outputs, 'truth table' and symbols.

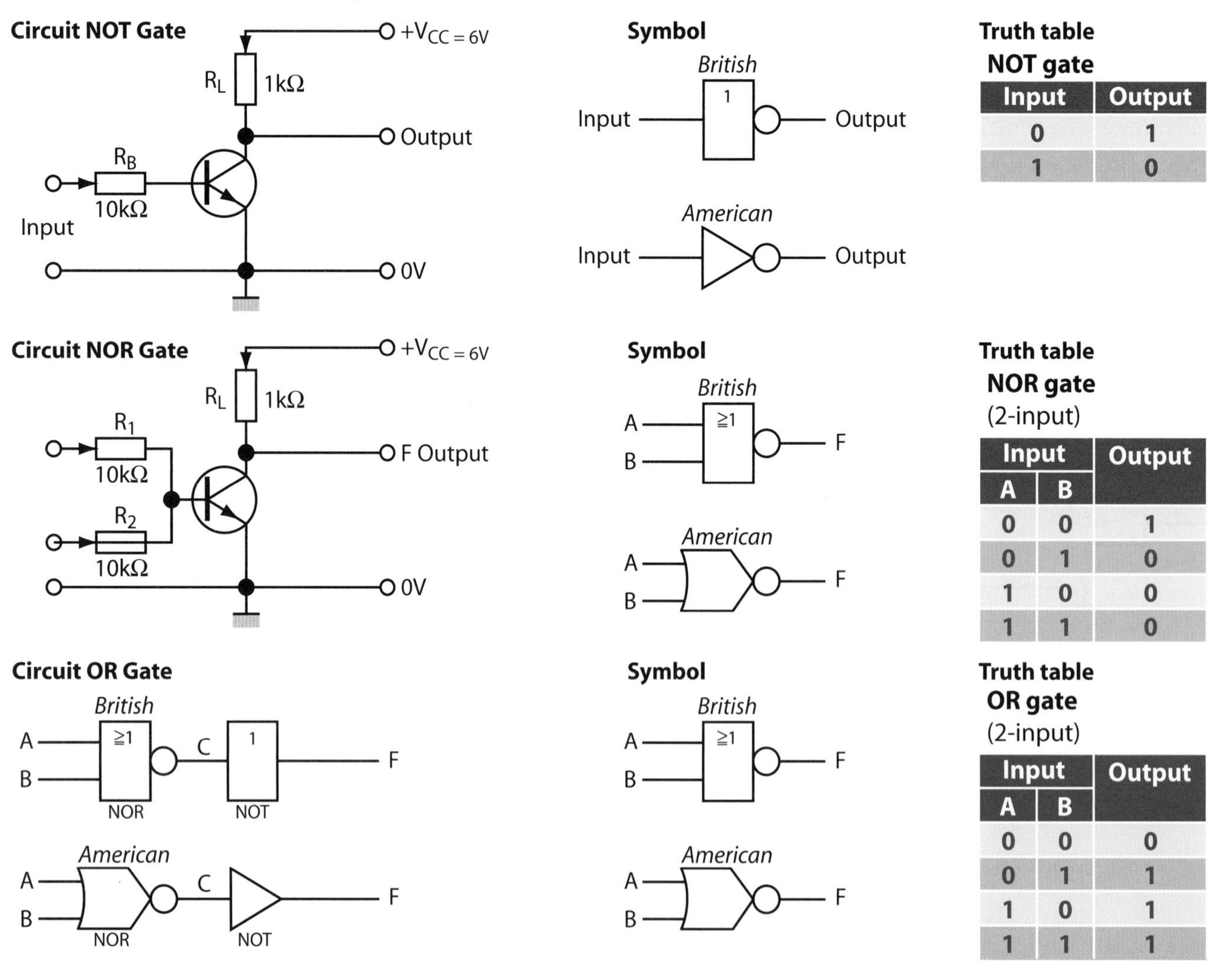

Input	Output
0	1
1	0

Input A	Input B	Output
0	0	1
0	1	0
1	0	0
1	1	0

Input A	Input B	Output
0	0	0
0	1	1
1	0	1
1	1	1

Figure 8.57 Basic logic gate circuits (continued on next page)

Circuit NAND Gate

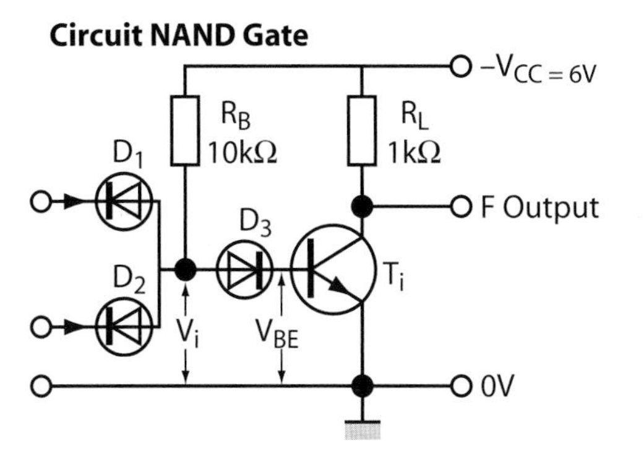

Symbol

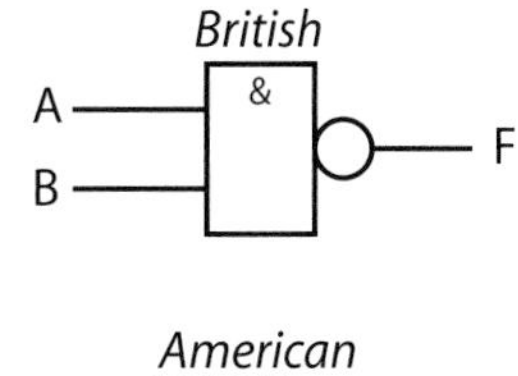

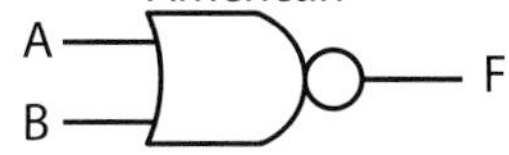

Truth table

NAND gate
(2-input)

A	B	F
0	0	1
0	1	1
1	0	1
1	1	0

Circuit AND Gate

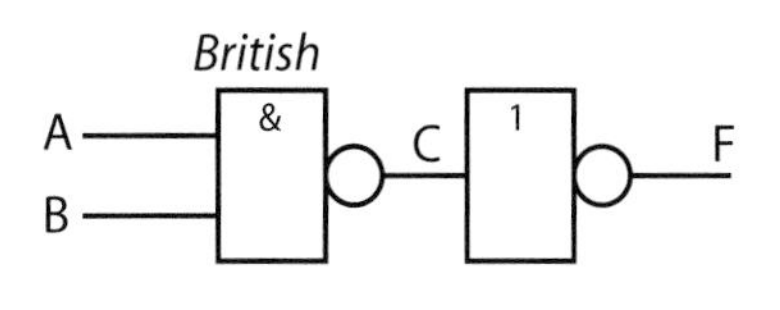

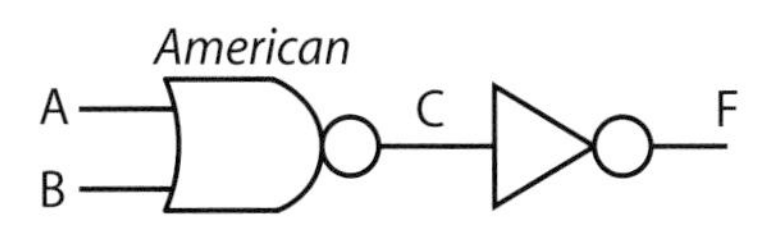

Symbol

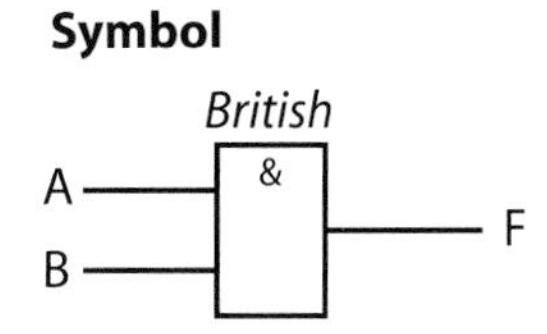

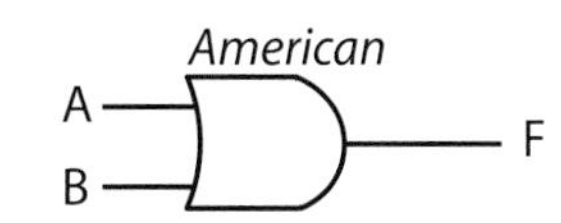

Truth table

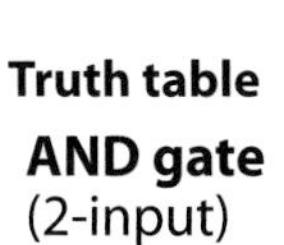

AND gate
(2-input)

A	B	C	F
0	0	1	0
0	1	1	0
1	0	1	0
1	1	0	1

Symbol

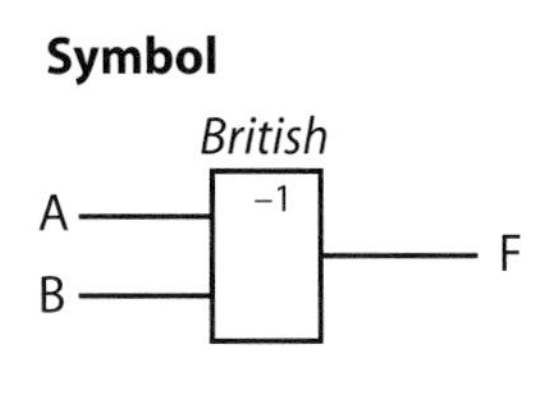

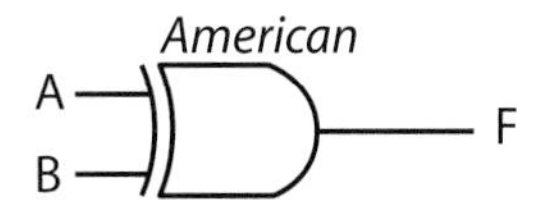

Truth table

Exclusive-OR gate

A	B	F
0	0	1
0	1	1
1	0	1
1	1	0

Figure 8.57 Basic logic gate circuits (continued)

There is also an Exclusive NOR which gives an output as indicated in Figure 8.58.

Symbol

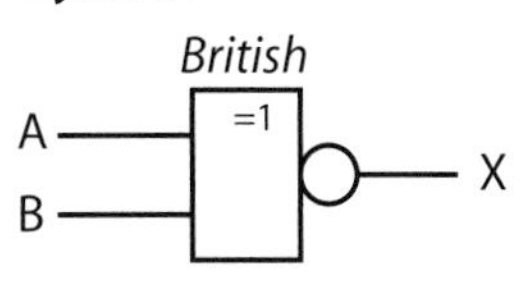

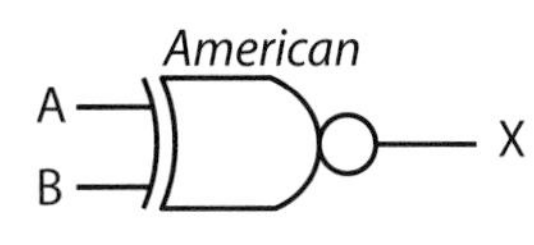

Truth table

Exclusive NOR gate

Input		Output
A	B	
0	0	0
0	1	1
1	0	1
1	1	1

Figure 8.58 Exclusive NOR gate

Testing transistors

As all transistors consist of either a npn or a pnp construction the testing of them is similar to diodes. Special meters with three terminals for testing transistors are available and many testing instruments have this facility. However, an ohmmeter can be used for testing a transistor to check if it is conducting correctly. The following results should be obtained from a transistor assuming that the red lead of an ohmmeter is positive.

Note: This is not always the case. With some older analogue meters, the battery connections internally are the opposite way round, so it is always good to check both ways across base and emitter as shown in Figures 8.60 and 8.61.

A good npn transistor will give the following readings:

- Red to base and black to collector or emitter will give a low resistance.
- However, if the connections are reversed it will result in a high resistance reading.
- Connections of any polarity between the collector and emitter will also give a high reading.

A good pnp transistor will give the following readings:

- Black to base and red to collector or the emitter will give a low resistance reading.
- However, if the connections are reversed a high resistance reading will be observed.
- Connections of either polarity between the collector and emitter will give a high resistance reading.

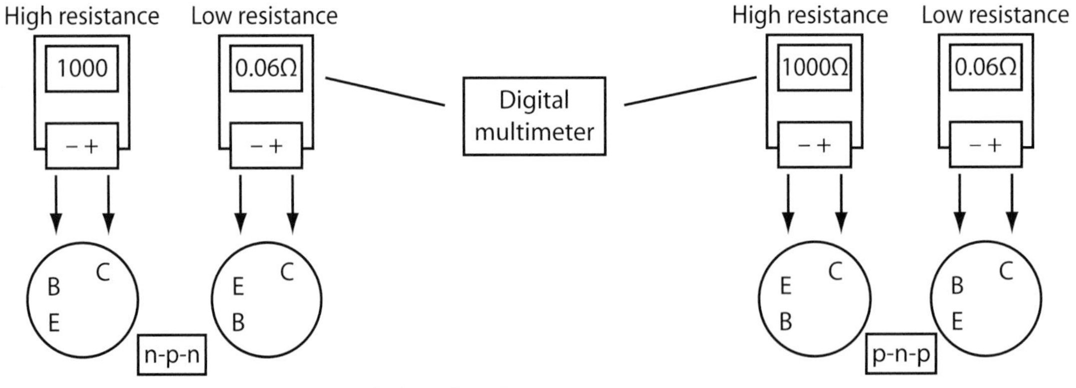

Figure 8.59 Testing transistors with digital multimeter

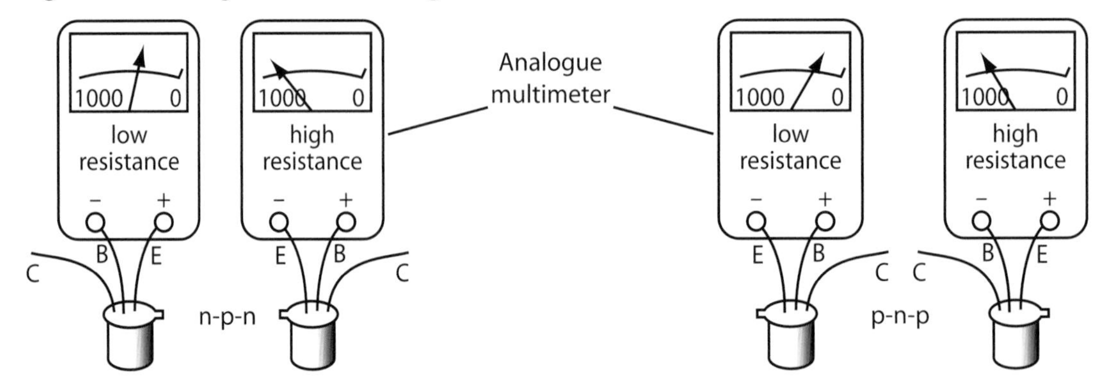

Figure 8.60 Testing transistors with analogue multimeter

Integrated circuits (ICs)

Integrated circuits are complete electronic circuits housed within a plastic case (known as the **black box**). The chip contains all the components required, which may include diodes, resistors, capacitors, transistors etc.

There are several categories, which include analogue, digital and memories. The basic layout is shown in Figure 8.61, which is an operational amplifier (dual in-line IC).

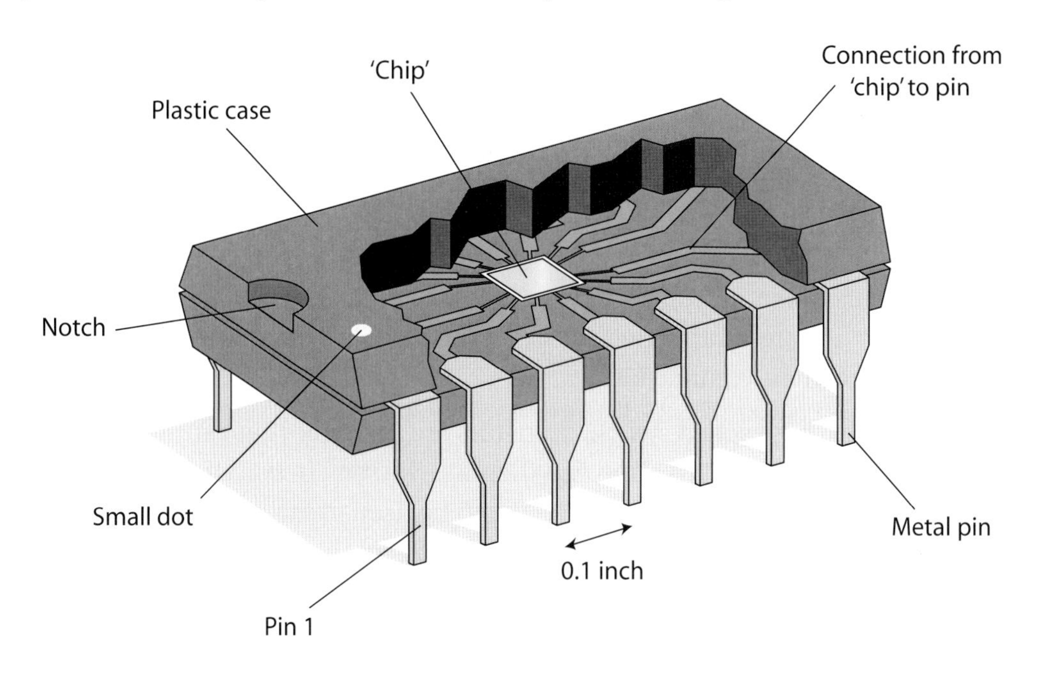

Figure 8.61 Operational amplifier

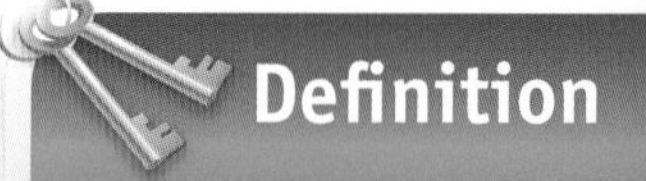

Dual in-line ICs – the types of IC with the pins lined up down each side

The plastic case has a notch at one end and, if you look at the back of the case with the notch at the top, Pin 1 is always the first one on the left hand side, sometimes noted with a small dot. The other pin numbers follow down the left hand side, 2, 3 and 4 then back up the right hand side from the bottom right to the top, 5, 6, 7 and 8. This is an 8-pin chip but you can get some chips with up to 32 pins or more.

Thyristors

Sometimes referred to as a 'silicon controlled rectifier' (SCR), the thyristor is a four-layer semiconducting device, with each layer consisting of an alternating n or p type material as shown in Figure 8.62.

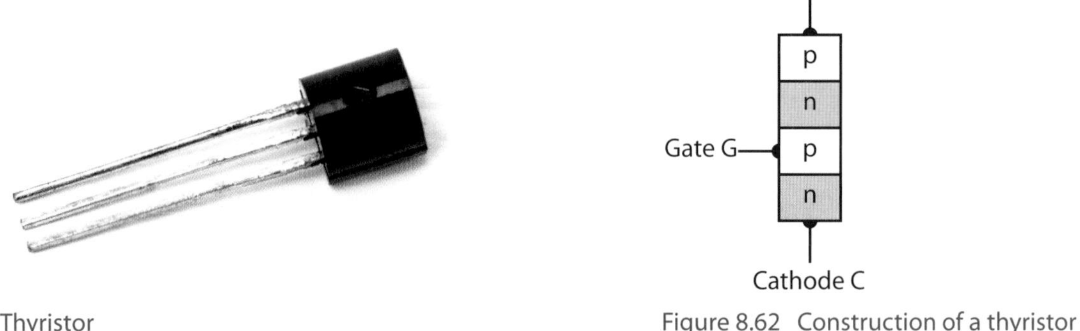

Thyristor

Figure 8.62 Construction of a thyristor

Figure 8.63 Thyristor circuit symbol

The main terminals (the anode and cathode) are across the full four layers, while the control terminal (the gate) is attached to one of the middle layers. The circuit symbol for a thyristor is shown in Figure 8.63.

Effectively acting as a high speed switch, thyristors are available that can switch large amounts of power (as high as MW) and can therefore be seen in use within High Voltage Direct Current (HVDC) systems. These can be used to interconnect two a.c. regions of a power-distribution grid, albeit the equipment needed to convert between a.c. and d.c. can add considerable cost. That said, above a certain distance (about 35 miles for undersea cables and 500 miles for overhead cables), the lower cost of the HVDC electrical conductors can outweigh the cost of the electronics required.

Principle of operation

Did you know?

The conversion from a.c. to d.c. is known as rectification and from d.c. to a.c. as inversion

A thyristor acts like a semiconductor version of a mechanical switch, having two states; in other words it is either 'on' or 'off' with nothing in between. This is how they gained their name from the Greek word thyra (which means door), the inference being something that is either open or closed.

The thyristor is very similar to a diode, with the exception that it has an extra terminal (the gate) which is used to activate it. Effectively in its normal or 'forward biased' state, the thyristor acts as an open-circuit between anode and cathode, thus preventing current flow through the device. This is known as the 'forward blocking' state.

However, the thyristor can allow current to flow through it by the application of a control (gate) current to the gate terminal. It is this concept that allows a small signal at the gate to control the switching of a higher power load. In this respect the thyristor is performing in a similar way to a relay (see Book 1 Electrical science).

Once activated, a thyristor doesn't require a control (gate) current to continue operating and will therefore continue to conduct until either the supply voltage is turned off, reversed or when a minimum 'holding' current is no longer maintained between the anode and cathode.

These concepts are shown in the following diagram:

In Figure 8.64, switch S_1 acts as a master isolator and no supply is present at either the thyristor or at switch S_2. Closing switch S_1 will allow a supply to be present at the thyristor, but there is no signal at the gate terminal as switch S_2 is open and therefore no current will flow to the indicator lamp. However, if we now close switch S_2, the gate will be energised and the thyristor will operate, thus allowing current to flow through it to the indicator lamp. The current at the anode would be large enough in this situation to allow the thyristor to continue operating, even if we opened switch S_2.

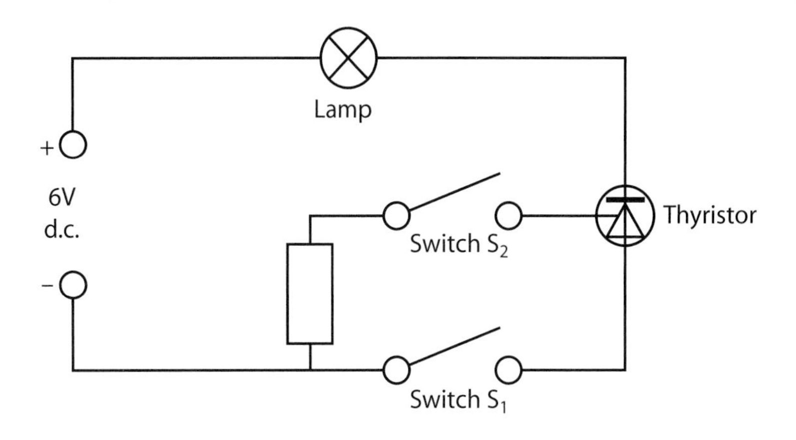

Figure 8.64 Circuit diagram of thyristor

The control of a.c. power can also be achieved with the thyristor by allowing current to be supplied to the load during part of each half cycle. If a gate pulse is applied automatically at a certain time during each positive half cycle of the input, then the thyristor will conduct during that period until it falls to zero for the negative half cycle.

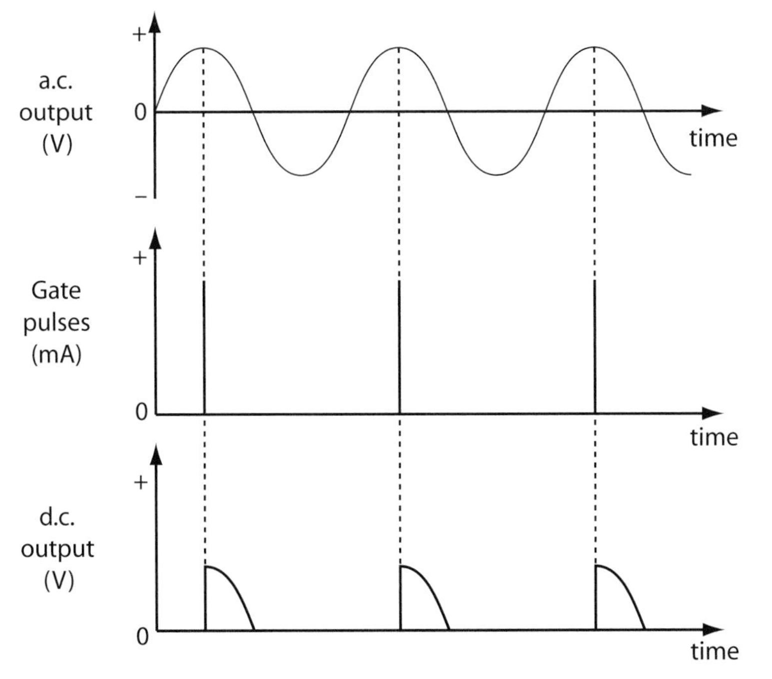

Figure 8.65 Supply of current to load during part of each half cycle

You will see from Figure 8.65 that the gate pulse (mA) occurs at the peak of the a.c. input (V). During negative half cycles the thyristor is reverse biased and will not conduct and will not conduct again until half way through the next positive half cycle. Current actually flows for only a quarter of the cycle, but by changing the timing of the gate pulses, this can be decreased further or increased. The power supplied to the load can be varied from zero to half wave rectified d.c.

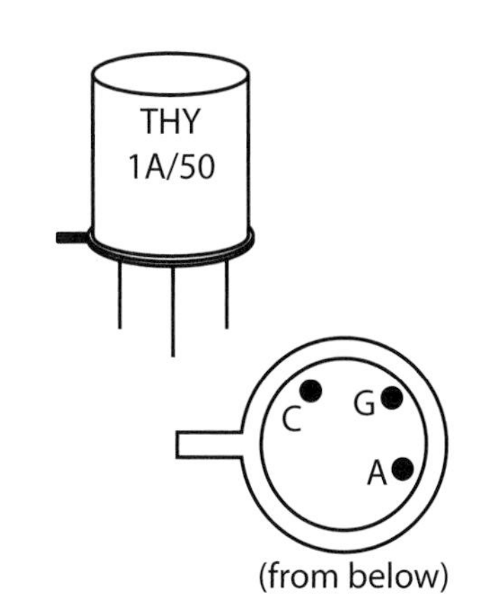

Figure 8.66 Typical small value thyristor

Thyristor testing

To test thyristors a simple circuit needs to be constructed as shown in Figure 8.67. When switch B only is closed the lamp will not light, but when switch A is closed the lamp lights to full brilliance. The lamp will remain illuminated even when switch A is opened. This shows that the thyristor is operating correctly. Once a voltage has been applied to the gate the thyristor becomes forward conducting like a diode and the gate loses control.

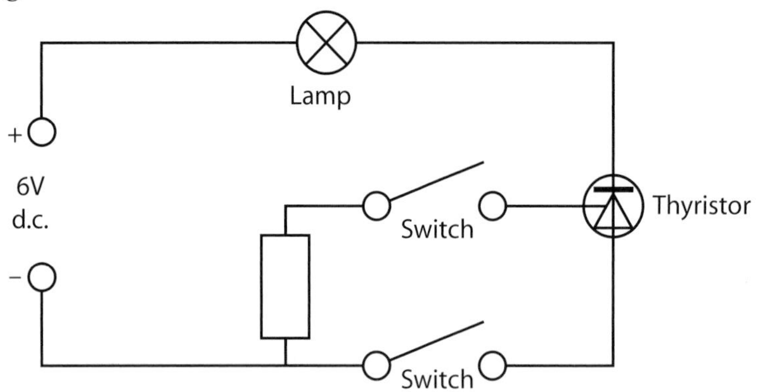

Figure 8.67 Circuit for testing thyristor

The triac

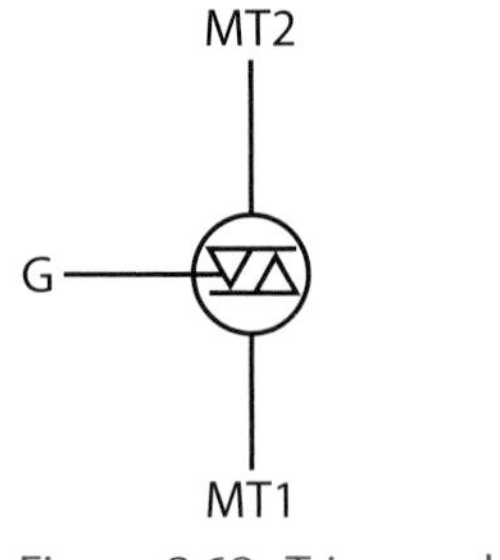

Figure 8.68 Triac symbol

The triac is a three terminal semiconductor for controlling current in either direction and the schematic symbol for a triac is shown in Figure 8.68.

If we look at Figure 8.68 more closely, we can see that the symbol looks like two thyristors that have been connected in parallel, albeit in opposite facing directions and with only one gate terminal. We refer to this type of arrangement as an inverse parallel connection.

The main power terminals on a triac are designated as MT1 (Main Terminal 1) and MT2. When the voltage on terminal MT2 is positive with regard to MT1, if we were to apply a positive voltage to the gate terminal the left thyristor would conduct. When the voltage is reversed and a negative voltage is applied to the gate, the right thyristor would conduct.

As with the thyristor generally, a minimum holding current must be maintained in order to keep a triac conducting. A triac therefore generally operates in the same way as the thyristor, but operating in both a forward and reverse direction. It is therefore sometimes referred to as a bidirectional thyristor, in that it can conduct electricity in both directions. One disadvantage of this is that triacs can require a fairly high current pulse to turn them on.

Did you know?

Triacs were originally developed for, and used extensively in, the consumer market. They are used in many low power control applications such as food mixers, electric drills and lamp dimmers, etc.

The diac

Before consideration is given to practical triac applications and circuits, it is necessary to examine the diac. This device is often used in triac triggering circuits because it, along with a resistor-capacitor network, produces an ideal pulse-style waveform. It does this without any sophisticated additional circuitry, due to its electrical characteristics. Also it provides a degree of protection against spurious triggering from electrical noise (voltage spikes).

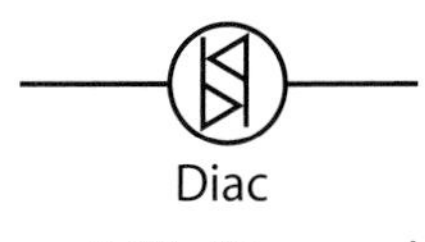

Figure 8.69 Diac symbol

The symbol is shown in Figure 8.69. The device operates like two breakdown (Zener) diodes connected in series, back to back. It acts as an open switch until the applied voltage reaches about 32–35 volts, when it will conduct.

Lamp dimmer circuit

Figure 8.70 shows a typical GLS lamp dimmer circuit.

The GLS lamp has a tungsten filament, which allows it to operate at about 2500°C and is wired in series with the triac. The variable resistor is part of a trigger network providing a variable voltage into the gate circuit, which contains a diac connected in series. Increasing the value of the resistor increases the time taken for the capacitor to reach its charge level to pass current into the diac circuit. Reducing the resistance allows the triac to switch on faster in each half cycle. By this adjustment the light output of the lamp can be controlled from zero to full brightness.

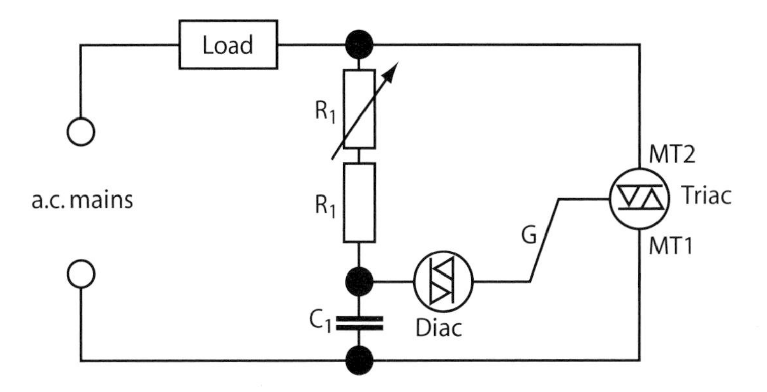

Figure 8.70 GLS lamp dimmer circuit

The capacitor is connected in series with the variable resistor. This combination is designed to produce a variable phase shift into the gate circuit of the diac. When the p.d. across the capacitor rises, enough current flows into the diac to switch on the triac.

The diac is a triggering device having a relatively high switch on voltage (32–35 volts) and acts as an open switch until the capacitor p.d. reaches the required voltage level.

The triac is a two-directional thyristor, which is triggered on both halves of each cycle. This allows it to conduct current in either direction of the a.c. supply. Its gate is in series with the diac, allowing it to receive positive and negative pulses.

A relatively high resistive value resistor R2 (100 Ω) is placed in series with a capacitor to reduce false triggering of the triac by mains voltage interference. The capacitor is of a low value (0.1 mF). This combination is known as the **snubber circuit**.

On the job: On-call

You are asked to look at a problem at a small private residential care home. When you arrive the warden explains that there is a problem with the nurse call system, in that it has very recently been installed but, as the electrical contractor that fitted it has gone into receivership, no one is quite sure of its operation. Additionally, they require another 'patient call' button to be installed in a further bedroom. The warden has a circuit diagram of the system, which is shown below.

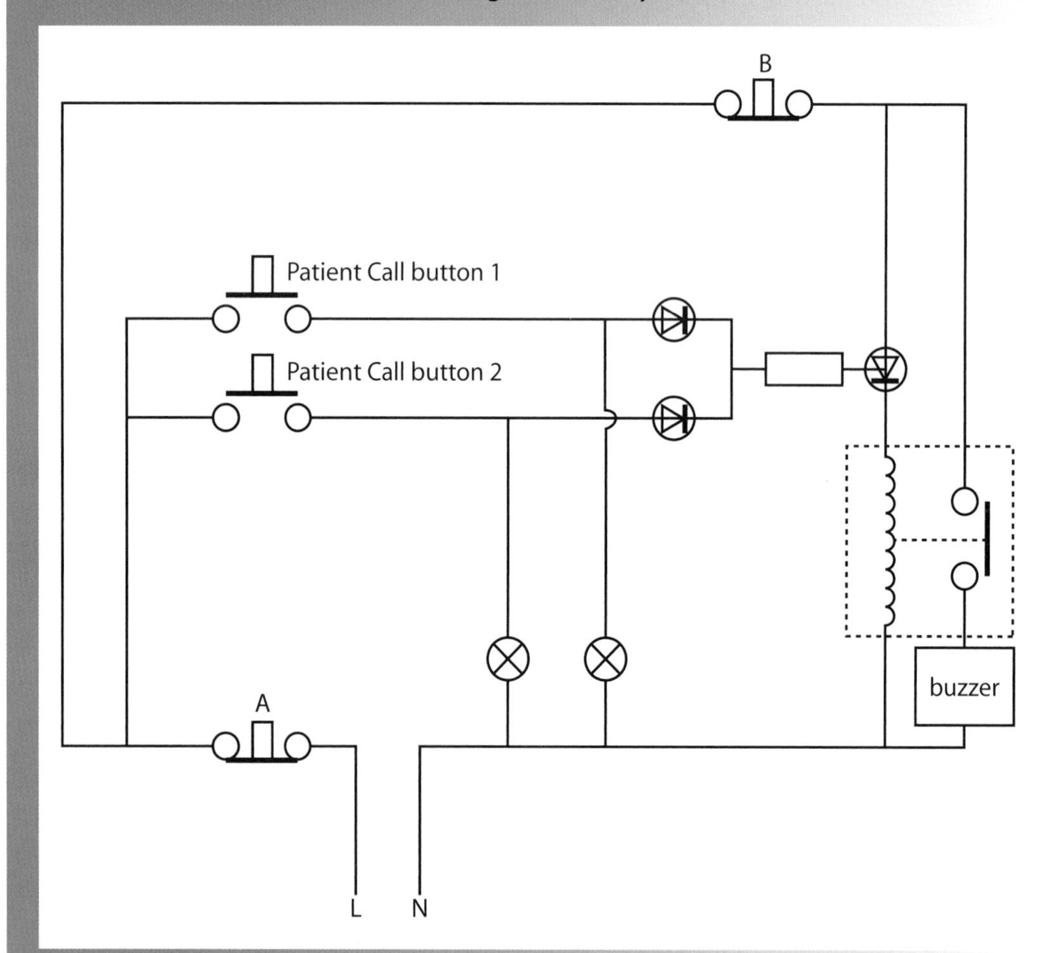

All components, with the exception of the patient call buttons, are located inside the nurse call panel, which is located at the nurses' station. Looking at the diagram, identify the components and prepare a written report for the warden, explaining in writing how the system operates. Now produce a revised circuit diagram to show how an additional patient could be added to the system. (Please assume that the values and ratings of any components for this exercise will be acceptable.) Your new drawing and report will then be held by the warden for future reference.

(A suggested answer can be found on page 316)

Field effect transistors (FETs)

Field effect transistor devices first appeared as separate (or discrete) transistors, but now the field effect concept is employed in the fabrication of large-scale integration arrays such as semiconductor memories, microprocessors, calculators and digital watches.

There are two types of field effect transistor: the junction gate field effect transistor, which is usually abbreviated to JUGFET, JFET or FET, and the metal oxide semiconductor field effect transistor known as the MOSFET. They differ significantly from the bipolar transistor in their characteristics, operation and construction.

The main advantages of an FET over a bipolar transistor are:

- Its operation depends on the flow of majority current carriers only. It is, therefore, often described as a unipolar transistor.
- It is simpler to fabricate and occupies less space in integrated form.
- Its input resistance is extremely high, typically above 10 MΩ especially for MOSFET devices. In practice, this is why voltage measuring devices such as oscilloscopes and digital voltmeters employ the FET in their input circuitry, so that the voltage being measured is not altered by the connection of the instrument.
- Electrical noise is the production of random minute voltages caused by the movement of current carriers through the transistor structure. Since the FET does not employ minority carriers, it therefore has the advantage of producing much lower noise levels compared with the bipolar transistor.
- Also due to its unipolar nature it is more stable during changes of temperature.

The main disadvantages of an FET over its bipolar counterpart are:

- Its very high input impedance renders it susceptible to internal damage from static electricity.
- Its voltage gain for a given bandwidth is lower. Although this may be a disadvantage at low frequencies (below 10 MHz), at high frequencies the low noise amplification that an FET achieves is highly desirable. This facet of FET operation, though, is usually only exploited in radio and TV applications, where very small high frequency signals need to be amplified.
- The FET cannot switch from its fully on to its fully off condition as fast as a bi-polar transistor. It is for this reason that digital logic circuits employing MOSFET technology are slower than bipolar equivalents, although even faster switching speeds are being achieved as FET production technology continues to advance.

Remember

The main reason why the FET has such a differing characteristic from the bipolar transistor is because current flow through the device is controlled by an electric field, which is not the case with the bipolar transistor. It is for this reason that the FET is considered to be a voltage operated device rather than current operated

Figure 8.71 illustrates the basic construction of the FET, which consists of a channel of n-type semiconductor material with two connections, source (S) and drain (D). A third connection is made at the gate (G), which is made of p-type material to control the n-channel current. The symbol is shown in Figure 8.72.

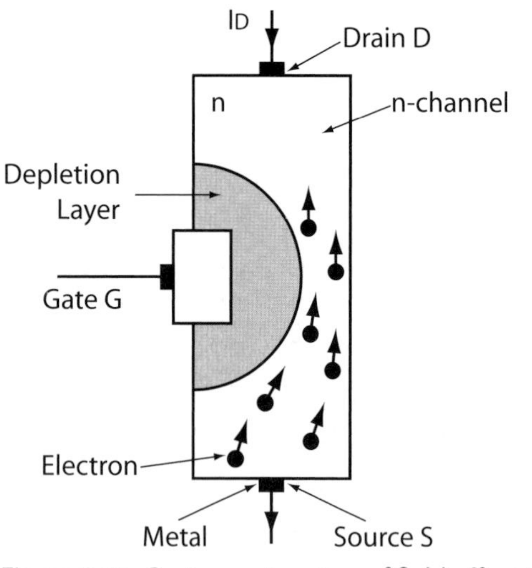

Figure 8.71 Basic construction of field effect transistor (FET)

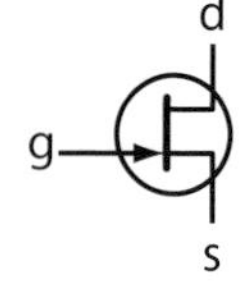

Figure 8.72 Field effect transistor (FET) symbol

In theory, the drain connection is made positive with respect to the source, and electrons are attracted towards the D terminal. If the gate is made negative there will be reverse bias between G and S, which will limit the number of electrons passing from S to D.

The gate and source are connected to a variable voltage supply, such as a potentiometer, and increasing or lowering the voltage makes G more or less negative, which in turn reduces or increases the drain current.

Component positional reference

As electronic diagrams become more complex a system called the component positional reference system is used. This system uses a simple grid reference to identify holes on a board on which components are installed. This is done by counting along the columns at the top of the board starting from the left and numbering them as you count. Then starting from the left and counting down rows, each row is given a letter in turn from the alphabet starting from A. For example the position reference point 7:J would be 7 holes from the left and 10 holes down.

Knowledge check

1. Name the four bands found in the colour code system for resistors and capacitors.
2. Give a rhyme for remembering the colours in the resistor colour code system.
3. What are the two ways that resistors can be used? Describe briefly these two ways.
4. List and describe the different types of dielectric found in capacitors and name the three factors that the value of capacitance depends on.
5. Capacitors of 10 μF, 12 μF and 40 μF are connected in series and then in parallel. Calculate the effective capacitance for each connection.
6. Describe the characteristics of a zener diode.
7. List the precautions to take with fibre optic cables.
8. Describe with the aid of diagrams the process for testing diodes.
9. Describe with the aid of diagrams voltage amplification in a transistor circuit.
10. Draw the lamp dimmer circuit and describe the operation of the individual components.

Suggested Answers

Answers to 'On the jobs'

Chapter 2 Earthing and protection

On the job: Intermittent fault page 83

The worst faults to find are intermittent ones and in an example such as this, there could be many reasons why the RCD may be tripping, most of which could never be covered by this book. However, we have stressed before that 'knowledge is power' and the customer has given us some clues that may help. Sometimes we have to think laterally to find faults, and the clues here, as a starting point for your investigation, are that the trips occur overnight and when it has been raining.

We know that there are security lights, and these will be controlled by PIR detectors, probably fitted as part of each light. The fact that the fault occurs through the night might lead us to think that one of the security lights has some kind of fault that is influenced by dampness. This fault in the light could only trip the RCD once it detects movement, which logically could only occur once it is dark. Following safe isolation procedures, an insulation-resistance test of the circuit may prove a good starting point. Also look for dead insects at terminations which may be causing the problem.

It is also possible that there is no single fault but an accumulation of problems. Immersion heaters, heating pumps and cooker elements are all known sources of small leakage (mA) to earth. Many appliances, including computer equipment, have built-in mains filtering in the form of capacitors between phase and earth and neutral and earth; computers have been recorded to leak as much as 3 mA.

Too much connected leakage current means that whereas from testing we know that a 30 mA RCD should not trip at 15 mA, a 'healthy' RCD may well trip at 16 mA if an extra problem is applied that 'pushes it over the edge'.

Another factor that is often overlooked is that a slight short between earth and neutral would not necessarily trip an RCD until a load was then applied to it. The amount of load would depend on the neutral-to-earth fault resistance. Therefore the operation of a microwave, a central-heating pump or a computer could be sufficient to trip the RCD, especially if there is already a continuous connected leakage as mentioned above.

Checking individual circuits neutral to earth can be difficult (e.g. earths all twisted together and put in one terminal) but should not be excluded from your investigation.

Chapter 3 Lighting

On the job: Fluorescent fault page 110

The discharge across the electrodes in a fluorescent fitting is extinguished about 100 times per second and this can cause a flickering effect. You normally can't see this, but when rotating machines are illuminated in this way, the stroboscopic effect can make machinery appear to slow down, change direction of rotation or even stop. You should therefore suggest the following to the factory manager. Either fit tungsten filament lamps locally to the machines, which will lessen the effect but not eliminate it completely, or, as the factory lighting will probably be split over three phases, connect adjacent fluorescent fittings to different phases of the supply. In a three-phase supply the phases are 120° out of phase with each other. Therefore, if the light falling on the machine arrives from two different sources, each will be flickering at a different time and they will interfere with one another, thus reducing the stroboscopic effect.

Chapter 4 Circuits and systems

On the job: Customer technical query page 118

You should always answer any customer enquiries politely and fully, sufficient to meet their needs. However, never mislead the customer and if you don't know the answer, say so, and promise either to find out or put them in touch with someone who does know. The basic answer would be to say that the thermostat controls the central-heating boiler by telling the programmer to switch it on when the temperature has fallen below a set level, or switch it off if it has risen above a set level.

If a more in-depth answer is required, you could say that electronic digital thermostats use a device called a thermistor to measure temperature. A thermistor is a resistor whose electrical resistance changes with temperature. The circuitry within the thermostat measures the resistance and converts it into a temperature reading.

Ideally, a room thermostat should be located in the part of the house where people spend the most time. It should be about 1.5 m above floor level and about 0.5 m away from an outside wall. It should not be exposed to any heat sources (other than the air in the room) such as sunlight, other appliances, heater vents, windows or hot-water pipes. Consequently, it is best located in a living room rather than the entrance hall (as is very commonly done), as the hall temperature can be lowered by the front door being used and the heating may, therefore, be running more often than is actually required.

Chapter 5 Inspection, testing and commissioning

On the job: Periodic inspection page 200

When you arrived at the premises, you should have fully explained the purpose and implications of your visit to someone in charge. Ideally, all aspects of your work should have been explained to the customer even before you arrived.

Whenever diagrams, charts or tables are not available, some exploratory work should be undertaken to make sure that you could carry out the inspection and testing safely and effectively. Check that alterations such as the extension have not affected the suitability of any of the wiring for its present load and method of installation. Remember that with periodic inspection and testing, inspection is the vital initial operation and that testing subsequently carried out is in support of that inspection. For safety it is necessary to carry out a visual inspection of the installation before beginning any tests or opening enclosures, removing covers etc.

Whenever electronic equipment such as a computer is in use, arrangements should also be made with the customer to have it disconnected for the period of the test. You should therefore try to make this arrangement now. However, remember that some IT systems use back-up power supplies, so check isolation carefully. Carry out your work and complete a Periodic Inspection & Test Certificate with the Schedule of Results sheet attached.

Chapter 6 Fault diagnosis and rectification

On the job: Fault diagnosis page 244

The secret of fault finding is to look logically for the differences between what is there and what should be there. So stay calm, don't panic and you are half way there! In a fictitious example such as this we can never be sure what the fault is, but let's start by reasoning out what should happen.

We know that the motor is only trying to start from the remote station. We also know that the motor should run when we take our finger off the start button at the remote station. The fact that it starts at all from the remote station indicates that a supply is present at the start button, but what does that start button actually do? Well, normally the start button in a remote station is a 'normally open' contact connected in parallel across the start button in the starter/isolator. The feed from the DB normally goes straight to the remote station start button, passing firstly through the 'normally closed' stop button inside the remote station. As the two start buttons are connected in parallel, one of the wires should create a permanent supply at the start button in the starter/isolator. When the remote start is pressed, normally the supply passes from one side of the button to the other, where that parallel wire goes to the other side of the start button in the starter/isolator to energise the coil. This energises the coil and the 'hold in' contact comes into play, thus maintaining the permanent supply to the coil when either start button is released.

We should therefore start our investigation by looking to see whether the parallel wire taking the permanent supply between start buttons has been disconnected or is broken.

Chapter 7 Motors, generators and motor control

On the job: Motor noise page 277

Noise indicates motor problems, but in itself does not cause damage. Noise, however, is usually accompanied by vibration, and vibration can cause damage in several ways. It can shake windings loose, damage insulation mechanically by cracking the material, cause brush sparking at commutators or current collector rings and speed up bearing failure.

Consequently, whenever noise or vibration is reported in an operating motor, the source should quickly be investigated and corrected. Noise and vibrations can be caused by a misaligned motor shaft or can be transmitted to the motor from the driven machine or power transmission system. They can also be the result of electrical imbalance in the motor.

So, having checked the motor shaft alignment, disconnect the motor from the driven load. If the motor then operates smoothly, look for the source of noise or vibration in the driven equipment. If the disconnected motor still vibrates, remove the supply from the motor. If the vibration stops, look for an electrical imbalance. If it continues as the motor coasts without power, look for a mechanical imbalance.

Electrical imbalance occurs when the magnetic attraction between stator and rotor is uneven around the periphery of the motor. This causes the shaft to deflect as it rotates, thus creating a mechanical imbalance. Electrical imbalance usually indicates an electrical failure such as an open stator or rotor winding, an open bar or ring in squirrel-cage motors or shorted field coils in synchronous motors. An uneven air gap, usually resulting from badly worn sleeve bearings, also produces electrical imbalance.

Chapter 8 Electronics

On the job: On-call page 328

The components used are five push buttons (N/O & N/C), three diodes, a thyristor, a resistor, three indicator lamps and a relay to control the buzzer.

Any patient pressing their button will send a signal to the gate of the thyristor which, as it has a permanent supply to it, will allow the supply to energise the coil of the relay, in turn activating the buzzer. The same signal from each push button will activate the respective indicator lamp to show which patient is calling, and the diodes prevent any signal 'backfeeding' onto any of the other buttons, thus preventing all of the indicators from being lit. The resistor limits the current flow in case more than one patient activates the system at the same time.

Push-button B then acts as an isolator to silence the sounder, but not the indicators, and button A isolates the entire system. Each individual patient's call button will be reset by the nurse once they have attended to the situation.

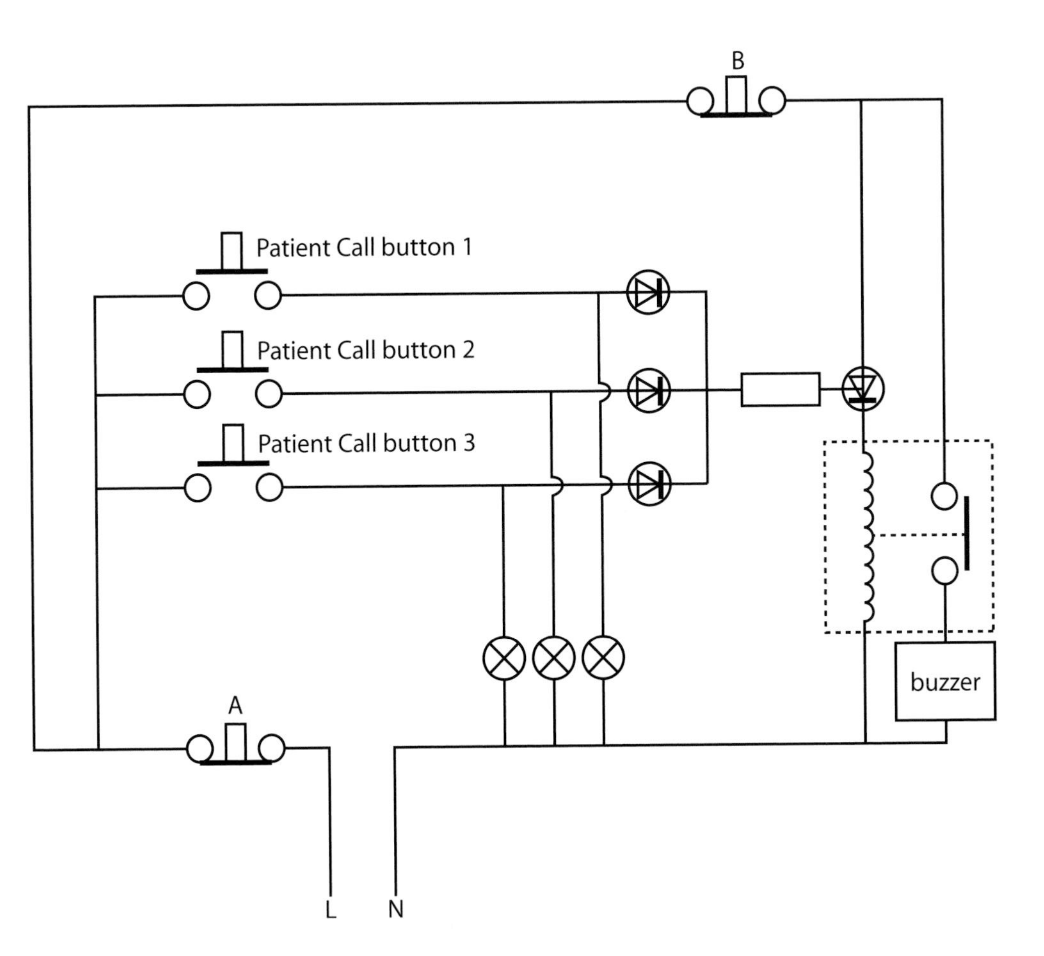

Glossary

Words in bold mean there is also a glossary entry for that term. *See also* **refers to a closely related term.**

a.c. Alternating current, but often used to describe anything that uses it, e.g. a.c. voltage, a.c. motor *(See also* **d.c.**, **sine wave**)

active or true power **Power** consumed by a resistor, dissipated as heat and not returned to the source *(See also* **apparent power**, **reactive power**)

adiabatic equation Used to check that a **conductor** will carry a **fault current** without overheating

adjacent In trigonometry, the side of a triangle adjacent to the angle under consideration and the right angle

ambient temperature Surrounding temperature within the area where an electric cable is installed

ammeter Instrument for measuring **current**; can be **analogue** or **digital**

analogue device Represents measurement by position of, for example, a needle on a calibrated scale *(See also* **digital device**)

apparent power Sum of **true power** and **reactive power** in an **a.c.** circuit

armature Moving part of electrical machine in which a **voltage** is induced by a magnetic field *(See also* **rotor**, **stator**)

asynchronous Used to describe motors in which the speed of the motor and of the rotating magnetic field are different *(See also* **synchronous**, **slip**)

average value For **sinewave a.c.** this equals either **peak value** or **voltage** × 0.637 *(See also* **r.m.s.**)

ballast (or choke) **Inductance** coil in fluorescent lights to limit variations of **a.c.** or alter its **phase**

black box Plastic case housing complete electronic circuit *(See also* **integrated circuit**)

capacitance Ability of circuit or device to store electric **charge**, measured in farads – generally microfarads (μF), nanofarads (nF) or picofarads (pF); symbol C *(See also* **capacitor**)

capacitive Circuit containing components that can store electric charge, i.e. have capacitance

capacitive reactance Opposition to flow of a.c. produced by a **capacitor**, measured in ohms (Ω); symbol X_C

capacitor Component with ability to store required quantity of electric **charge** if **voltage** applied to it, which can be returned to a circuit *(See also* **capacitance**, **dielectric**)

CCD Camera system used in **CCTV** system, more expensive than **CMOS** but images clear and sharp

CCTV Closed circuit television, used in security systems

charge Quantity of electrons, e.g. as stored in a **capacitor**, measured in **coulombs**; symbol Q

choke (or ballast) **Inductance** coil in fluorescent lights to limit variations of **a.c.** or alter its **phase**

coercivity Due to **hysteresis**, reverse magnetic **field strength** that must be applied to reduce **flux density** to zero *(See also* **remanence**)

CMOS Camera system used in **CCTV** system, cheap but images not clear or sharp *(See also* **CCD**)

commutator Device for connecting a rotating **current** carrying coil to a **d.c.** supply

conductor Any material allowing electric charges to flow freely, including metal pipes, metal structures of buildings, salt water and ionised gases *(See also* **semiconductor**, **insulator**)

contactor Widely used in electrical installations, using **solenoid** effect to make or break contacts

control panel Programmable 'brains' of a system to which all parts are connected

cosine In a right-angle triangle, ratio of length of side **adjacent** to an angle to length of the longest side (**hypotenuse**), abbreviated to 'cos' *(See also* **power factor**)

coulomb SI unit of electric charge, the quantity of electricity transported in one second by **current** of one ampere; symbol C

cpc Circuit protective conductor

csa Cross-sectional area, frequently used when describing **conductors**

current Flow of free electrons in a **conductor**, measured in amperes (A); symbol I; direction is opposite to actual electron flow

d.c. Direct current, but often used to describe anything that uses one, e.g. d.c. voltage, d.c. motor *(See also* **a.c.**)

delta connection Triangular arrangement of electrical three-phase windings

dielectric Type of **insulator**, especially used as thin film in **capacitors**, in which charges are displaced when subjected to a **potential difference** but do not flow

digital device Represents measurement as numbers (liquid crystal or **LED** display) rather than a needle on a scale as in **analogue device**

diode **Semiconductor** device that allows current to flow in one direction only

directly proportional If two properties are directly proportional then an increase in one will cause a corresponding increase in the other *(See also* **indirectly proportional**)

discrimination Correct arrangement of fuses in sequence to protect parts of a circuit

DOL Direct-on-line, type of motor starter arrangement

dual-in-line IC Integrated circuit with pins in line down each side

EEBADOS Earthed equipotential bonding and automatic disconnection of the supply

e.m.f. *See* **electromotive force**

electromotive force Greatest potential difference that can be generated by a current source, measured in volts *(See also* **potential difference** and **voltage**)

fault current Current due to a fault in a circuit, including earth-fault current

FET Field-effect transistor, type of transistor common in integrated circuits

field strength Force field around a magnet or current carrying **conductor**, measured in teslas (T); symbol H *(See also* **flux density**)

Fleming's left-hand (motor) rule If the first finger, the second finger and the thumb of the left hand are held at right angles to each other, then the first finger pointing in the direction of the Field (N to S) and the second finger pointing in the direction of the current in the conductor, the thumb will now be indicating the direction in which the conductor tends to move

flux density Quantity of magnetism taking account of magnetic **field strength** and its extent, measured in webers; symbol B

former Block of material, normally cylindrical or cuboid, around which wire is wound in a helical fashion to produce coils or windings

frequency (f) Rate at which **a.c.** goes from zero to positive maximum, to negative maximum and back to zero (one cycle), measured in hertz (Hz)

fused spur **Spur** connected to ring final circuit through fused connection

fusing factor Ratio of minimum current causing a fuse or circuit breaker to trip and stated current that either can sustain without blowing

HBC High breaking capacity, BS 88 sophisticated variant of a cartridge fuse

HSE Health and Safety Executive

hypotenuse Side in a right-angle triangle opposite to the right angle *(See also* **sine**, **cosine**, **Pythagoras' theorem**)

hysteresis Lagging of an effect behind its cause, e.g. change in magnetism of a body which lags behind changes in magnetic field *(See also* **coercivity**, **remanence**)

IC *See* **integrated circuit**

illuminance Measure of visible light reaching a surface, measured in lumens per square metre (lux); symbol lx

impedance Total opposition to flow of **a.c.** in a circuit, combining **resistance**, **inductance** and **capacitance**, measured in ohms (Ω); symbol Z

indirectly proportional If two properties are indirectly proportional then an increase in one will cause a corresponding decrease in the other *(See also* **directly proportional**)

inductance Opposition created by a changing **current** in a magnetic field which induces a **voltage** to oppose change in current, either within a circuit (self-inductance) or a neighbouring circuit in the same magnetic field (mutual inductance), measured in henrys (H); symbol L *(See also* **mutual inductance**)

inductive circuit or **load** Containing components with windings, e.g. motor, generator or transformer, which have **inductance**

inductive reactance Opposition to flow of **a.c.** produced by an inductor, measured in ohms (Ω); symbol X_L

inductor Component introduced into a circuit to provide required amount of inductance

infrared Electromagnetic radiation below the wavelength of visible light that we sense as heat, used in sensors for intruder alarm systems *(See also* **PIR**)

in phase Where **current** and **voltage** alternate at the same time *(See also* **out of phase**)

insulator Poor conductor of electric charges (i.e. a **current**) or heat *(See also* **conductor**, **dielectric**, **semiconductor**)

integrated circuit Miniature electronic circuit produced on a single **semiconductor** crystal offering low cost, bulk, reliability and high speed

Invar Nickel-iron alloy with lowest known rate of expansion from room temperature to 230°C, widely used in instruments, oven thermostats etc.

isolator Mechanical switch to separate an installation from all sources of electrical energy

IT Information technology, generally computers and computer equipment

JIB Joint Industry Board for the Electrical Contracting Industry: sets procedures, particularly for safety

LED Light emitting **diode**: **semiconductor** device that emits light when **current** flows through it, often used in instrument displays

line current **Current** flowing in any one **phase** of a three-phase circuit between source and load

line voltage **Potential difference** between any two-phase conductors between source and load in a three-phase electrical circuit

MCB *See* **miniature circuit breaker**

MCCB Moulded case circuit breaker, protective device usually integral to switchgear

M_i Mineral-insulated cable

miniature circuit breaker automatic switch which opens when excess **current** flows and can be closed again when current returns to normal

multimeter Single instrument (**analogue** or **digital**) combining **ammeter**, **voltmeter**, **ohmmeter**, **wattmeter** etc.

mutual inductance Production of **e.m.f.** in a circuit by a change in **current** in an adjacent circuit *(See also* **inductance**)

no volts protection Prevention of automatic restarting should a motor stop because of supply failure

non-fused spur Usually directly connected to a circuit at the terminal of socket outlets *(See also* **spur**, **fused spur**)

ohmmeter Instrument for measuring **resistance**, can be **analogue** or **digital**

Ohm's Law **Current** flowing in a circuit is directly proportional to the **voltage** applied and indirectly proportional to circuit **resistance**, provided temperature remains constant (I=V/R or V=IR or R=V/I)

oil dashpot Used in **MCB** construction to achieve time-delay tripping

out of phase Where **current** and **voltage** are alternating at different times *(See also* **in phase**)

overcurrent Current exceeding rated value or current-carrying capacity of a **conductor**

overload current Overcurrent occurring in a circuit that is electrically sound

p.d. *See* **potential difference**

panel *See* **control panel**

peak value Maximum value of **a.c.** or **d.c.** wave form, positive or negative, value from maximum to minimum is known as 'peak to peak'

PELV Protective extra-low voltage system, electrically separate from earth and other systems to protect against direct and indirect contact (also **SELV**)

PEN Protective earth neutral

phase *See* **in phase** and **out of phase**

phase angle Difference in **phase** between two sinusoidally varying quantities *(See also* **power factor**); symbol Ø

phase current **Current** through any one component comprising a three-phase source or load

phase voltage **Voltage** measured across a single component in a three-phase source or load

phasor Straight line with length representing size of an **a.c.** quantity and direction representing relationship between **voltage** and **current** *(See also* **phase angle**)

PIR Passive infra red, commonly found as PIR sensor in alarm systems *(See also* **infra red**)

PME Protective multiple earth, the most commonly used earthing system (others are **TN-S**, **TN-C-S**, **TT**)

pole pair A magnet always comprises a north and south pole; hence we always calculate using 'pole pairs'

potential difference Difference in electrical **charge** between two points in a circuit; or force available to push **current** around a circuit; measured in volts, symbol V *(See also* **voltage**, **e.m.f.**)

power Energy used per second, measured in watts (W); symbol P *(See also* **power factor** and **reactive power**)

power factor Number less than 1.0, ratio of apparent to true power in a circuit; equals cos Ø *(See also* **cosine**, **phase angle**)

Pythagoras' theorem In a right-angle triangle, if you square the lengths of the two short sides and add them, the sum equals the square of the length of the long side (**hypotenuse**)

reactive current Current in a **capacitive** or **inductive circuit**, with no resistance and no dissipation of energy

reactive power Power 'consumed' by a **capacitor** or **inductor** is not dissipated as heat but returned to source when **current** reverses *(See also* **apparent power**, **true or active power**)

remanence Due to **hysteresis**, when an applied magnetic **field strength** reduces to zero some **flux density** remains *(See also* **coercivity**)

r.m.s. Root mean square value is the effective **voltage** or **current**, defined as the **a.c.** equivalent of a **d.c.** quantity delivering the same average power to the same resistor, equals 0.707 X peak current or voltage *(See also* **peak value**, **average value**)

RCBO Residual current device with overload protection, which combines the capabilities of an **RCD** and an **MCB**

RCD *See* **residual current device**

reactance *See* **capacitive reactance**, **inductive reactance**

reactive power Power returned to source by reactive components in a circuit (**capacitors**, **inductors**), measured in volt-amperes-reactive; symbol V_{AR}

remanence **Flux density** remaining in a magnetic material after the magnetic field has reduced to zero *(See also* **hysteresis**)

residual current device Monitors **current** flowing in phase and neutral **conductors** and trips if an earth-fault creates a difference above a predetermined value *(See also* **fault current**)

resistance Opposition to flow of electrons (**current**) in a circuit, measured in amperes (A); symbol R

resistive circuit Circuit containing no **capacitance** or **inductance** (in practice there is always some), whereby voltage and current are in phase

resistor Component in a circuit to give a desired amount of **resistance**

rotor Part of a motor or generator that rotates in a magnetic field; it may carry conductors or the magnetic field system *(See also* **stator**, **armature**)

SELV Separate extra-low voltage system, electrically separate from earth and other systems to protect against direct and indirect contact (also **PELV**)

semiconductor Material (commonly silicon, germanium) that will conduct under certain conditions, allowing electrons (negative charge) or 'holes' (positive charge) to flow *(See also* **diode**, **transistor semiconductor**)

short-circuit current Overcurrent resulting from a fault that creates negligible **impedance** between live conductors *(See also* **overcurrent**)

simmerstat Energy regulator to control temperature of electric cooker plates using a bi-metal strip

sine In a right-angle triangle, the ratio of the length of the side opposite to an angle to the length of the longest side (**hypotenuse**), abbreviated 'sin' *(See also* **sine wave**)

sine wave Path traced on a graph by a pure **a.c. current** or **voltage** going from zero to positive maximum, to negative maximum and back to zero (one cycle) *(See also* **sine**, **frequency**)

slip In a motor or generator, the difference between the speed of the rotor and the speed of the rotating magnetic field, measured as a fraction (s) or percentage (S)

solenoid Coil of wire on a cylindrical former which, when **current** is passed through it, creates a magnetic field parallel to the axis to operate a plunger or other device *(See also* **contactor**)

spur Radial branch from a ring final circuit, perhaps to feed a socket outlet

star connection Y-shaped arrangement of three-phase electrical windings *(See also* **delta connection**)

star-delta System sometimes used in induction motors, whereby a **star connection** switches to a **delta connection**

static Electric charge caused by an excess of electrons (negative charge) or a deficiency (positive charge) on a conducting or non-conducting surface, creating a **potential difference** (volts)

stator Stationary part of a motor or generator which rotates in a magnetic field; it may carry **conductors** or the magnetic field system *(See also* **stator, armature**)

stroboscopic Effect when the flicker rate in a light coincides with the rotation of an object and causes it to look as if it has slowed down or stopped

synchronous Used to describe motors in which the speed of the motor and of the rotating magnetic field are the same *(See also* **asynchronous, slip**)

tan Abbreviation for tangent, in a right-angle triangle the ratio of the length of the side opposite an angle to the length of the **adjacent** side

thermistor Device in which **resistance** quickly increases (PTC type) or decreases (NTC type) as temperature rises

thermostat Temperature controller for ovens, central heating etc.

thyristor **Semiconductor** device enabling high-speed switching, formerly known as a silicon-controlled rectifier (SCR)

TN-C-S system Type of earthing system (others are **TT, TN- S, PME**)

TN-S system Type of earthing system (others are **TT, TN-C-S, PME**)

torque Turning force, such as that produced by an electric motor on its rotating shaft; measured as force X distance from its point of rotation

transformer Transfers electrical energy from one **a.c.** circuit to another by **mutual inductance** between two stationary coils wound onto a former; output to input values (e.g. voltage) are determined by the relative number of turns in each coil

transistor semiconductor Device that allows a larger **current** to be controlled by a smaller one, allowing **voltage** or **current** amplification or as a switch

trimmer capacitor Flat metal leaves separated by plastic film, screwed inwards to vary **capacitance** by a small amount *(See also* **tuning capacitor**)

true or active power Power consumed by a resistor, dissipated as heat and not returned to the source *(See also* **apparent power, reactive power**)

TT system Type of earthing system (others are **TN-S, TN-C-S, PME**)

tuning capacitor Interleaved metal plates, one set fixed, one movable; **capacitance** controlled by area overlapping *(See also* **trimming capacitor**)

UPS Uninterruptible power supply, used especially with computer systems

voltage Difference in electrical **charge** between two points in a circuit; or force available to push **current** around a circuit; measured in volts, symbol V *(See also* **potential difference** and **electromotive force**)

voltage drop Loss of **voltage** (electrical pressure) due to **resistance** in **conductor** and components in a circuit

voltmeter Instrument for measuring voltage

wattmeter Instrument for measuring power

windage Physical resistance experienced by rotating parts of a motor or generator created by air within the casing

Index

S

T

U

W

Z